Theory and Applications of the Poincaré Group

Fundamental Theories of Physics

A New International Book Series on The Fundamental Theories of Physics: Their Clarification, Development and Application

Editor: **ALWYN VAN DER MERWE**
University of Denver, U.S.A.

Editorial Advisory Board:

ASIM BARUT, *University of Colorado, U.S.A.*
HERMANN BONDI, *University of Cambridge, U.K.*
BRIAN D. JOSEPHSON, *University of Cambridge, U.K.*
CLIVE KILMISTER, *University of London, U.K.*
GÜNTER LUDWIG, *Philipps-Universität, Marburg, F.R.G.*
NATHAN ROSEN, *Israel Institute of Technology, Israel*
MENDEL SACHS, *State University of New York at Buffalo, U.S.A.*
ABDUS SALAM, *International Centre for Theoretical Physics, Trieste, Italy*
HANS-JÜRGEN TREDER, *Zentralinstitut für Astrophysik der Akademie der Wissenschaften, D.D.R.*

Theory and Applications of the Poincaré Group

by

Y. S. Kim
*Department of Physics and Astronomy,
University of Maryland, U.S.A.*

Marilyn E. Noz
*Department of Radiology,
New York University, U.S.A.*

D. Reidel Publishing Company

A MEMBER OF THE KLUWER ACADEMIC PUBLISHERS GROUP

Dordrecht / Boston / Lancaster / Tokyo

Library of Congress Cataloging in Publication Data

Kim, Y. S.
  Theory and applications of the Poincaré group.

  (Fundamental theories of physics)
  Bibliography: p.
  Includes index.
   1. Groups, Theory of. 2. Poincaré series. 3. Harmonic oscillators. 4. Hadrons. I. Noz, Marilyn E. II. Title. III. Series.
QC174.17.G7K56 1986   512'.22   86-3280
ISBN 90-277-2141-6

Published by D. Reidel Publishing Company,
P.O. Box 17, 3300 AA Dordrecht, Holland.

Sold and distributed in the U.S.A. and Canada
by Kluwer Academic Publishers,
190 Old Derby Street, Hingham, MA 02043, U.S.A.

In all other countries, sold and distributed
by Kluwer Academic Publishers Group,
P.O. Box 322, 3300 AH Dordrecht, Holland.

All Rights Reserved

© 1986 by D. Reidel Publishing Company, Dordrecht, Holland
No part of the material protected by this copyright notice may be reproduced or utilized in any form or by any means, electronic or mechanical including photocopying, recording or by any information storage and retrieval system, without written permission from the copyright owner

Printed in The Netherlands

Table of Contents

Preface	ix
Introduction	xi
Chapter I: Elements of Group Theory	1
1. Definition of a Group	2
2. Subgroups, Cosets, and Invariant Subgroups	4
3. Equivalence Classes, Orbits, and Little Groups	7
4. Representations and Representation Spaces	9
5. Properties of Matrices	13
6. Schur's Lemma	15
7. Exercises and Problems	17
Chapter II: Lie Groups and Lie Algebras	25
1. Basic Concepts of Lie Groups	26
2. Basic Theorems Concerning Lie Groups	32
3. Properties of Lie Algebras	36
4. Properties of Lie Groups	40
5. Further Theorems of Lie Groups	43
6. Exercises and Problems	45
Chapter III: Theory of the Poincaré Group	50
1. Group of Lorentz Transformations	51
2. Orbits and Little Groups of the Proper Lorentz Group	55
3. Representations of the Poincaré Group	60
4. Lorentz Transformations of Wave Functions	64
5. Lorentz Transformations of Free Fields	67
6. Discrete Symmetry Operations	71
7. Exercises and Problems	74
Chapter IV: Theory of Spinors	79
1. $SL(2,c)$ as the Covering Group of the Lorentz Group	80

2. Subgroups of $SL(2, c)$ 82
  3. $SU(2)$ 86
  4. $SL(2, c)$ Spinors and Four-Vectors 91
  5. Symmetries of the Dirac Equation 95
  6. Exercises and Problems 101

Chapter V: Covariant Harmonic Oscillator Formalism 107
  1. Covariant Harmonic Oscillator Differential Equations 109
  2. Normalizable Solutions of the Relativistic Oscillator Equation 110
  3. Irreducible Unitary Representations of the Poincaré Group 115
  4. Transformation Properties of Harmonic Oscillator Wave Functions 117
  5. Harmonic Oscillators in the Four-Dimensional Euclidean Space 122
  6. Moving $O(4)$ Coordinate System 126
  7. Exercises and Problems 127

Chapter VI: Dirac's Form of Relativistic Quantum Mechanics 135
  1. C-Number Time-Energy Uncertainty Relation 137
  2. Dirac's Form of Relativistic Theory of "Atom" 143
  3. Dirac's Light-Cone Coordinate System 147
  4. Harmonic Oscillators in the Light-Cone Coordinate System 151
  5. Lorentz-Invariant Uncertainty Relations 152
  6. Exercises and Problems 155

Chapter VII: Massless Particles 159
  1. What is the $E(2)$ Group? 161
  2. $E(2)$-like Little Group for Photons 166
  3. Transformation Properties of Photon Polarization Vectors 170
  4. Unitary Transformation of Photon Polarization Vectors 174
  5. Massless Particles with Spin 1/2 176
  6. Harmonic Oscillator Wave Functions for Massless Composite Particles 178
  7. Exercises and Problems 184

Chapter VIII: Group Contractions 189
  1. $SE(2)$ Group as a Contraction of $SO(3)$ 190
  2. $E(2)$-like Little Group as an Infinite-momentum/zero-mass Limit of the $O(3)$-like Little Group for Massive Particles 193
  3. Large-momentum/zero-mass Limit of the Dirac Equation 196
  4. Finite-dimensional Non-unitary Representations of the $SE(2)$ Group 198
  5. Polarization Vectors for Massless Particles with Integer Spin 202
  6. Lorentz and Galilei Transformations 204
  7. Group Contractions and Unitary Representations of $SE(2)$ 207
  8. Exercises and Problems 209

Table of Contents vii

Chapter IX: $SO(2, 1)$ and $SU(1, 1)$ 214
1. Geometry of $SL(2, r)$ and $Sp(2)$ 216
2. Finite-dimensional Representations of $SO(2, 1)$ 221
3. Complex Angular Momentum   225
4. Unitary Representations of $SU(1, 1)$ 228
5. Exercises and Problems   232

Chapter X: Homogeneous Lorentz Group   236
1. Statement of the Problem   237
2. Finite-dimensional Representations of the Homogeneous Lorentz Group   238
3. Transformation Properties of Electric and Magnetic Fields   241
4. Pseudo-unitary Representations for Dirac Spinors   244
5. Harmonic Oscillator Wave Functions in the Lorentz Coordinate System   245
6. Further Properties of the Homogeneous Lorentz Group   249
7. Concluding Remarks   252

Chapter XI: Hadronic Mass Spectra   255
1. Quark Model   256
2. Three-particle Symmetry Classifications According to the Method of Dirac   260
3. Construction of Symmetrized Wave Functions   262
4. Symmetrized Products of Symmetrized Wave Functions   263
5. Spin Wave Functions for the Three-Quark System   267
6. Three-quark Unitary Spin and $SU(6)$ Wave Functions   268
7. Three-body Spatial Wave Functions   271
8. Totally Symmetric Baryonic Wave Functions   273
9. Baryonic Mass Spectra   275
10. Mesons   279
11. Exercises and Problems   280

Chapter XII: Lorentz-Dirac Deformation in High-Energy Physics   286
1. Lorentz-Dirac Deformation of Hadronic Wave Functions   288
2. Form Factors of Nucleons   291
3. Calculation of the Form Factors   296
4. Scaling Phenomenon and the Parton Picture   300
5. Covariant Harmonic Oscillators and the Parton Picture   305
6. Calculation of the Parton Distribution Function for the Proton   310
7. Jet Phenomenon   313
8. Exercises and Problems   316

References   320

Index   327

Preface

Special relativity and quantum mechanics, formulated early in the twentieth century, are the two most important scientific languages and are likely to remain so for many years to come. In the 1920's, when quantum mechanics was developed, the most pressing theoretical problem was how to make it consistent with special relativity. In the 1980's, this is still the most pressing problem. The only difference is that the situation is more urgent now than before, because of the significant quantity of experimental data which need to be explained in terms of both quantum mechanics and special relativity.

In unifying the concepts and algorithms of quantum mechanics and special relativity, it is important to realize that the underlying scientific language for both disciplines is that of group theory. The role of group theory in quantum mechanics is well known. The same is true for special relativity. Therefore, the most effective approach to the problem of unifying these two important theories is to develop a group theory which can accommodate both special relativity and quantum mechanics.

As is well known, Eugene P. Wigner is one of the pioneers in developing group theoretical approaches to relativistic quantum mechanics. His 1939 paper on the inhomogeneous Lorentz group laid the foundation for this important research line. It is generally agreed that this paper was somewhat ahead of its time in 1939, and that contemporary physicists must continue to make real efforts to appreciate fully the content of this classic work.

Wigner's 1939 paper is also a fundamental contribution in mathematics. Since 1939, in order to achieve a better understanding of Wigner's work, mathematicians have developed many concepts and tools, including little groups, orbits, groups containing Abelian invariant subgroups, induced representations, group extensions, group contractions and expansions. These concepts are widely discussed in many of the monographs and textbooks in mathematics [Segal (1963), Gel'fand et al. (1966), Hermann (1966), Gilmore (1974), Mackey (1978), and many others].

Indeed, the mathematical research along this line has been extensive. It is therefore fair to say that there is at present an imbalance between mathematics and physics in the sense that there are not enough physical examples to enrich the theorems mathematicians have developed. The main purpose of this book is to reduce the gap between mathematical theorems and physical examples.

This book combines in a systematic manner numerous articles published by the authors primarily in the *American Journal of Physics* and lecture notes prepared by the authors over the past several years. It is intended mainly as a teaching tool directed toward those who desire a deeper understanding of group theory in terms of examples applicable to the physical world and/or of the physical world in terms of the symmetry properties which can best be formulated in terms of group theory. Both graduate students and others interested in the relationship between group theory and physics will find it instructive. In particular, those engaged in high-energy physics and foundations of quantum mechanics will find this book rich in illustrative examples of relativistic quantum mechanics.

For numerous discussions, comments, and criticisms while the manuscript was being prepared, the authors would like to thank S. T. Ali, L. C. Biedenharn, J. A. Brooke, W. E. Caswell, J. F. Carinena, A. Das, D. Dimitroyannis, P. A. M. Dirac, G. N. Fleming, H. P. W. Gottlieb, O. W. Greenberg, M. Haberman, M. Hamermesh, D. Han, W. J. Holman, T. Hubsch, P. E. Hussar, S. Ishida, P. B. James, T. J. Karr, S. K. Kim, W. Klink, R. Lipsman, G. Q. Maguire, V. I. Man'ko, M. Markov, S. H. Oh, S. Oneda, E. F. Redish, M. J. Ruiz, G. A. Snow, D. Son, L. J. Swank, K. C. Tripathy, A. van der Merwe, D. Wasson, A. S. Wightman, E. P. Wigner, and W. W. Zachary. The chapters of this book on massless particles are largely based on the series of papers written by one of the authors (YSK) in collaboration with D. Han and D. Son.

Finally, the authors would like to express their sincere gratitude to H. J. Laster for encouraging their collaboration. One of them (YSK) wishes to thank J. S. Toll for providing key advice at the critical stages of his research life.

Introduction

One of the most fruitful and still promising approaches to unifying quantum mechanics and special relativity has been and still is the covariant formulation of quantum field theory. The role of Wigner's work on the Poincaré group in quantum field theory is nicely summarized in the fourth paragraph of an article by V. Bargmann *et al.* in the commemorative issue of *Reviews of Modern Physics* in honor of Wigner's 60th birthday [*Rev. Mod. Phys.* **34**, 587 (1962)], which concludes with the sentences:

Those who had carefully read the preface of Wigner's great 1939 paper on relativistic invariance and had understood the physical ideas in his 1931 book on group theory and atomic spectra were not surprised by the turn of events in quantum field theory in the 1950's. A fair part of what happened was merely a matter of whipping quantum field theory into line with the insights achieved by Wigner in 1939.

It is important to realize that quantum field theory has not been and is not at the present the only theoretical machine with which physicists attempt to unify quantum mechanics and special relativity. Paul A. M. Dirac devoted much of his professional life to this important task. In his attempt to construct a "relativistic dynamics of atom" using "Poisson brackets" contained in the commemorative issue of *Reviews of Modern Physics* in honor of Einstein's 70th Birthday (1949), Dirac emphasizes that the task of constructing a relativistic dynamics is equivalent to constructing a representation of the inhomogeneous Lorentz group.

Dirac's form of relativistic quantum mechanics had been overshadowed by the success of quantum field theory throughout the 1950's and 1960's. However, in the 1970's, when it was necessary to deal with quarks confined permanently inside hadrons, the limitations of the present form of quantum field theory became apparent. Currently, there are two different opinions on the difficulty of using field theory in dealing with bound-state problems or systems of confined quarks. One of these regards the present difficulty merely as a complication in calculation. According to this view, we should continue developing mathematical techniques which will someday enable us to formulate a bound-state problem with satisfactory solutions within the framework of the existing form of quantum field theory. The opposing opinion is that quantum field theory is a model that can handle only scattering problems in which all particles can be brought to free-particle asymptotic states. According to this view, we have to make a fresh start for relativistic bound-state problems, possibly starting from Dirac's 1949 paper.

We contend that these two opposing views are not mutually exclusive. *Bound-state models developed in these two different approaches should have the same space-time symmetry.* It is quite possible that independent bound-state models, if successful in explaining what we see in the real world, will eventually complement field theory. One of the purposes of this book is to discuss a relativistic bound-state model built in accordance with the principles laid out by Wigner (1939) and Dirac (1949), which can explain basic hadronic features observed in high-energy laboratories.

Another important development in modern physics is the extensive use of gauge transformations in connection with massless particles and their interactions. Wigner's 1939 paper has the original discussion of space-time symmetries of massless particles. However, it was only recently recognized that gauge-dependent electromagnetic four-potentials form the basis for a finite-dimensional non-unitary representation of the little group of the Poincaré group. This enables us to associate gauge degrees of freedom with the degrees of freedom left unexplained in Wigner's work. Hence it is possible to impose a gauge condition on the electromagnetic four-potential to construct a unitary representation of the photon polarization vectors.

The organization of this book is identical to that of Wigner's original paper, but the emphasis will be different. In discussing representations of the Poincaré group for free particles, we use the method of little groups, as is summarized in Table 1.

Wigner observed in 1939 that Dirac's electron has an $SU(2)$-like internal space-time symmetry. However, quarks and hadrons were unknown at that time. By discussing Dirac's form of relativistic bound-state quantum

TABLE 1
Wigner's little groups discussed in this book.

P: four momentum	Subgroup of $O(3, 1)$	Subgroup of $SL(2, c)$
Massive: $M^2 > 0$	$O(3)$-like subgroup of $O(3, 1)$: *hadrons*	$SU(2)$-like subgroup of $SL(2, c)$: *electrons*
Massless: $M^2 = 0$	$E(2)$-like subgroup of $O(3, 1)$: *photons*	$E(2)$-like subgroup of $SL(2, c)$: *neutrinos*
Spacelike $M^2 < 0$	$O(2, 1)$-like subgroup of $O(3, 1)$	$Sp(2)$-like subgroup of $SL(2, c)$
$P = 0$	$O(3, 1)$	$SL(2, c)$

Introduction

mechanics, which starts from the representations of the Poincaré group, it is possible to study the $O(3)$-like little group for massive particles. Since Dirac's form leads to hadronic wave functions which can describe fairly accurately the distribution of quarks inside hadrons, a substantial portion of hadronic physics can be incorporated into the $O(3)$-like little group for massive particles.

As for massless particles, Wigner showed that their internal space-time symmetry is locally isomorphic to the Euclidean group in two-dimensional space. However, Wigner did not explore the content of this isomorphism, because the physics of the translation-like transformations of this little group was unknown in 1939. Neutrinos were known only as "Dirac electrons without mass," although photons were known to have spins either parallel or antiparallel to their respective momenta. We now know the physics of the degrees of freedom left unexplained in Wigner's paper. Much more is also known about neutrinos today than in 1939. For instance, it is firmly established that neutrinos and anti-neutrinos are left and right handed respectively. Therefore, it is possible to discuss internal space-time symmetries of massless particles starting from Wigner's $E(2)$-like little group.

The $O(2, 1)$-like little group could explain internal space-time symmetries of particles which move faster than light. Since these particles are not observable, this little group is not of immediate physical interest. The story is the same for the $O(3, 1)$-like little group, since it is difficult to observe particles with vanishing four-momentum. However, the mathematics of these groups has been and is still being discussed extensively in the literature. We shall discuss the mathematical aspect of these two little groups in this book. It is of interest to note in particular that $O(2, 1)$ is isomorphic to the two-dimensional symplectic group or $Sp(2)$, which is playing an increasingly important role in all branches of physics.

Also included in this book are discussions of hadronic phenomenology. The above-mentioned $O(3)$-like little group for hadrons and the bound-state model based on this concept will be meaningful only if they can describe the real world. We shall discuss hard experimental data and curves describing mass spectra, form factors, the parton model, and the jet phenomenon. We shall then show that a simple harmonic oscillator model developed along the line suggested by Wigner (1939) and Dirac (1945, 1949) can produce results which can be compared with the relevant experimental data.

In Chapters I and II, we start from the basic concept of group, and concentrate our discussion on continuous groups. The treatment given in these chapters is not meant to be complete. However, we organize the material in such a way that the reader can acquire a basic introduction to and understanding of group theory. Chapters III and IV contain a pedagogical elaboration of Wigner's original work on the Poincaré group. Little groups, forms of the Casimir operators, and the relation between them are discussed in detail.

In Chapter V, the covariant harmonic oscillator formalism is discussed as a mathematical device useful in constructing space-time solutions of the commutator equations for Dirac's relativistic bound-state quantum mechanics. It is pointed out that the oscillator formalism is useful also in explaining basic hadronic features we observe in the real world. Chapter VI shows that the oscillator formalism indeed satisfies all the requirements for Dirac's form of relativistic quantum mechanics, and therefore that the formalism is consistent with the established rules of quantum mechanics and special relativity.

Chapters VII and VIII deal with the $E(2)$-like little group for massless particles. Both finite and infinite-dimensional representations of the $E(2)$ group are considered. The content of the isomorphism between the little group and the two-dimensional Euclidean group is discussed in detail. The concept of group contraction is introduced to obtain the $E(2)$-like little group for massless particles from the $O(3)$-like little group in the infinite-momentum/zero-mass limit.

In Chapters IX and X, we discuss representations of the $O(2, 1)$ and $O(3, 1)$ groups. Although the physics of these little groups is not well understood, constructing their representations has been a challenging problem in mathematics, since the appearance of the papers of Bargmann (1947) and of Harish-Chandra (1947). In particular, $SO(2, 1)$ is locally isomorphic to $Sp(2)$, $SU(1, 1)$, $SL(2, r)$, and has a rich mathematical content. The homogeneous Lorentz group plays the central role in studying Lorentz transformation properties of quantum mechanical state vectors and operators. We shall study $O(3, 1)$ not as a little group but as the symmetry group for the process of orbit completion.

Chapters XI and XII deal with various applications to hadronic phenomenology of the harmonic oscillator formalism developed in Chapter V and Chapter VI. Since Hofstadter's discovery in 1955, it has been known that the proton or hadron is not a point particle, but has a space-time extension. This idea is compatible with the basic concept of the quark model in which hadrons are quantum bound states of quarks having space-time extensions. At present, this concept is totally consistent with all qualitative features of hadronic phenomenology. It is widely believed that quark motions inside the hadron generate Rydberg-like mass spectra. It is also believed that fast-moving extended hadrons are Lorentz-deformed, and that this deformation is responsible for peculiarities observed in high-energy experiments such as the parton and jet phenomena. The question is how to describe all these covariantly.

In Chapter XI, we show first that the $O(3)$-like little group is the correct language for describing covariantly the observed mass spectra and the space-time symmetry of confined quarks, and then show that the harmonic oscillator model describes the mass spectra observed in the real world. In Chapter XII, we point out first that the space-time extension of the hadron and its

Lorentz deformation are responsible for behavior of hadronic form factors. We present a comprehensive review of theoretical models constructed along this line, and compare the calculated form factors with experimental data.

Second, we discuss in detail the peculiarities in Feynman's parton picture which are universally observed in high-energy hadronic experiments. According to the parton model, the hadron consisting of a finite number of quarks appears as a collection of an infinite number of partons when the hadron moves very rapidly. Since partons appear to have properties which are different from those of quarks, the question arises whether quarks are partons. While this question cannot be answered within the framework of the present form of quantum field theory, the harmonic oscillator formalism provides a satisfactory resolution of this paradoxical problem.

Third, the proton structure function is calculated from the Lorentz-boosted harmonic oscillator wave function, and is compared with experimental data collected from electron and neutrino scattering experiments. Detailed numerical analyses are presented. Fourth, it is shown that Lorentz deformation is responsible also for formation of hadronic jets in high-energy experiments in which many hadrons are produced in the final state. Both qualitative and quantitative discussions are presented.

Chapter I

Elements of Group Theory

Since there are already many excellent textbooks and lecture notes on group theory, it is not necessary to give another full-fledged treatment of group theory in this book. We are interested here in only those aspects of group theory which are essential for understanding the theory and applications of the Poincaré group. We shall not present proofs of theorems if they are readily available in standard textbooks [Hamermesh (1962), Pontryagin (1966), Boerner (1979), Miller (1972), Gilmore (1974), and many others]. Physicists learn group theory not by proving theorems but by working out examples. The purpose of this Chapter and of the entire book is to present a systematic selection of examples from which the reader can formulate his/her own concepts.

Physicists' first exposure to group theory takes place through the three-dimensional rotation group in quantum mechanics. Since 1960, group theory has become an indispensable tool in theoretical physics in connection with the quark model. During the past twenty years, the trend has been toward more abstract group theory, with emphasis on constructing unitary representations of compact Lie groups which are simple or semisimple in connection with constructing multiplet schemes for quarks and other fundamental particles.

However, the recent trend, as is manifested in the study of supersymmetry and the Kaluza-Klein theory, is that we are becoming more interested in studying space-time coordinate transformations and in constructing explicit representations of noncompact groups, particularly those which are neither simple nor semisimple. Because it describes the fundamental space-time symmetries in the four-dimensional Minkowskian space, the Poincaré group occupies an important place in this recent trend. The purpose of this book is to discuss the representations of the Poincaré group which is noncompact and which is neither simple nor semisimple.

The mathematical theorems connected with noncompact groups are somewhat beyond the grasp of most physics students. Fortunately, however, calculations involved in the study of the Poincaré group are simple enough to carry out with only a basic knowledge of group theory. The examples selected in this book are aimed at building a bridge between the abstract concepts and explicit calculations.

It is assumed that the reader is already familiar with the three-dimensional rotation group. This group will therefore be used throughout this book to illustrate the abstract concepts. Among other examples contained in this chapter and in Chapter II, the two-dimensional Euclidean gorup or $E(2)$ is used extensively, for the following reasons.

(a) Transformations of the $E(2)$ group are mathematically simple. They can be explained in terms of the two-dimensional geometry which can be sketched on a piece of paper.

(b) Like the Poincaré group, $E(2)$ is noncompact and contains an Abelian invariant subgroup.

(c) The $E(2)$ group is isomorphic to the little group of the Poincaré group for massless particles, i.e. it is neither simple nor semisimple.

(d) Like the three-dimensional rotation group, $E(2)$ is a three-parameter Lie group. It can be obtained as a contraction of $O(3)$.

(e) Among the standard textbooks available today, Gilmore's book (Gilmore, 1974) contains a very comprehensive coverage of Lie groups, and is thus one of the most popular books among students in group theory. As an illustrative example, Gilmore uses in his book the two-parameter Lie group consisting of multiplications and additions of real numbers. This group is very similar to the $E(2)$ group. For this reason, by using the $E(2)$ group, we can establish a bridge between Gilmore's textbook and what we do in this Chapter and also in Chapter II.

In Section 1, we give definitions of standard terms used in group theory. In Section 2, subgroups are discussed. In Section 3, vector spaces and representations are discussed. In Section 4, equivalence classes, orbits, and little groups are discussed. In Section 5, the properties of matrices are given, and Section 6 contains a discussion of Schur's Lemma. In Section 7, we list further examples in the form of exercises and problems.

1. Definition of a Group

A set of elements g forms a group G if they have a multiplication (i.e. a group operation) defined for any two elements a and b of G according to the group axioms:

(1) Closure: for any a and b in G, $a \cdot b$ is in G, where $\cdot$ is the group operation generally referred to as group multiplication.
(2) Associative law: $(a \cdot b) \cdot c = a \cdot (b \cdot c)$, where c is also in G.
(3) Identity: there exists a unique identity element e such that $e \cdot a = a \cdot e = a$ for all a in G.
(4) Inverse: for every a in G, there exists an inverse element, denoted a^{-1}, such that $a^{-1} \cdot a = a \cdot a^{-1} = e$.

The third axiom is a consequence of the other three.

The number of elements in the group is called the *order* of G. The order can be either finite or infinite. It is assumed that the reader is familiar with some examples of groups, such as the group S_3 which consists of permutations of three objects. The group most familiar to physicists is $SO(3)$, namely the three-dimensional rotation group without space inversions. The order of S_3 is six, while the order of the three-dimensional rotation group is infinite.

Groups in which the commutative law for group multiplication holds are called Abelian groups. The group consisting of rotations around the origin on the xy plane or $SO(2)$ is Abelian. The group of translations in three-dimensional space is also Abelian. S_3 and $SO(3)$ are not Abelian groups.

Two groups are *isomorphic* if there is a one-to-one mapping F from one group (G) to the other (H) that preserves group multiplication, i.e. $F(g_1 g_2) = F(g_1)\ F(g_2)$, where $h_1 = F(g_1)$ and $h_2 = F(g_2)$. S_3 is isomorphic to the symmetry group of an equilateral triangle. This group consists of three rotations (by 0°, 120°, and 240°) and three flips each of which results in interchange of two vertices. The groups are *homomorphic* if the mapping is onto but not one-to-one. The group $SU(2)$ consisting of two-by-two unitary unimodular matrices is homomorphic to the three-dimensional rotation group.

There are many examples of various groups discussed in standard textbooks. One example which will be useful in this book is the "four group", consisting of four elements, e, a, b, c, where e is the unit element. This group can have two distinct multiplication laws:

$$\text{(A)}\quad a^2 = b, \quad ab = c = a^3, \quad a^4 = b^2 = e, \tag{1.1}$$

with the multiplication table:

e	a	b	c
a	b	c	e
b	c	e	a
c	e	a	b

$$\text{(B)}\quad a^2 = b^2 = c^2 = e, \quad ab = c, \quad ac = b, \quad bc = a, \tag{1.2}$$

with the table:

e	a	b	c
a	e	c	b
b	c	e	a
c	b	a	e

It is clear that type A is cyclic and is isomorphic to the group consisting of 1, i, -1, and $-i$. The multiplication law for the four group of type B is identical to that for the Pauli spin matrices up to the phase factor. This gorup will be useful in understanding space-time reflection properties (Wigner, 1962a).

Another example which will play an important role throughout this book is the two-dimensional Euclidean group. This group, which is often called $SE(2)$, consists of rotations and translations in the xy plane, resulting in the linear transformation:

$$x = x_0 \cos \theta - y_0 \sin \theta + u,$$
$$y = x_0 \sin \theta + y_0 \cos \theta + v. \tag{1.3}$$

This is the rotation of the coordinate system by angle θ followed by the translation of the origin to (u, v). In matrix notation, this transformation takes the form

$$\begin{bmatrix} x \\ y \\ 1 \end{bmatrix} = \begin{bmatrix} \cos \theta & -\sin \theta & u \\ \sin \theta & \cos \theta & v \\ 0 & 0 & 1 \end{bmatrix} \begin{bmatrix} x_0 \\ y_0 \\ 1 \end{bmatrix}. \tag{1.4}$$

The determinant of the above transformation matrix is 1, and the inverse transformation takes the form

$$\begin{bmatrix} x_0 \\ y_0 \\ 1 \end{bmatrix} = \begin{bmatrix} \cos \theta & \sin \theta & -u' \\ -\sin \theta & \cos \theta & -v' \\ 0 & 0 & 1 \end{bmatrix} \begin{bmatrix} x \\ y \\ 1 \end{bmatrix}, \tag{1.5}$$

where

$$u' = u \cos \theta + v \sin \theta,$$
$$v' = -u \sin \theta + v \cos \theta.$$

Like matrices representing the three-dimensional rotation group, the matrix in Equation (1.4) is also three-by-three and contains three parameters. While similar to the rotation group in some aspects, $SE(2)$ also shares many characteristics with the inhomogeneous Lorentz group. Indeed, this group will play a very important role both in illustrating mathematical theorems and in constructing representations of the Poincaré group.

2. Subgroups, Cosets, and Invariant Subgroups

A set of elements H, contained in a group G, forms a subgroup of G, if all elements in H satisfy group axioms. Even permutations in S_3 form a subgroup. Rotations around the z axis form a subgroup of the rotation group.

If H is a subgroup of G, the set gH with g in G is called the *left coset* of H.

Similarly, the set Hg is called the *right coset* of H. The number of right cosets and the number of left cosets are either both infinite, or both finite and equal. The number of cosets is called the index of H in G. If the order of G is finite, it is the product of the order and the index of H. In other words, if the order of G is a, and the order and index of H are b and c respectively, then $a = bc$. Two cosets are either identical or disjoint. The group G therefore is a union of all left or right cosets, or can be expanded in terms of cosets. This leads us to the concept of *coset space* in which cosets form the entire group. The coset space is usually written as G/H.

S_3 is the union of the two cosets of the even permutation subgroup. One coset is the even permutation subgroup itself, and the other is the set of even permutations followed or preceded by a transposition of two objects resulting in odd permutations. The coset space consists of the set of even permutations and the set of odd permutations. The order of the subgroup in this case is 3, and the index is 2, while the order of S_3 is 6.

Rotations around the z axis form a subgroup of the three-dimensional rotation group. Cosets in this case are rotations of the z axis followed or preceded by the rotation around the z axis. The coset space consists of all possible directions the rotation axis can take or the surface of a unit sphere. In this case, the order of G, the order of H, and the index are all infinite.

A right coset of H in general need not be identical to the left coset. There are certain subgroups N in which every right coset is a left coset, satisfying

$$gN = Ng \quad \text{or} \quad gNg^{-1} = N \quad \text{for all} \quad g \text{ in } G. \tag{2.1}$$

Subgroups N having this property are called normal or *invariant* subgroups. The set of even permutations in S_3 is an invariant subgroup. The group of rotations around the z axis is not an invariant subgroup of the three-dimensional rotation group.

If one multiples two cosets of an invariant subgroup N, the result itself is a coset. For, from $Nn = nN = N$ for every n in N, $NN = N$. For g_1 and g_2 in G,

$$\begin{aligned}(Ng_1)(Ng_2) &= N(g_1N)g_2 \\ &= N(Ng_1)g_2 = (NN)(g_1g_2) \\ &= N(g_1g_2).\end{aligned} \tag{2.2}$$

The set of cosets of a invariant subgroup is therefore a group, with N as the identity element. This group is called the *quotient group* of G with respect to N and is written G/N.

If a group does not contain invariant subgroups, it is called a *simple group*. $SO(3)$ is a simple group. If a group contains invariant subgroups which are not Abelian, it is called a *semisimple* group. S_3 is a semisimple group. $SE(2)$ contains an Abelian invariant subgroup, and is therefore neither simple nor semisimple.

Let us look at $SE(2)$ closely. This group has two subgroups. One of them is the rotation subgroup consisting of matrices:

$$R(\theta) = \begin{bmatrix} \cos\theta & -\sin\theta & 0 \\ \sin\theta & \cos\theta & 0 \\ 0 & 0 & 1 \end{bmatrix}, \tag{2.3}$$

and the other is the translation subgroup represented by:

$$T(u, v) = \begin{bmatrix} 1 & 0 & u \\ 0 & 1 & v \\ 0 & 0 & 1 \end{bmatrix}. \tag{2.4}$$

If we use the notation $F(u, v, \theta)$ for the three-by-three matrix of Equation (1.4), then

$$F(u, v, \theta) = T(u, v) R(\theta). \tag{2.5}$$

Both T and R are closed under their respective group multiplications. They are Abelian subgroups. The three-by-three matrix in Equation (1.4) is a product of the two matrices: $T(u, v) R(\theta)$. It is easy to verify that T is an invariant subgroup, while R is not.

As we can see in S_3 and $SE(2)$, a group can be generated by two subgroups H and K:

$$G = KH. \tag{2.6}$$

This means that an element of H is to be multiplied from left by an element of K. This is usually written as

$$g = kh = (k, h). \tag{2.7}$$

The multiplication law for g is then

$$\begin{aligned} g_2 g_1 &= (k_2 h_2)(k_1 h_1) \\ &= k_2 (h_2 k_1 h_2^{-1}) h_2 h_1. \end{aligned} \tag{2.8}$$

If $(h_2 k_1 h_2^{-1})$ belongs to K, the above product can be written as

$$g_2 g_1 = (k_2 h_2 k_1 h_2^{-1}, h_2 h_1). \tag{2.9}$$

The multiplication of two groups given in (2.9) is called the *semi-direct product*. The subgroup K of G in this case has to be an invariant subgroup in order to satisfy the requirement of Equation (2.7).

If every element in H commutes with every element in K, then we say that K and H commute with each other. In that case, the product becomes

$$\begin{aligned} g_2 g_1 &= (k_2 h_2)(k_1 h_1) \\ &= (k_2 k_1, h_2 h_1). \end{aligned} \tag{2.10}$$

Elements of Group Theory

If this multiplication law is satisfied, it is said that G is a *direct product* of H and K. The addition of orbital and spin angular momenta in quantum mechanics is an example of a direct product of the $SO(3)$ and $SU(2)$ groups.

S_3 is a semi-direct product of the even permutation group and the two-element group consisting of the identity and a transposition of the first two elements. $SE(2)$ is a semi-direct product of the translation and rotation subgroups. Because the translation subgroup is an invariant subgroup, T and R can still be separated as

$$F(u_2, v_2, \theta_2) F(u_1, v_1, \theta_1) = T(u', v') R(\theta_2 + \theta_1), \tag{2.11}$$

where

$$u' = u_2 + u_1 \cos \theta_2 - v_1 \sin \theta_2,$$
$$v' = v_2 + u_1 \sin \theta_2 + v_1 \cos \theta_2.$$

Indeed, this property will play an important role in constructing representations of the Poincaré group in later chapters.

3. Equivalence Classes, Orbits, and Little Groups

An element a of G is said to be *conjugate* to the element b if there exists an element g in G such that $b = gag^{-1}$. It is easy to show that conjugacy is an equivalence relation, i.e. $a \sim a$ (reflexive), (2) $a \sim b$ implies $b \sim a$ (symmetric), and (3) $a \sim b$ and $b \sim c$ implies $a \sim c$ (transitive). For this reason, a and b are said to belong to the same *equivalence class*. Thus the element of G can be divided into equivalent classes of mutually conjugate elements.

The class containing the identity element consists of just one element, because $geg^{-1} = e$ for all g in G. The even and odd permutations in S_3 form separate classes. All rotations by the same angle around the different axes going through the origin in three-dimensional space belong to the same equivalence class in the rotation group.

The subgroup H of G is said to be *conjugate* to the subgroup K if there is an element g such that $K = gHg^{-1}$. If H is an invariant subgroup, then it is conjugate to itself. As was noted in Section 2, even permutations form an invariant subgroup in S_3. In the three-dimensional rotation group, the subgroup of rotations around any given axis is conjugate to the $O(2)$-like subgroup consisting of rotations around the z axis.

Let p be a point of the vector space X. The maximal subgroup $G^{(p)}$ of G which leaves p invariant, i.e., $G^{(p)}p = p$, is called the *little group* of G at p. In the three-dimensional x, y, z space, rotations around the z axis form the $O(2)$-like little group of $SO(3)$ at $(0, 0, 1)$. This little group is conjugate to the little group at $(1, 0, 0)$ which consists of rotations around the x axis.

The set of points V that can be reached through the application of G on a single point p in X is called the *orbit* of G at p. Two orbits are either identical or disjoint. The orbit of $SO(3)$ at $(0, 0, R)$ is the surface of the sphere with radius R centered around the origin. The orbit of $SO(2)$ at the point $(x = a, y = 0)$ on the xy plane is the circumference of a circle with radius a.

The coset space $G/G^{(p)}$ is therefore identical to orbit V. As was noted in Section 2, the coset space $SO(3)/SO(2)$ is the surface of a unit sphere in three-dimensional space.

Let us see how these concepts are helpful in understanding the $SE(2)$ group. The little group of $SE(2)$ at $x = y = 0$ is the rotation subgroup represented by Equation (2.3) satisfying the relation

$$\begin{bmatrix} 0 \\ 0 \\ 1 \end{bmatrix} = \begin{bmatrix} \cos\theta & -\sin\theta & 0 \\ \sin\theta & \cos\theta & 0 \\ 0 & 0 & 1 \end{bmatrix} \begin{bmatrix} 0 \\ 0 \\ 1 \end{bmatrix}. \qquad (3.1)$$

The orbit of $SE(2)$ at the origin is the entire plane, because every point on the plane can be reached through a translation of the origin:

$$\begin{bmatrix} u \\ v \\ 1 \end{bmatrix} = \begin{bmatrix} 1 & 0 & u \\ 0 & 1 & v \\ 0 & 0 & 1 \end{bmatrix} \begin{bmatrix} 0 \\ 0 \\ 1 \end{bmatrix}. \qquad (3.2)$$

The coset space $SE(2)/SO(2)$ is therefore the entire plane.

Rotations by the same angle around two different points on the plane belong to the same equivalence class in the $SE(2)$ group. For instance, the rotation around the origin is represented by the matrix of Equation (3.1). The rotation around (u, v) is

$$\begin{bmatrix} 1 & 0 & u \\ 0 & 1 & v \\ 0 & 0 & 1 \end{bmatrix} \begin{bmatrix} \cos\theta & -\sin\theta & 0 \\ \sin\theta & \cos\theta & 0 \\ 0 & 0 & 1 \end{bmatrix} \begin{bmatrix} 1 & 0 & -u \\ 0 & 1 & -v \\ 0 & 0 & 1 \end{bmatrix}$$

$$= \begin{bmatrix} \cos\theta & -\sin\theta & -u\cos\theta - v\sin\theta + u \\ \sin\theta & \cos\theta & -u\sin\theta - v\cos\theta - v \\ 0 & 0 & 1 \end{bmatrix}. \qquad (3.3)$$

Clearly, the rotation around the point (u, v) is conjugate to the rotation around the origin. In matrix language, the above operation is known as a *similarity* transformation.

Let us consider another example. The group consisting of four-by-four Lorentz transformation matrices is called the Lorentz group. We can now consider a four-vector P and the maximal subgroup which leaves this four-vector invariant. This subgroup is the little group of the Lorentz group at P.

Elements of Group Theory

If P is the four-momentum of a free massive particle, and if this particle is at rest, the little group is the three-dimensional rotation group. If we perform Lorentz transformations on this, then the four-momentum traces the hyperbolic surface:

$$P_0^2 - |\mathbf{P}|^2 = M^2, \tag{3.4}$$

where M, P_0, and $\mathbf{P}$ are the mass, energy, and momentum of the particle respectively. This hyperbolic surface is the orbit. The matrices of the little group are Lorentz-boosted rotation matrices. We shall study the little groups and orbits of the Lorentz group in more detail in Chapters III and IV.

4. Representations and Representation Spaces

A set of linear transformation matrices homomorphic to the group multiplication of G is called a *representation* of the group. The word "homomorphic" is appropriate here because the matrices need not have a one-to-one correspondence with the group elements. In addition, there can be more than one set of matrices forming a representation of the group. For instance, the three-dimensional rotation group can be represented by two-by-two, three-by-three, or n-by-n matrices, where n is an arbitrary integer.

$GL(n, c)$ and $GL(n, r)$ are the groups of non-singular n-by-n complex and real matrices respectively. $SL(n, c)$ is an invariant subgroup of $GL(n,c)$ with determinant 1, and is called the *unimodular group*. These matrices perform linear transformations on a linear vector space V containing vectors of the form:

$$X = (x_1, x_2, x_3, \ldots, x_n). \tag{4.1}$$

$U(n)$ is the unitary subgroup of $GL(n, c)$ which leaves the norm $= (|x_1|^2 + |x_2|^2 + \cdots + |x_n|^2)^{1/2}$ invariant. $O(n)$ is a subgroup of $U(n)$ in which all elements are real, and $SO(n)$ is the unimodular subgroup of $O(n)$.

$U(n, m)$ is the *pseudo-unitary group* applicable to the $(n + m)$-dimensional vector space which leaves the quantity $[(|x_1|^2 + |x_2|^2 + \cdots + |x_n|^2) - (|y_1|^2 + |y_2|^2 + \cdots + |y_m|^2)]$ invariant. $O(n, m)$ is the real subgroup of $U(n, m)$. The group of four-by-four Lorentz transformation matrices is $O(3, 1)$.

In addition to the above mentioned homogeneous linear transformations, there are inhomogeneous linear transformations. The $SE(2)$ transformation defined in Equation (1.3) is an inhomogeneous linear transformation in that the u and v variables are added to x and y respectively after a rotation. Inhomogeneous linear transformations can also be represented by matrices as can be seen in the case of $SE(2)$. Properties of this kind of matrix transformation are not well known to physicists. While avoiding general theorems on this subject, we shall study some special cases of the inhomogeneous transformations, including those of the inhomogeneous Lorentz transformations.

We shall use the symbol L to denote a representation of G. L therefore consists of matrices acting on the vector space V. A closed subspace W of V will be said to be *invariant* if the action of L on any element in W for all elements in W transforms into an element belonging to W. If we restrict the operations of L to W, we obtain a new representation L^W whose vector space is W. We call L^W a subrepresentation of L. If a representation does not contain subrepresentations, it is said to be *irreducible*.

Let us consider the $n = 2$ state hydrogen atom consisting of spinless proton and electron. This energy state has four degenerate states with two different values of the angular momentum quantum number ℓ. There is only one state for $\ell = 0$ which is usually called the s state. There are three different states for the p state with $\ell = 1$. Under rotations, the s state remains invariant, and the p state wave functions undergo homogeneous linear transformations. The p state wave functions never mix with the s state. Thus we say that the $n = 2$ hydrogen wave functions can be divided into two invariant subspaces, and the rotation matrices become reduced to the three-by-three matrix for the p state and a trivial one-by-one matrix for the s state.

In addition to vectors, there are tensors. For example, the second rank tensor is formed from the *direct product* of two vector spaces. Let V and V' be two invariant vector spaces with n and m components respectively, and let L and L' be representations of the same group. The direct product of these two spaces results in a set of $x_i y_j$, where $i = 1, \ldots, n$ and $j = 1, 2, \ldots, m$. The question of whether these elements form an nm-dimensional vector space to which nm-by-nm matrices are applicable, whether these matrices are homomorphic to the multiplication law of the group G, and whether this matrix is reducible is one of the prime issues in representation theory. We are already familiar with some aspects of this through our experience with the rotation group.

Let us consider two spin-1/2 particles. Because each particle can have two different spin states, the dimension of the resulting space is 4. However, this space can be divided into one corresponding to spin-0 state and three for the spin-1 states. The rotation matrix for each of the spinors is two-by-two. The direct product of the two two-by-two matrices result in a four-by-four matrices which can be reduced to a block diagonal form consisting of one three-by-three matrix and one trivial one-by-one matrix. In quantum mechanics, this procedure is known as the calculation of Clebsch-Gordon coefficients. Calculations of Clebsch-Gordan coefficients occupy a very important part of mathematical physics, and the literature on this subject is indeed extensive. For this reason, we shall not elaborate further on this problem.

If L consists of finite-dimensional matrices acting on finite-dimensional vector spaces, such as those of $GL(n, c)$ and its subgroups, the representation is said to be finite-dimensional. If L is reducible, the matrices can be brought to a block diagonal form. In addition to finite-dimensional matrices, we have to consider the possibility of the size of the representation matrices becoming infinite. In this case, the representation is called infinite-dimensional. The technique of handling infinite-dimensional mtarices is far more complicated than that for the finite-dimensional case. We shall discuss some of the infinite-dimensional matrices in later chapters without going into full-fledged representation theory.

The ultimate purpose of physics is to calculate numbers which can be measured in laboratories. Therefore we have to convert the concept of abstract group theory into measurable numbers. This is only possible through construction of representations or matrices. The most common practice in constructing such matrices is solving the eigenvalue equations. Solving the time-independent Schrödinger equation is a process of constructing representations. Wave functions correspond to vector spaces. Construction of vector spaces often precedes that of the matrices representing the group. Therefore we often call this *representation space*. The spherical harmonics with a given value of ℓ form a representation space for $(2\ell + 1)$-by-$(2\ell + 1)$ matrices representing the three-dimensional rotation group.

There are many different methods of constructing representations, and many of them are based on the techniques of deriving new representations of the same or a different group starting from known representations. The most common practice is the above-mentioned direct product of two representations. As we demonstrated in the case of $SE(2)$, the semi-direct product is also used often in physics, especially in the study of the inhomogeneous Lorentz group. We shall elaborate on this in later chapters.

It is also quite common to use the exponentiation of matrices. Let us consider a set of matrices A. Its exponentiation results in another set of matrices of the form

$$B = e^A = \sum_{n=0}^{\infty} A^n/n!. \tag{4.2}$$

This method is quite convenient when the matrices A take a simpler form than that of B. This point will be studied in greater detail in Chapter II.

The use of complex numbers is often helpful in constructing representations. Unitary or orthogonal matrices are much easier to deal with than other forms of matrices. Therefore we sometimes study a pseudo-unitary matrix by converting it to a unitary matrix. The best known example is the group of Lorentz transformations. Lorentz transformation matrices constitute a four-by-four pseudo-orthogonal representation of $O(3, 1)$. However, if we change the time variable t to $i(t)$, the transformation matrix becomes orthogonal. We shall discuss this method in Chapter IV.

Another useful method in constructing representations is the method of group contraction. The surface of the earth appears flat most of the time.

However, it is a spherical surface. We are therefore led to the idea that the $SE(2)$ group which deals with transformations on a flat surface is an approximation of those on a spherical surface. Therefore, $SE(2)$ may be a limiting case of the rotation group. This process and its physical relevance are discussed in Chapter VIII.

Most of the matrix calculations are straightforward. However, sometimes, it is easier to do the same calculation with different mathematics. For instance, the matrix algebra of the $SE(2)$ group which we carried out using three-by-three matrices throughout this chapter can be translated into the algebra of the complex variable $z = x + iy$. Another examnple is the $SL(2, c)$ group. This group consists of two-by-two matrices of the form

$$\begin{bmatrix} a & b \\ c & d \end{bmatrix}, \tag{4.3}$$

with the condition: $ad - bc = 1$. The algebra of this matrix can be that of the conformal transformation (Bargmann, 1947):

$$w = \frac{az + b}{cz + d}. \tag{4.4}$$

Quite often in physics, we use the differential form for operators. For instance, momentum and angular momentum operators are written in terms of differential operators. Where do these operators stand in group theory? This question will be addressed in Chapter II on Lie groups.

When we construct representations, we are constructing matrices. Then it is important to know to what vector or representation space these matrices are applied. Indeed, the construction of a representation cannot be separated from the construction of its representation space. In dealing with this vector space, it is convenient to use the *basis vectors*. Every vector in the vector space can be written as a linear combination of the basis vectors. The basis vectors in the three-dimensional Cartesian system are usually discussed in Freshmen physics. The spherical harmonics constitute the basis vectors for a given integer ℓ representation of the rotation group.

There are many other forms for basis vectors. Let us for example consider an arbitrary polynomial in the complex variable z:

$$F_n(z) = a_0 + a_1 z^1 + a_2 z^2 + \cdots a_n z^n. \tag{4.5}$$

This is a linear combination of z^k, with $k = 0, 1, 2, \ldots, n$. Thus each z^k can be regarded as a basis vector. Thus a linear transformation on this representation space should result in rearrangement of the coefficients $a_0, a_1, \ldots, a_n$. This point is discussed in detail in Chapter IV.

In terms of the basis vectors, a similarity transformation is a rearrangement of the basis vectors. In the three-dimensional rotation group, a similarity transformation is a rotation of the coordinate system. In the

Elements of Group Theory

two-dimensional Euclidean space, a similarity transformation with translation matrix results in a translation of the origin. Again in the three-dimensional rotation group, it is more convenient to use the combinations $(x + iy)$, $(x - iy)$ and z, instead of x, y, and z coordinate system. This is also a similarity transformation.

5. Properties of Matrices

Matrices are in general rectangular, i.e., the number of rows does not have to be the same as that of columns. We shall in fact be using one of them in Chapter II. However, in order to represent a group, every matrix in the set has to have its inverse. Therefore, matrices representing a group have to be square and non-singular. This means that the determinant of the matrix cannot vanish.

It is easy to study a matrix if it is of diagonal form. Therefore it is of prime importance to see whether any given matrix can be diagonalized by a similarity transformation or a conjugate operation. Any matrix M defined over a complex vector space can be brought to diagonalized form by a similarity transformation provided that M commutes with its Hermitian conjugate, i.e. $MM^\dagger = M^\dagger M$. Diagonal elements of the diagonalized matrices are the eigenvalues of the matrices.

Those matrices with which physicists are most familiar are unitary and Hermitian matrices. Unitary matrices U satisfy the condition:

$$U^\dagger = U^{-1}, \quad \text{or} \quad U^\dagger U = I. \tag{5.1}$$

Hermitian matrices are of the type:

$$H = H^\dagger. \tag{5.2}$$

Both unitary and Hermitian matrices can be brought to a diagonal form through a similarity transformation:

$$S H S^{-1} = \text{Diagonal}, \tag{5.3}$$

$$S U S^{-1} = \text{Diagonal}. \tag{5.4}$$

S in this case is a unitary matrix. The eigenvalues of unitary matrices have unit modulus. The eigenvalues of Hermitian matrices are real. It is possible to write a unitary matrix as

$$U = \exp(-iH). \tag{5.5}$$

There are also anti-Hermitian matrices:

$$A = -A^\dagger. \tag{5.6}$$

The exponentiation of the anti-Hermitian matrix is not unitary. However, an anti-Hermitian matrix commutes with its Hermitian conjugate, and therefore can be diagonalized by a similarity transformation.

In dealing with inhomogenous linear transformations, we often use triangular matrices. For example, if we choose the bases to be $x + iy$ and $x - iy$ on the two-dimensional Euclidean plane as discussed in Section 2, then the rotation matrix becomes diagonal:

$$R(\theta) = \begin{bmatrix} e^{-i\theta} & 0 & 0 \\ 0 & e^{i\theta} & 0 \\ 0 & 0 & 1 \end{bmatrix}, \tag{5.7}$$

and the translation matrix becomes

$$T(u, v) = \begin{bmatrix} 1 & 0 & u + iv \\ 0 & 1 & u - iv \\ 0 & 0 & 1 \end{bmatrix}. \tag{5.8}$$

This translation matrix is an upper triangular matrix in the sense that it has only zero elements below the main diagonal. The semi-direct product of the rotation and translation becomes the triangular matrix:

$$T(u, v) R(\theta) = \begin{bmatrix} e^{-i\theta} & 0 & u + iv \\ 0 & e^{i\theta} & u - iv \\ 0 & 0 & 1 \end{bmatrix}. \tag{5.9}$$

A lower triangular matrix has only zero elements above the main diagonal. Throughout this book, by triangular matrix, we shall mean an upper triangular matrix unless otherwise specified.

Matrix multiplication of two triangular matrices yields a triangular matrix. Thus there can be groups of triangular matrices, provided an inverse exists for each matrix.

Unlike the above-mentioned unitary, Hermitian and anti-Hermitian matrices, triangular matrices cannot be brought to diagonal form by a similarity transformation. However, every non-zero matrix B defined over a complex vector space is unitarily equivalent (similarity transformation with the transformation matrix being unitary) to an upper triangular matrix if not a diagonal matrix. The diagonal elements of this triangular matrix are the eigenvalues of B.

Among triangular matrices, there are those with zero diagonal elements, such as

$$T_0(3) = \begin{bmatrix} 0 & a & b \\ 0 & 0 & c \\ 0 & 0 & 0 \end{bmatrix}, \tag{5.10}$$

Elements of Group Theory

This matrix has the property that

$$[T_0(3)]^3 = 0. \tag{5.11}$$

In general, for an n-by-n triangular matrix with zero diagonal elements,

$$[T_0(n)]^n = 0. \tag{5.12}$$

Therefore, the exponentiation of the T_0 matrix yields

$$\exp(T_0) = \sum_{k=0}^{n-1} \frac{1}{k!} (T_0)^k. \tag{5.13}$$

The series truncates!

In physics, the size of the matrix is an important factor in carrying out calculations, and also in illustrating mathematical theorems. As the size of the matrix becomes larger, the problem becomes more complicated. For this reason, we seek constantly the smallest matrix with which we can do our calculations. There are two important methods which are most commonly employed. One is to seek whether the matrices can be reduced to block diagonal form. This method is called the construction of irreducible representation which we shall discuss in Section 6. Another method commonly used in physics is the method of covering group. The best known example of this method is to use the two-by-two unitary unimodular matrices or $SU(2)$ for the rotation group. We shall discuss this problem in more detail in later chapters.

In addition, there are infinite-by-infinite matrices applicable to infinite-dimensional vector spaces. This kind of matrix is quite common in physics. Fourier transformations, which cannot be separated from quantum mechanics, are applications of the infinite-dimensional matrix. The so-called S matrix which relates the initial and final states in scattering processes is also an infinite-dimensional matrix. As we shall see in later chapters, certain representations of the Lorentz group are infinite-dimensional.

6. Schur's Lemma

We discussed in Section 4 the fact that a set of matrices or linear transformations on a vector space V_m, homomorphic to the group multiplication of g, constitutes a representation of the group. The dimension of the vector space is denoted by m. We shall denote this set of matrices or linear transformations as $L(s)$ where the parameter s may have a domain which is finite or infinite, denumerable or continuous. For the permutation group S_3, the domain of s is denumerable and finite, while the domain is continuous and infinite for the three-dimensional rotation group.

If we consider $L'(s)$ to be another set of matrices which also forms a representation of the same group over another vector space V'_n, then the two representations are said to be equivalent if V_m can be mapped onto V'_n by a one-to-one linear transformation. This means that any two vectors related by the transformation $L(s)$, e.g., x and $x' = L(s)\,x$, map into related vectors y and $y' = L'(s)\,y$ with the same value of s. Then m must equal n, otherwise there would not be any linear one-to-one mapping from V_m to V'_n. The equivalence mapping is accomplished as usual by a non-singular matrix S:

$$x = S\,y, \tag{6.1}$$

and hence

$$L'(s) = S^{-1} L(s)\, S. \tag{6.2}$$

In Section 4 we noted also that the set of matrices or linear transformation $L(s)$ is called reducible if there exists an invariant subspace W of V, i.e., if every vector of W maps into a vector of W under each of the transformations L. If there exists no invariant subspace of V except V and the null vector, the set $L(s)$ is said to be irreducible. The set is reducible if it can be decomposed into several irreducible systems. The corresponding vector space V is then a direct sum of irreducible invariant linear subspaces.

The construction of nonequivalent irreducible representations is one of the fundamental problems in the representation theory of groups. In addition, a practical method must be devised for decomposing a reducible representation into irreducible representations. Schur's Lemma, which consists of the following two theorems, yields via the second theorem, a practical method for determining if a given representation is irreducible.

The *first theorem* consists of the following. Let $L(s)$ and $L'(s)$ be two irreducible representations of the group G defined on the vector spaces V_m and V'_n respectively. Let S be a (rectangular) matrix of n rows and m columns, mapping V into V', i.e. for every s,

$$S\,L(s) = L'(s)\,S. \tag{6.3}$$

Then either $S = 0$ or S is nonsingular. If S is nonsingular, $m = n$ and $L(s)$ and $L'(s)$ are equivalent. That this is true can be seen from the following.

Consider the matrix S of rank r. Then all the transformations Sx where x is a vector in V form a linear subspace y in V' [$y = Sx$] of dimension r which by our theorem is invariant under $L'(s)$. Since $L'(s)$ is irreducible, either $y = 0$ (then $r = 0$ and $S = 0$) or $y = V'$ (then $r = m$ and $n \geq m$). The vectors x of V for which $0 = Sx$ constitute a linear subspace x_1 of V of dimension $(n - r)$ which is invariant under $L(s)$ by Equation (6.3). Since $L(s)$ is irreducible, either $x = V$ (then $r = 0$ and $S = 0$) or $x = 0$ (then $r = n$ and $m \geq n$). Thus either $S = 0$ or $r = n = m$, and S is nonsingular. Hence $L(s)$ and $L'(s)$ are equivalent. Since we are free to choose the coordinate systems as we like, we choose that coordinate system in which $L(s) = L'(s)$. We use this form of the equivalence statement in what follows.

Elements of Group Theory

The theorem which constitutes the second part of Schur's Lemma is extremely useful, because it provides a practical method for determining if a group representation is irreducible. This *second theorem* states that, if $L(s)$ is a representation of the group G on a complex vector space V, $L(s)$ is irreducible if and only if the only transformation $SV \to V$ such that

$$L(s) S = S L(s) \qquad (6.4)$$

for all x in V are those for which S is a multiple of the identity. This says that $S = \lambda I$, where λ is a complex number and I is the identity matrix on V.

We know from linear algebra that a linear operator, such as matrix, operating on a finite-dimensional complex vector space always has at least one eigenvalue. Suppose that λ is that eigenvalue of S which satisfies Equation (6.4). Then there is a subspace C_λ of V consisting of vectors which satisfy

$$S \xi = \lambda \xi, \qquad (6.5)$$

for all ξ in C_λ. This means that C_λ is a subspace of V with dimension greater than 0. Also, C_λ is invariant under $L(s)$ because

$$S L(s) \xi = L(s) S \xi = \lambda L(s) \xi. \qquad (6.6)$$

If $L(s)$ is ireducible, then there is a parameter s such that $x = [L(s) \xi]$ for x in V. Thus the subspace $C_\lambda = V$.

Conversely, suppose that $L(s)$ is reducible. Then we know that V can be decomposed into a direct sum of invariant subspaces. Call one of these V_1. Any vector x in V can also be written uniquely as a direct sum of its vector components in each invariant subspace. Call x_1 the component in V_1. It is always possible to define a projection operator P on V such that $Px = x_1$ in V_1. Then $P L(s) x = L(s) P x = L(s) x_1$. Since, however, P cannot be a multiple of identity, $L(s)$ must be irreducible.

7. Exercises and Problems

Exercise 1. Construct the maximal set of commuting operators for S_3.

For the system of three similar objects labeled as 1, 2, 3 respectively, we can perform six permutations. First, there are three permutations of the form

$$(12), \quad (23), \quad (31), \qquad (7.1)$$

where each number is replaced by the succeeding number in parentheses, while the first one goes to the last position. In addition, there are two permutations of the form

$$(123), \quad (132). \qquad (7.2)$$

The above five permutations together with the identity form the six permutations which can be performed on the three objects.

As Dirac did in his Equation (13) of Section 55 in his book entitled *Principles of Quantum Mechanics* (1958), we construct the following operators for the three-body system:

$$X_1 = I,$$
$$X_2 = [(12) + (23) + (31)]/3, \qquad (7.3)$$
$$X_3 = [(123) + (132)]/2,$$

where I is the identity operator. X_1 can also be written as

$$X_1 = [I + (12) + (23) + (31) + (123) + (132)]/6. \qquad (7.4)$$

The above operators commute with every permutation, and therefore with one another. They form the maximal set of commuting operators, the eigenvalues of these operators specify the representation.

Exercise 2. S_3 deals with three objects. Therefore it must be possible to construct three-by-three matrices performing permutations. Construct the multiplication table.

The identity matrix is

$$I = \begin{bmatrix} 1 & 0 & 0 \\ 0 & 1 & 0 \\ 0 & 0 & 1 \end{bmatrix}. \qquad (7.5)$$

The even permutations (123) and (132) are respectively

$$A = \begin{bmatrix} 0 & 0 & 1 \\ 1 & 0 & 0 \\ 0 & 1 & 0 \end{bmatrix}, \quad B = \begin{bmatrix} 0 & 1 & 0 \\ 0 & 0 & 1 \\ 1 & 0 & 0 \end{bmatrix}. \qquad (7.6)$$

The odd permutations (23), (13) and (12) are

$$C = \begin{bmatrix} 1 & 0 & 0 \\ 0 & 0 & 1 \\ 0 & 1 & 0 \end{bmatrix}, \quad D = \begin{bmatrix} 0 & 0 & 1 \\ 0 & 1 & 0 \\ 1 & 0 & 0 \end{bmatrix}, \quad E = \begin{bmatrix} 0 & 1 & 0 \\ 1 & 0 & 0 \\ 0 & 0 & 1 \end{bmatrix}. \qquad (7.7)$$

Using these matrices, we can construct the multiplication table of Table 7.1. It is clear from this table that S_3 is closed under group operation, and that I, A and B form a subgroup.

Elements of Group Theory

TABLE 7.1
Group Multiplication Table for S_3. If we multiply one
element by another, we end up with an element in the group.
I, A and B form a subgroup.

	A	B	C	D	E
A	B	I	D	E	C
B	I	A	E	C	D
C	E	D	I	B	A
D	C	E	A	I	B
E	D	C	B	A	I

Exercise 3. Using Schur's Lemma check whether the above three-by-three representation of S_3 is irreducible.

We can start with the most general three-by-three matrix:

$$F = \begin{bmatrix} a & b & c \\ d & e & g \\ g & h & k \end{bmatrix}, \tag{7.8}$$

and impose the condition that

$$AF = FA, \quad BF = FB, \ldots,$$

for all five non-trivial matrices given in Equations (7.6) and (7.7). Then the result is that F is a multiple of I or

$$G = \begin{bmatrix} 1 & 1 & 1 \\ 1 & 1 & 1 \\ 1 & 1 & 1 \end{bmatrix}, \quad \text{or} \quad H = \begin{bmatrix} 0 & 1 & 1 \\ 1 & 0 & 1 \\ 1 & 1 & 0 \end{bmatrix}. \tag{7.9}$$

We could have calculated these matrices using the expressions for X_2 and X_3 of Equation (7.3). The above matrices commute with all six permutations, but are not diagonal. What went wrong? The answer to this question is simple. H in the above expression is a linear combination of I and G. G is a singular matrix. Therefore, the only non-singular matrix which commutates with all the elements is the unit matrix or a multiple thereof. Therefore, according to Schur's lemma, the matrices given in Exercise 2 form an irreducible representation. In Chapter XI, we shall discuss representations for which X_2 and X_3 are not singular.

Exercise 4. We noted in Section 5 that a multiplication of two triangular matrices leads to another triangular matrix. Is it possible to construct a triangular matrix by taking a product of diagonalizable non-triangular matrices?

We shall point out that this is possible by giving a concrete example. Let us consider the following three matrices:

$$A = \begin{bmatrix} \cosh(\eta/2) & -\sinh(\eta/2) \\ -\sinh(\eta/2) & \cosh(\eta/2) \end{bmatrix}. \tag{7.10}$$

$$B = [1/\cosh(\eta)]^{1/2} \begin{bmatrix} \cosh(\eta/2) & -\sinh(\eta/2) \\ \sinh(\eta/2) & \cosh(\eta/2) \end{bmatrix}. \tag{7.11}$$

$$C = \begin{bmatrix} [\cosh \eta]^{1/2} & 0 \\ 0 & [\cosh \eta]^{-1/2} \end{bmatrix}. \tag{7.12}$$

If we take the product of these matrices:

$$D = ABC, \tag{7.13}$$

then the result is

$$D = \begin{bmatrix} 1 & -\tanh \eta \\ 0 & 1 \end{bmatrix}. \tag{7.14}$$

In mathematics, this operation is called the Iwasawa decomposition (see Iwasawa, 1949; Hermann, 1966). The physics of the above form is discussed in Chapter VII.

Exercise 5. One of the most common representations of the three-dimensional rotation group consists of three-by-three orthogonal matrices applicable to a three-component column vector consisting of three real numbers.

Show this three-component vector can be represented by a two-by-two unimodular Hermitian matrix consisting also of three-independent real numbers.

Let us consider the Pauli spin matrices. In addition to their well known Hermiticity and commutativity properties, they satisfy the orthogonality condition:

$$\tfrac{1}{2}\mathrm{Tr}(\sigma_i\sigma_j) = \sigma_{ij}. \tag{7.15}$$

Therefore, the Pauli spin matrices can serve as three basis vectors for the rotation group, and a vector **A** with three real components A_1, A_2, and A_3, can be written as

$$[A] = \begin{bmatrix} A_3 & A_1 - iA_2 \\ A_1 + iA_2 & -A_3 \end{bmatrix}, \tag{7.16}$$

with

$$\tfrac{1}{2}\mathrm{Tr}([A]^\dagger[B]) = A_1 B_1 + A_2 B_2 + A_3 B_3. \tag{7.17}$$

The rotation of the above matrix is performed by a unitary matrix $U(\theta, \phi, \alpha)$:

$$[A'] = U[A]U^\dagger, \tag{7.18}$$

where

$$U(\theta, \phi, \alpha) = \begin{bmatrix} e^{-i(\phi+\alpha)/2}\cos\dfrac{\theta}{2} & e^{-i(\phi-\alpha)/2}\sin\dfrac{\theta}{2} \\ -e^{i(\phi-\alpha)/2}\sin\dfrac{\theta}{2} & e^{i(\phi+\alpha)/2}\cos\dfrac{\theta}{2} \end{bmatrix}.$$

It is quite clear that the trace of Equation (7.17), which represents the inner product, is invariant under the unitary transformation given in Equation (7.18).

The matrix form for a vector is quite common in physics. If this procedure is generalized to $SU(3)$, the above procedure is the "eightfold-way" representation of elementary particles [Gell-Mann (1961 and 1962), Ne'eman (1961)]. This procedure can also be extended to four-vectors in Minkowskian space, as we shall discuss in Chapter IV.

Problem 1. Write out explicitly the multiplication table associated with the rotation through angles of 0°, 120° and 240° and reflections at each vertex of an equilateral triangle. From the multiplication table, determine the number of subgroups.

Problem 2. Using Schur's lemma, show that the following six matrices constitute an irreducible representation of S_3 (Wigner, 1959).

$$\begin{bmatrix} 1 & 0 \\ 0 & 1 \end{bmatrix}, \quad \begin{bmatrix} 1 & 0 \\ 0 & -1 \end{bmatrix}, \quad \begin{bmatrix} -1/2 & \sqrt{3}/2 \\ \sqrt{3}/2 & 1/2 \end{bmatrix},$$

$$\begin{bmatrix} -1/2 & -\sqrt{3}/2 \\ -\sqrt{3}/2 & 1/2 \end{bmatrix}, \quad \begin{bmatrix} -1/2 & \sqrt{3}/2 \\ -\sqrt{3}/2 & -1/2 \end{bmatrix},$$

$$\begin{bmatrix} -1/2 & -\sqrt{3}/2 \\ \sqrt{3}/2 & -1/2 \end{bmatrix}. \tag{7.19}$$

Problem 3. The $SL(2, c)$ group consists of complex two-by-two matrices of the form given in Equation (4.3). Show how this group corresponds to the conformal transformation of Equation (4.4). Do they have the same algebraic property? Is the correspondence one-to-one?

Problem 4. Show that $SL(2, r)$ consisting of two-by-two real unimodular matrices is unitarily equivalent to the subgroup of $SL(2, c)$ of the form

$$\begin{bmatrix} a & b \\ b^* & a^* \end{bmatrix}. \tag{7.20}$$

Problem 5. Assume that $c = [c_{ij}]$ is any square matrix over a complex vector space. Show that C can be written as $A + iB$, where A and B are Hermitian.

Problem 6. If we take a direct product of two $SU(2)$ spinors, the product is reducible to the symmetric component with spin-1 and the spin-0 antisymmetric state. Construct rotation matrices for this reducible representation. Show that the rotation matrix for the spin-0 state is 1, and that those for the spin-1 states are equivalent to the three-by-three rotation matrix. See Section 3 of Chapter IV.

Problem 7. In Equation (4.5), we noted that an arbitrary polynomial in the complex variable z:

$$F_n(z) = a_0 + a_1 z^1 + a_2 z^2 + \ldots a_n z^n.$$

can be regarded as a vector with the basis vectors z^k. Show that the replacement of z by $(z - z_0)$ constitutes a similarity transformation. Calculate the matrix which performs this similarity transformation when $n = 2$. Explain why this matrix has to be triangular. Calculate the inverse of the transforma-

Elements of Group Theory

tion matrix. Is the inverse matrix also triangular? See discussions in Chapter VIII.

Problem 8. Repeat Problem 7 when the transformation is

$$z' = \alpha z - z_0. \tag{7.21}$$

See Gilmore (1974).

Problem 9. Let us consider a quadratic form x defined for the interval $-1 < x < 1$:

$$G(x) = a_0 + a_1 x^1 + a_2 x^2. \tag{7.22}$$

This form can also be expressed as a linear expansion in terms of the Legendre polynomials:

$$G(x) = b_0 P_0(x) + b_1 P_1(x) + b_2 P_2(x). \tag{7.23}$$

Now the coefficients b_1, b_2 and b_3 are linear combinations of a's. Show that this is a similarity transformation and calculate the transformation matrix.

Problem 10. Problem 9 is an example of the procedure generally known as the *Gramm-Schmidt* orthonormalization (Gilmore, 1974). Calculate two lowest non-trivial order Hermite and Laguerre polynomials using the Gramm-Schmidt procedure.

Problem 11. Show that $SE(2)$ discussed throughout this chapter is isomorphic to the group of Galilei transformations in two-dimensional space. See Section 6 of Chapter VIII.

Problem 12. We are familiar with the complex z ($= x + iy$) plane. For a given complex number z, show that $SE(2)$ is isomorphic to the group consisting of multiplication by a factor of unit modulus and addition by a complex number.

Problem 13. Show that the matrix of $E(2)$ given in Equation (1.4) has the same algebraic property as that of

$$\begin{bmatrix} e^{-i\theta/2} & (u - iv)e^{i\theta/2} \\ 0 & e^{i\theta/2} \end{bmatrix}, \tag{7.24}$$

or

$$\begin{bmatrix} e^{-i\theta/2} & 0 \\ (u + iv)e^{-i\theta/2} & e^{i\theta/2} \end{bmatrix}. \tag{7.25}$$

Problem 14. Let us go back to Equation (4.2) where two matrices A and B are related by

$$B = e^A. \tag{7.26}$$

Show that the determinant of B is $e^{\mathrm{Tr}(A)}$.

Problem 15. Addition of numbers can be converted to multiplication if we use an exponential function. Show that the mtarix of the form:

$$\begin{bmatrix} 1 & a \\ 0 & 1 \end{bmatrix} \tag{7.27}$$

also converts addition into a matrix multiplication.

Problem 16. Let A, B be two n-by-n matrices. We call

$$[A, B] = AB - BA \tag{7.28}$$

the commutator of the two matrices. Show that

$$[[A, B], C] + [[B, C], A] + [[C, A], B] = 0. \tag{7.29}$$

This is often called the Jacobi identity.

Chapter II

Lie Groups and Lie Algebras

This chapter is a continuation of Chapter I on the basic elements of group theory, and is devoted to continuous groups whose elements depend on a given number of continuous parameters. We are already familiar with some aspects of continuous groups from our experience with the three-dimensional rotation group.

Because the parameters are continuous, we ae led to ask whether there exists differential and integral calculus for groups with continuous parameters. The question then is whether it is possible to develop the mathematics of converting a continuous group into a theory of infinitesimal transformations from which the original group can be reconstructed. If this process is possible, the group is called a Lie group.

Because most of the groups in modern physics are Lie groups, there are already many excellent textbooks on Lie groups for physicists, and it is not necessary to prove all the theorems. As in Chapter I, we shall discuss a selected set of examples in order to elucidate the basic concepts of Lie groups. The Lie group in which we are most interested in this book is the inhomogeneous Lorentz group which is often called the Poincaré group. Fortunately, the rotation group is a subgroup of the Poincaré group, and we can expand our knowledge from what we know about the three-dimensional rotation group. The Poincaré group has many properties which are shared by the two-dimensional Euclidean group. For this reason, we continue to use the $E(2)$ group as the prime example.

In Section 1, we explain the basic concepts of Lie groups using the $E(2)$ group as a specific example. The concept of generators is introduced. In Section 2, we give Lie's theorems which enable us to construct a Lie group by using only its generators. It is pointed out that the commutation relations for the generators, which are called Lie algebras, determine the multiplication properties of the group. Section 3 states group properties, such as subgroups and their invariance in terms of the Lie algebra. Section 4 explains how the properties of the Lie algebra are translated into those of the group.

In addition, there are many other theorems which we intend to use without proof. In Section 5, we list some of them. Indeed, the purpose of this book is to give concrete illustrations for those mathematical theorems the average physicist does not wish to prove. Section 6 contains exercises and problems which will further serve as illustrative examples for Lie groups.

1. Basic Concepts of Lie Groups

Many of the groups in physics are continuous groups. The rotation and translation groups are continuous groups. The Lorentz and Poincaré groups which we intend to study in this book are also continuous groups. Continuous groups depend on a set of continuous parameters. The number of parameters which dictates the form of the transformation matrix is not in general equal to the dimension of the vector space to which the matrix is applied. For example, the spinor rotation matrices have three parameters while acting on a two-dimensional space.

A Lie group is a connected component of a continuous group which can be continuously connected with the identity element. $O(3)$ is only a continuous group while $SO(3)$ is a Lie group. In studying Lie groups, we are interested not only in linear transformations on coordinates, but also in transformations of functions which depend on the coordinate variables. For example, the rotation of the $\ell = 1$ spherical harmonics $Y_1^m(\theta, \phi)$ can be regarded as a coordinate transformation, but the rotation of Y_ℓ^m with $\ell > 1$ is the same transformation of a function depending on the coordinate variables.

We are interested first in linear transformations on a set of n variables x_0^i ($i = 1, \ldots, n$), which may be regarded as the coordinates of a point in a certain space. Consider now the set of equations

$$x^i = f^i(x_0^1, \ldots, x_0^n; \alpha^1, \ldots, \alpha^r), \tag{1.1}$$

in which α^ρ appear as a set of r independent parameters. By omitting indices, we shall write this and similar relations in the form

$$x = f(x_0; \alpha). \tag{1.2}$$

Linear transformation are either homogeneous or inhomogeneous. Linear coordinate transformations in $SO(3)$ are homogeneous, and those in $SE(2)$ are inhomogeneous.

Let us go back to the coordinate transformations of Equation (1.1). We shall assume that the f^i have all the required derivatives, and that r is the smallest number of parameters needed to specify the transformations completely and uniquely. The set of transformations f^i will form a group if they obey the following two conditions.

(i) The result of performing successively any two transformations of the set is another transformation belonging to this set. Formally if $x = f(x_0; \alpha)$ and $x' = f(x; \beta)$, then there exists a set of parameters

Lie Groups and Lie Algebras

$$\gamma^\rho = \phi^\rho(\alpha; \beta) \tag{1.3}$$

such that

$$x' = f(x; \beta) = f(f(x_0; \alpha); \beta) = f(x_0; \phi(\alpha; \beta)). \tag{1.4}$$

(ii) Corresponding to every transformation, there exists a unique inverse, which also belongs to the set: Given Equation (1.1), there exists a set of parameters α^{-1} such that

$$x_0 = f(x, \alpha^{-1}). \tag{1.5}$$

Transforming x_0 onto x and then inversely back to x_0, we obtain according to (i) a transformation which belongs to the group and is characterized by the set of parameters α_0.

$$x_0 = f(x_0; \alpha_0). \tag{1.6}$$

This is the identity transformation. Since it imposes no restriction on x_0 or on α_0, we shall take

$$\alpha_0^\rho = 0, \quad \rho = 1, \ldots, r, \tag{1.7}$$

so that

$$f(x; 0) = x, \tag{1.8}$$

and

$$\phi(\alpha; 0) = \alpha. \tag{1.9}$$

The coordinate transformations given in Equations (1.1) and (1.6) can be written respectively as

$$x = f(x_0, \alpha), \tag{1.10}$$

and

$$x = f(x, 0). \tag{1.11}$$

Corresponding to these, there are two ways of expressing a transformation such that the new components of x differ infinitesimally from the old ones:

$$x + dx = f(x_0; \alpha + d\alpha), \tag{1.12}$$

or

$$x + dx = f(x, \delta\alpha), \tag{1.13}$$

where $d\alpha$ and $\delta\alpha$ are the infinitesimal increments in the parameter space from α and from the origin $\alpha = 0$ respectively. From the above equations, we derive

$$dx = \frac{\partial f(x_0, \alpha)}{\partial \alpha^\sigma} d\alpha^\sigma, \tag{1.14}$$

or

$$dx = \left.\frac{\partial f(x, \alpha)}{\partial \alpha^\sigma}\right|_{\alpha=0} \delta\alpha^\sigma. \tag{1.15}$$

The basic idea of the Lie group is to derive all the group properties from the behavior of the functions near the origin in the parameter space. We are thus led to look at the second equation more closely. Equation (1.15) can be written as

$$dx^i = u^i_\sigma(x)\, \delta\alpha^\sigma, \tag{1.16}$$

where

$$u^i_\sigma(x) = \left.\frac{\partial f^i(x;\alpha)}{\partial \alpha^\sigma}\right|_{\alpha=0}. \tag{1.17}$$

Next, let us consider transformations of a function depending on the coordinate variables, i.e., the same function on the same point only in terms of a different coordinate system,

$$F(f(x_0)) = F_0(x_0),$$

or

$$\begin{aligned} F(x) &= F_0(x_0) \\ &= F_0(f(x; \alpha^{-1})). \end{aligned} \tag{1.18}$$

This is a reflection of the fact that the transformation of the function is achieved through the inverse transformation of its arguments. We are interested in the operator which changes $F_0(x)$ to $F(x)$:

$$F(x) = S_\alpha F_0(x). \tag{1.19}$$

For example, consider a function:

$$F_0(x_0, y_0) = [x_0 + iy_0]^2, \tag{1.20}$$

defined on the two-dimensional $x_0 y_0$ plane, and an inhomogeneous linear transformation

$$\begin{aligned} x &= x_0 + u, \\ y &= y_0 + v. \end{aligned} \tag{1.21}$$

The original function of Equation (1.20) can be described in the form

$$F(x, y) = F_0(x_0, y_0) = [(x - u) + i(y - v)]^2. \tag{1.22}$$

Thus

$$F(x, y) - F_0(x, y) = -2(x + iy)(u + iv) + (u + iv)^2. \tag{1.23}$$

Lie Groups and Lie Algebras

$F(x)$ is clearly different from $F_0(x)$, and this difference is the effect of the S_α operation defined in Equation (1.19). The corresponding coordinate transformation is given in Equation (1.21).

The transformation of the coordinate point through Equation (1.10) or Equation (1.21) is called the *active transformation*. The transformation of the coordinate system through which the function is transformed as is given in Equation (1.18) or Equation (1.22) is called the *passive transformation*. From a purely calculational point of view, the active and passive transformations are the inverse of each other [See Exercise 2 in Section 6].

The infinitesimal transformation of the function F_0 corresponding to the coordinate transformation of Equation (1.12) or Equation (1.13) is

$$dF_0(x) = F(x) - F_0(x) = F_0(x_0) - F_0(x)$$
$$= F_0(x - dx) - F_0(x). \tag{1.24}$$

Thus

$$dF_0(x) = -\frac{\partial F_0}{\partial x^i} dx^i = -\delta a^\sigma u^i_\sigma(x) \frac{\partial F_0}{\partial x^i} = -i\delta a^\sigma X_\sigma(x) F_0(x). \tag{1.25}$$

where

$$X_\sigma(x) \equiv -i u^i_\sigma(x) \frac{\partial}{\partial x^i}. \tag{1.26}$$

The operators $X_\sigma(x)$ are called the *generators* of the Lie group.

Let us carry out explicit calculations for the $SE(2)$ group. There are two coordinate variables, namely

$$x^1 = x, \quad x^2 = y, \tag{1.27}$$

and three parameters:

$$a^1 = \theta, \quad a^2 = u, \quad a^3 = v. \tag{1.28}$$

For small values of the parameters,

$$\begin{bmatrix} x + dx \\ y + dy \\ 1 \end{bmatrix} = \begin{bmatrix} 1 & -\delta\theta & \delta u \\ \delta\theta & 1 & \delta v \\ 0 & 1 & 1 \end{bmatrix} \begin{bmatrix} x \\ y \\ 1 \end{bmatrix}, \tag{1.29}$$

and Equation (1.16) in this case can be written as

$$dx = \left(\frac{\partial x}{\partial \theta}\right)\bigg|_{\theta=0} \delta\theta + \left(\frac{\partial x}{\partial u}\right)\bigg|_{u=0} \delta u + \left(\frac{\partial x}{\partial v}\right)\bigg|_{v=0} \delta v$$
$$= -y\,\delta\theta + \delta u. \tag{1.30}$$

and

$$dy = \left(\frac{\partial y}{\partial \theta}\right)\bigg|_{\theta=0} \delta\theta + \left(\frac{\partial y}{\partial u}\right)\bigg|_{u=0} \delta u + \left(\frac{\partial y}{\partial v}\right)\bigg|_{v=0} \delta v$$

$$= x\,\delta\theta + \delta v.$$

The generators are

$$X_\theta(x) = -i\left[\left(\frac{\partial x}{\partial \theta}\right)\bigg|_{\theta=0}\frac{\partial}{\partial x} + \left(\frac{\partial y}{\partial \theta}\right)\bigg|_{\theta=0}\frac{\partial}{\partial y}\right]$$

$$= -i\left(x\frac{\partial}{\partial y} - y\frac{\partial}{\partial x}\right),$$

$$X_u(x) = -i\left[\left(\frac{\partial x}{\partial u}\right)\bigg|_{u=0}\frac{\partial}{\partial x} + \left(\frac{\partial y}{\partial u}\right)\bigg|_{u=0}\frac{\partial}{\partial y}\right]$$

$$= -i\frac{\partial}{\partial x}, \qquad (1.31)$$

and

$$X_v(x) = -i\left[\left(\frac{\partial x}{\partial v}\right)\bigg|_{v=0}\frac{\partial}{\partial x} + \left(\frac{\partial y}{\partial v}\right)\bigg|_{v=0}\frac{\partial}{\partial y}\right]$$

$$= -i\frac{\partial}{\partial y}.$$

The expressions given in Equations (1.30) and Equation (1.31) are Equation (1.16) and Equation (1.26) applied to the $SE(2)$ group respectively. Both forms need the matrix $u^i_\sigma(x)$, which can be written out as

$$[u^i_\sigma(x)] = \begin{bmatrix} -y & -1 & 0 \\ x & 0 & -1 \end{bmatrix}. \qquad (1.32)$$

Since $SE(2)$ is a three-parameter group operating on a two-dimensional geometrical space, the above matrix is two-by-three. This matrix has no inverse. The generators of $SE(2)$ given in Equation (1.31) satisfy the commutation relations:

$$[X_\theta(x), X_u(x)] = iX_v(x),$$
$$[X_\theta(x), X_v(x)] = -iX_u(x), \qquad (1.33)$$
$$[X_u(x), X_v(x)] = 0.$$

Lie Groups and Lie Algebras

Let us next consider the transformation of functions defined in the parameter space specified by Equation (1.3). Following the same procedure as that for deriving Equations (1.14) and (1.15), we get

$$\alpha + d\alpha = \alpha + \left.\frac{\partial \phi(\alpha; \beta)}{\partial \beta^\tau}\right|_{\beta=0} \delta\alpha^\tau. \tag{1.34}$$

Thus $d\alpha$ is a linear combination of the $\delta\alpha$:

$$d\alpha^\rho = \Theta^\rho_\tau(\alpha)\, \delta\alpha^\tau, \tag{1.35}$$

where

$$\Theta^\rho_\tau(\alpha) = \left.\frac{\partial \phi^\rho(\alpha; \beta)}{\partial \beta^\tau}\right|_{\beta=0}. \tag{1.36}$$

Following the same logic as that which led to Equation (1.25) for $dF_0(x)$, we can write for a function $\Phi(\alpha)$:

$$d\Phi(\alpha) = -i\delta\alpha\, \Theta^\rho_\sigma(\alpha)\, \frac{\partial \Phi_0}{\partial \alpha^\rho}$$

$$= -i\delta\alpha^\sigma X_\sigma(\alpha)\, \Phi_0(\alpha), \tag{1.37}$$

where $\Theta^\rho_\sigma(\alpha)$ is given in Equation (1.36). Consequently

$$X_\sigma(\alpha) = -i\Theta^\rho_\sigma(\alpha)\, \frac{\partial}{\partial \alpha^\rho}. \tag{1.38}$$

Unlike the case for $u^i_\sigma(x)$, the $\Theta^\rho_\sigma(\alpha)$ matrix is a square matrix. Since there are the minimum necessary number of α parameters, this matrix is expected to have an inverse:

$$\Psi^\sigma_\rho \Theta^\rho_\tau = \delta^\sigma_\tau, \tag{1.39}$$

such that

$$\delta\alpha^\sigma = \Psi^\sigma_\tau\, d\alpha^\tau. \tag{1.40}$$

Let us carry out explicit calculations for the $SE(2)$ case [See also Exercise 1 in Section 6]. The Θ^ρ_σ and Ψ^σ_τ matrices take the form

$$[\Theta^\rho_\sigma(\alpha)] = \begin{bmatrix} 1 & 0 & 0 \\ -v & 1 & 0 \\ u & 0 & 1 \end{bmatrix}. \tag{1.41}$$

$$[\Psi^\sigma_\rho(\alpha)] = \begin{bmatrix} 1 & 0 & 0 \\ v & 1 & 0 \\ -u & 0 & 1 \end{bmatrix}. \tag{1.42}$$

From these expressions, we can derive

$$X_\theta(a) = -i\left(\frac{\partial}{\partial \theta} + u\frac{\partial}{\partial v} - v\frac{\partial}{\partial u}\right),$$

$$X_u(a) = -i\frac{\partial}{\partial u}, \qquad (1.43)$$

$$X_v(a) = -i\frac{\partial}{\partial v}.$$

These operators satisfy the commutation relations:

$$[X_\theta(a), X_u(a)] = iX_v(a),$$
$$[X_\theta(a), X_v(a)] = -iX_u(a), \qquad (1.44)$$
$$[X_u(a), X_v(a)] = 0.$$

These commutation relations are identical to those given in Equation (1.33). Is this an accidental coincidence or a manifestation of a more fundamental theorem? We shall study this point in more detail in Section 2.

2. Basic Theorems Concerning Lie Groups

Now combining Equation (1.16) and Equation (1.40), we arrive at

$$dx^i = u^i_\rho(x)\,\Psi^\rho_\tau(a)\,da^\tau,$$

or

$$\frac{\partial x^i}{\partial a^\sigma} = u^i_\rho(x)\,\Psi^\rho_\sigma(a). \qquad (2.1)$$

This is known as Lie's first theorem.

The commutation relations of Equation (1.33) and Equation (1.44) for the $SE(2)$ case strongly suggest some fundamental theorems concerning the generators. In order to see this possibility, we observe first that the necessary and sufficient condition for the differential equation of Equation (2.1) to have a unique solution with a given initial condition is that all mixed derivatives be equal:

$$\frac{\partial^2 x^i}{\partial a^\sigma \partial a^\lambda} = \frac{\partial^2 x^i}{\partial a^\lambda \partial a^\sigma}. \qquad (2.2)$$

This condition is often called the integrability condition.

Let us then apply this condition to Equation (2.1):

$$\frac{\partial}{\partial a^\sigma}(u^i_\rho(x)\,\Psi^\rho_\lambda(a)) = \frac{\partial}{\partial a^\lambda}(u^i_\rho(x)\,\Psi^\rho_\sigma(a)). \qquad (2.3)$$

The calculation of this equation will require the derivatives of the $u(x)$ function with respect to the α parameters:

$$\frac{\partial u^i_\rho(x)}{\partial a^\sigma} = \frac{\partial u^i_\rho(x)}{\partial x^j} \frac{\partial x^j}{\partial a^\sigma} = \Psi^\lambda_\sigma(a) \, u^j_\lambda(x) \frac{\partial u^i_\rho(x)}{\partial x^j}. \quad (2.4)$$

Then Equation (2.3) becomes

$$\Psi^\tau_\sigma(a) \Psi^\sigma_\lambda(a) \left\{ u^j_\rho(x) \frac{\partial u^i_\tau(x)}{\partial x^j} - u^j_\tau(x) \frac{\partial u^i_\rho(x)}{\partial x^j} \right\}$$

$$= \left\{ \frac{\partial \Psi^\rho_\tau(a)}{\partial a^\sigma} - \frac{\partial \Psi^\rho_\sigma(a)}{\partial a^\tau} \right\} u^i_\rho(x). \quad (2.5)$$

Because the Ψ matix has an inverse,

$$u^j_\nu(x) \frac{\partial u^i_\mu(x)}{\partial x^j} - u^j_\mu(x) \frac{\partial u^i_\nu(x)}{\partial x^j}$$

$$= \left\{ \Theta^\sigma_\mu(a) \Theta^\lambda_\nu(a) \left(\frac{\partial \Psi^\rho_\tau(a)}{\partial a^\sigma} - \frac{\partial \Psi^\rho_\sigma(a)}{\partial a^\tau} \right) \right\} u^i_\rho(x). \quad (2.6)$$

The left-hand side of the above expression depends only on the x variables. The quantity inside the curly bracket on the right-hand side depends only on the α variables. Furthermore, this equation cannot be completely separated because the $u^i_\rho(x)$ matrix is not a square matrix and does not have an inverse.

This, however, does not discourage us from asking whether the quantity inside the square bracket is constant. In order to determine this, let us replace $f^i(x; \alpha)$ by $\phi^\mu(\beta; \alpha)$. Then Equation (2.6) can be rewritten as

$$\Theta^\rho_\nu(\beta) \frac{\partial \Theta^\sigma_\mu(\beta)}{\partial \beta^\rho} - \Theta^\rho_\mu(\beta) \frac{\partial \Theta^\sigma_\nu(\beta)}{\partial \beta^\rho}$$

$$= \left\{ \Theta^\sigma_\mu(a) \Theta^\lambda_\nu(a) \left(\frac{\partial \Psi^\rho_\tau(a)}{\partial a^\sigma} - \frac{\partial \Psi^\rho_\sigma(a)}{\partial a^\tau} \right) \right\} \Theta^\sigma_\rho(\beta). \quad (2.7)$$

Now the $\Theta^\sigma_\rho(\beta)$ on the right-hand side of the above equation can be moved to the left-hand side. Then the separation of the variables becomes complete. The quantity in the curly bracket is indeed a constant. This allows us to write

$$u^j_\rho(x) \frac{\partial u^i_\tau(x)}{\partial x^j} - u^j_\tau(x) \frac{\partial u^i_\sigma(x)}{\partial x^j} = iC^\rho_{\tau\sigma} u^i_\rho(x), \quad (2.8)$$

and

$$\Theta^\nu_\rho(\alpha) \frac{\partial \Theta^\mu_\tau(\alpha)}{\partial \alpha^\nu} - \Theta^\nu_\tau(\alpha) \frac{\partial \Theta^\mu_\rho(\alpha)}{\partial \alpha^\nu} = iC^\sigma_{\rho\tau} \Theta^\mu_\sigma(\alpha), \tag{2.9}$$

where the coefficients $C^\mu_{\sigma\tau}$ are constants.

The immediate consequence of the above relations is that the generators of the Lie group $X_\sigma(x)$ defined in Equation (1.26) and in Equation (1.38) satisfy the commutation relations:

$$[X_\rho(x), X_\sigma(x)] = iC^\tau_{\rho\sigma} X_\tau(x), \tag{2.10}$$

$$[X_\rho(\alpha), X_\sigma(\alpha)] = iC^\tau_{\rho\sigma} X_\tau(\alpha). \tag{2.11}$$

The constants $C^\tau_{\rho\sigma}$ are commonly called *the structure constants*. This result is known as Lie's second theorem.

Evidently, from Equations (2.9) and (2.10),

$$C^\tau_{\rho\sigma} = -C^\tau_{\sigma\rho}, \tag{2.12}$$

and the generators satisfy the Jacobi identity:

$$[[X_\rho, X_\sigma], X_\tau] + [[X_\sigma, X_\tau], X_\rho] + [[X_\tau, X_\rho], X_\sigma] = 0, \tag{2.13}$$

resulting in

$$C^\mu_{\rho\sigma} C^\nu_{\mu\tau} + C^\mu_{\sigma\tau} C^\nu_{\mu\rho} + C^\mu_{\tau\rho} C^\nu_{\mu\sigma} = 0. \tag{2.14}$$

The properties of the structure constants given in Equations (2.12) and (2.14) constitute *Lie's third theorem*.

As stated in Section 1, the fundamental idea of the theory of Lie groups is to consider only that part of the group which lies near the identity. This is equivalent to considering only infinitesimal transformations. In terms of the generators, an infinitesimal transformation of the group takes the form

$$S_{\delta\alpha} = I - iX_\sigma \, \delta\alpha^\sigma, \tag{2.15}$$

where I is the identity operator, and $\delta\alpha^\sigma$ is an infinitesimal quantity defined to be of first order. The commutation relations for the generators given in Equations (2.10) and (2.11) determine the algebraic properties of the Lie group. These commutation relations are of course determined by the structure constants.

Let us see how these infinitesimal transformations lead to transformations with finite parameters. Consider

$$S_N = S_N[(S_{N-1})^{-1} S_{N-1}] [(S_{N-2})^{-1} S_{N-2}] \ldots [(S_1)^{-1} S_1] I$$
$$= [S_N(S_{N-1})^{-1}] [S_{N-1}(S_{N-2})^{-1}] \ldots [S_2(S_1)^{-1}] S_1 I. \tag{2.16}$$

The condition (2.16) defines the multiplication law of the gorup.

Let us consider the parameterization $\alpha^\mu t$, where t is allowed to vary along the real axis while the α^μ are a fixed set of numbers. The transformations

Lie Groups and Lie Algebras

with different values of t form a one-parameter Abelian subgroup. We are particularly interested in the interval $0 \leq t \leq 1$. This interval can be divided into

$$t = k/N, \tag{2.17}$$

with

$$k = 0, 1, 2, \ldots, N,$$

where N can be an arbitrarily large integer. On this interval, the transformation S_N satisfies the multiplication law:

$$S_k S_m = S_{k+m}, \tag{2.18}$$

and

$$S_N = [S_1]^N. \tag{2.19}$$

S_N is therefore determined from S_1 which describes the transformation for small values of t.

$$S_\alpha = \exp(-i\alpha^\rho X_\rho) = \sum_{k=0}^{\infty} \frac{1}{k!} (-i\alpha^\rho X_\rho)^k. \tag{2.20}$$

In Equation (2.15), an infinitesimal transformation of the group S was written as the sum of the identity plus an infinitesimal transformation, where $\delta\alpha^\sigma$ is an infinitesimal quantity defined to be of first order. If two such transformations are considered, we get

$$\begin{aligned} S_\alpha S_\beta &= (I - i\delta\alpha^\rho X_\rho)(I - i\delta\beta^\sigma X_\sigma) \\ &= I - i\delta\alpha^\rho X_\rho - i\delta\beta^\sigma X_\sigma, \end{aligned} \tag{2.21}$$

where the first non-vanishing infinitesimal terms have been written. Thus the operation of multiplication in S corresponds to addition in the infinitesimal group of S. If the first-order quantities vanish, then higher-order quantities must be used. But in order for the u^i in Equation (2.1) to be the vector space of a transformation such as is given in Equation (1.1), Equation (2.1) must be completely integrable. The integrability of Equation (2.1) implies that in Equation (2.21) we never need go beyond the second order infinitesimals [second partial derivatives must exist and be continuous in order for Equation (2.1) to be integrable]. Thus we need only consider "commutators" which are expressions of the form $S_\alpha S_\beta S_\alpha^{-1} S_\beta^{-1}$ and to ask that the corresponding infinitesimal operators of second order,

$$\delta\alpha^\rho \delta\beta^\sigma [X_\rho, X_\sigma] \tag{2.22}$$

be continuous in the linear manifold of the infinitesimal operators. Thus the structure constants given in Equations (2.10) and (2.11) come in second order because they are given by the commutator alone, and it is the structure constants which determine the group.

3. Properties of Lie Algebras

As we have seen in Sections 1 and 2, the properties of Lie groups are determined by the closed algebra of the generators. We call this Lie algebra. Let us rewrite the commutation relations for the generators:

$$[X_\rho, X_\sigma] = iC_{\rho\sigma}^\tau X_\tau, \tag{3.1}$$

where

$$\rho, \sigma = 1, 2, \ldots, r.$$

If all the generators commute and the structure constants vanish, we say that the algebra is Abelian.

If a subset of generators X_i satisfy the closed algebraic relation

$$[X_i, X_j] = iC_{ij}^k X_k \tag{3.2}$$

where

$$i, j = 1, \ldots, s < r,$$

we say that X_i's form a *subalgebra* generating a subgroup. If the subset of the algebra has the property that the commutator of any member of the subset with any member of the algebra produces a member of the subset:

$$[X_\rho, X_i] = iC_{\rho i}^k X_k, \tag{3.3}$$

then the subalgebra of X_i is *invariant*, and the subgroup generated by this subalgebra is an invariant subgroup discussed in Section 2 of Chapter I. If all the X_i's commute:

$$[X_i, X_j] = 0, \tag{3.4}$$

we say that X_i's form an *Abelian invariant subalgebra* generating an Abelian invariant subgroup.

Algebras with no invariant subalgebra are called *simple algebras*, as in the case of simple groups. Algebras with no Abelian invariant subalgebras are called semi-simple Lie algebras. For example, the generators of the rotation group form a simple algebra. The Lie algebra of $SE(2)$ given in Equations (1.33) and (1.44) form an algebra which is neither simple nor semi-simple.

Next, let us discuss the isomorphism between a Lie algebra A and another algebra A'. We say that the two algebras A and A' are isomorphic to each other if their generators satisfy the same set of commutation relations. This means that to each element X of A, there is an element $F(X)$ of A' such that

$$F([X_\rho, X_\sigma]) = [F(X_\rho), F(X_\sigma)]. \tag{3.5}$$

Lie groups having the same Lie algebra are said to be *locally isomorphic*. The best known example of this isomorphism is the relation between the three-dimensional rotation group and the $SU(2)$ group generated by the Pauli spinors.

It is important to note that the generators of $SE(2)$ given in Equation (1.31) are not the only ones satisfying the commutation relations of Equation (1.33). We can consider the two-by-two matrices:

$$Y_\theta = \frac{1}{2}\begin{bmatrix} 1 & 0 \\ 0 & -1 \end{bmatrix}, \qquad (3.6)$$

$$Y_u = \begin{bmatrix} 0 & -i \\ 0 & 0 \end{bmatrix}, \quad Y_v = \begin{bmatrix} 0 & 1 \\ 0 & 0 \end{bmatrix}.$$

These matrices also satisfy the commutation relations for $SE(2)$ given in Equation (1.33). Since Wigner was the first one (Wigner, 1939) to recognize this, we take the liberty of calling the group generated by the above matrices "$SW(2)$." $SE(2)$ is locally isomorphic to $SW(2)$.

Once the commutation relations for the generators are written, we can construct sets of matrices which satisfy the commutation relations:

$$M(X_\rho)\,M(X_\sigma) - M(X_\sigma)\,M(X_\rho) = iC^\tau_{\rho\sigma}\,M(X_\tau), \qquad (3.7)$$

where $M(X_\rho)\,M(X_\sigma)$ implies matrix multiplication. Each such set is called a *representation* of the Lie algebra. In this case, each generator of the algebra can be represented by a square matrix of rank k, in which each element of every matrix is an explicit number. The rank of matrix k is also the dimension of the representation. If k is a finite integer, we say that the representation is *finite-dimensional*. If k is infinite, the representation is called an *infinite-dimensional representation*.

One representation which is an immediate consequence of the Jacobi identity given in Equations (2.13) and (2.14) is the set of r matrices consisting of structure constants, defined as

$$(M_\rho)^\sigma_\tau = C^\sigma_{\rho\tau}, \qquad (3.8)$$

then

$$(M_\mu)^\sigma_\tau (M_\nu)^\tau_\rho - (M_\nu)^\sigma_\tau (M_\mu)^\tau_\rho = iC^\tau_{\mu\nu}(M_\tau)^\sigma_\rho. \qquad (3.9)$$

This representation is called the regular representation of the Lie algebra.

We are quite familiar with finite-dimensional representations of various groups. All representations of the three-dimensional rotation group are finite-dimensional. The three three-by-three matrices which generate the rotation of the three-dimensional coordinate system form the regular representation of the Lie algebra of the rotation group.

The three-by-three matrices describing the $SE(2)$ transformation in Section 1 constitute a finite-dimensional representation of the $SE(2)$ group, generated by

$$X_\theta = \begin{bmatrix} 0 & -i & 0 \\ i & 0 & 0 \\ 0 & 0 & 0 \end{bmatrix}, \tag{3.10}$$

$$X_u = \begin{bmatrix} 0 & 0 & i \\ 0 & 0 & 0 \\ 0 & 0 & 0 \end{bmatrix}, \quad X_v = \begin{bmatrix} 0 & 0 & 0 \\ 0 & 0 & i \\ 0 & 0 & 0 \end{bmatrix}.$$

These matrices satisfy the commutation relations for the $SE(2)$ group given in Equation (1.33). Furthermore, it is easy to show that they form the regular representation of the Lie algebra for $SE(2)$.

Since X_u and X_v for $SE(2)$ commute with each other, they form an Abelian subalgebra. Since the commutation of X_θ with X_u or X_v leads to X_u or X_v, the subalgebra of X_u and X_v is also an invariant subalgebra. The Lie algebra of the $SE(2)$ group is neither simple nor semi-simple. In addition to finite-dimensional representations, $SE(2)$ has infinite-dimensional representations which we shall study in Chapter V.

If the generators are Hermitian, the resulting representation of the Lie group is unitary, as is evident from the equation

$$(\exp[-i\alpha^\rho X_\rho])^\dagger = \exp[i\alpha^\rho (X_\rho)^\dagger]. \tag{3.11}$$

The generators of the rotation group are Hermitian and all the representations of the rotation gorup are unitary representations. On the other hand, in the case of $SE(2)$, X_θ is Hermitian while X_u and X_v are not. Therefore, the representation generated by the three matrices of Equation (3.10) is not a unitary representation. However, this does not rule out the possibility of the differential form of the generators given in Equation (1.31) being Hermitian when they are applied to a suitable Hilbert space. This possibility will also be discussed in detail in Chapter V.

When we construct representations of the group starting from the generators, we use the exponential expansion. We know how to do this in the case of rotation group. We know also that

$$\exp\{-i(\alpha^1 X_1 + \alpha^2 X_2)\} \tag{3.12}$$

is not necessarily equal to

$$\exp(-i\alpha^1 X_1) \exp(-i\alpha^2 X_2), \tag{3.13}$$

unless X_1 and X_2 commute with each other.

Lie Groups and Lie Algebras

As for the $SE(2)$ case, the matrix $R(\theta)$ defined in Equation (2.3) can be obtained from

$$\exp(-i\theta M(X_\theta)) = \sum_{k=0}^{\infty} \frac{(-i\theta)^k}{n!} (M(X_\theta))^k \tag{3.14}$$

as in the case of the rotation group. However, for X_u and X_v, the series truncates, and

$$\exp(-iuM(X_u)) = I - iuM(X_u),$$
$$\exp(-ivM(X_v)) = I - ivM(X_v). \tag{3.15}$$

This is because $[M(X_u)]^2 = [M(X_v)]^2 = 0$. In general, if any power of a non-zero matrix M vanishes:

$$(M)^n = 0, \tag{3.16}$$

for an integer n, then this matrix is said to be *nilpotent*. The triangular matrices with vanishing diagonal elements which we discussed in Chapter I are nilpotent matrices. We shall discuss further nilpotent matrices in Chapter VII.

In constructing representations, as in many other mathematical procedures in physics, it is convenient to find a set of variables which remain invariant under group transformations. In the language of Lie algebras, we should find first the maximal set of operators which commute with all the generators of a given Lie group. These operators are often called the *Casimir operators*. Then by Schur's lemma, the necessary and sufficient condition for the representation to be irreducible is that the Casimir operators be multiples of unity. The eigenvalues of the Casimir operators therefore specify the representation.

The Casimir operator for the rotation group is the square of the total angular momentum. It is straightforward to show that the Casimir operator for the $SE(2)$ group is the sum of the squares of the generators of translations. As is well known to many physicists, the $SU(3)$ group has two Casimir operators. As we shall see in Chatper III, the homogeneous and inhomogeneous Lorentz groups have two Casimir operators.

In addition to the explicit forms of the matrices, we can construct representations by finding differential operators and the Hilbert space to which the operators are applicable. This procedure applied to $SO(3)$ is well known. The Hilbert space to which the rotation operators are applicable is formed by the spherical harmonics. As for the $SE(2)$ group, we have already constructed the differential operators. We shall discuss in Chapter VII the Hilbert space to which the $SE(2)$ operators are applicable.

After constructing the Casimir operators, it is convenient to construct representations which are also diagonal in a commuting set of generators. In the case of the rotation group, we can construct representations which are simultaneously diagonal in the Casimir operator and one of the generators. These representations are not diagonal in the rest of the generators. In the case of $SE(2)$, we can choose the representations which are diagonal in the Casimir operator and the two commuting translation operators *or* in the Casimir operator and the rotation operators.

Let X_ρ and Y_ρ be two different representations of the same Lie algebra such that

$$[X_\rho, Y_\sigma] = 0, \tag{3.17}$$

we can then consider another representation Z_ρ such that

$$Z_\rho = X_\rho + Y_\rho. \tag{3.18}$$

The operators Z_ρ satisfy the same Lie algebra. We say in this case that the representation Z is a direct sum of X and Y. The Casimir operators for each of the X and Y representations remain as the Casimir operators for the resulting Z representation. However, the Z representation can have new Casimir operators. The new Casimir operators commute with every Z_ρ, but do not necessarily commute with X_ρ or Y_ρ. For example, in the rotation group, the total angular momentum is the direct sum of the generators of the $SO(3)$ and $SU(2)$ groups, which are the orbital and spin angular momenta respectively. The total orbital angular momentum and the total spin remain as good quantum numbers even if the two angular momenta are coupled. However, the third component of the orbital angular momentum does not commute with the square of the total angular momentum.

The procedure for constructing representations of a semi-direct product of two groups is much more delicate than the case for direct products. As was pointed out before, the $SE(2)$ group is a semi-direct product of the rotation and translation groups in a two-dimensional plane. As we shall see in Chapter III, the Poincaré group is a semi-direct product of the Lorentz and space-time translation groups.

The techniques of constructing representations of semi-direct products have been discussed extensively in the mathematical literature. However, for a physicist, it is much more profitable to work out concrete examples before attempting to prove theorems. The $SE(2)$ and Poincaré groups are good examples for both physicists and mathematicians.

4. Properties of Lie Groups

It is not difficult to translate the properties of Lie algebras to those of Lie groups. A Lie group generated by a given Lie algebra is Abelian if the Lie algebra is Abelian. A subalgebra of a Lie algebra generates a subgroup. A Lie group is simple or semi-simple if its Lie algebra is simple or semi-simple.

Lie Groups and Lie Algebras

The basic difference between Lie groups and Lie algebras is that, in the case of Lie groups, we have to deal with finite values of the group parameters. For this purpose, it is convenient to consider a Euclidean space spanned by the group parameters. This space is often called the *manifold*. The manifold for a one-parameter group is a straight line. The manifold for the group of translations on a two-dimensional plane, as discussed in Chapter I, is a two-dimensional plane. Both the $SO(3)$ and $SE(2)$ groups have three-dimensional manifolds.

The Lie algebra of a given Lie group is a description of the group in the neighborhood of the origin in the manifold which corresponds to the identity element of the Lie group. Every element of the group is specified by its coordinate position in the manifold. Because this element can be obtained through the process of continuously increasing the parameters from the identity element, there is a continuous line between the origin and the coordinate point for the group. Every point on this line corresponds to an element of the Lie group.

If all elements of a given group can be specified by the coordinate points confined within an finite region in the manifold, this group is said to be *compact*. If a group has no finite boundary in the manifold, it is said to be *non-compact*. As is specified in Figure 4.1, $SO(3)$ is compact while $SE(2)$ is non-compact.

A given Lie algebra may generate different Lie groups having different boundaries in the manifold. For example, in the case of a rotation group, the manifold for $SO(3)$ is a sphere with radius π, while that for $SU(2)$ is a sphere with radius 2π (Hamermesh, 1962).

If any closed loop in the manifold can be continuously deformed to a point, we say that the group is *simply connected*. Among all possible Lie groups sharing the same Lie algebra, there is only one simply connected group. This group is often called the *universal covering group*. The fact that

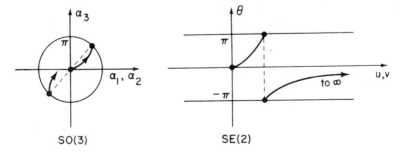

Fig. 4.1. Manifolds for $SO(3)$ and $SE(2)$. $SO(3)$ is a compact group because all the elements of $SO(3)$ are bounded by the surface of a sphere with radius π. Another way of stating this is to say that every geodesic through the identity returns to the vicinity of the identity. Such is not true for $SE(2)$. The translation variables u and v are not bounded. Hence it is not required for every translation through the identity to return to the vicinity of the identity.

$SU(2)$ is the universal covering group for the three-dimensional rotation group is discussed in many textbooks on group theory.

Let us study the manifold structure and the universal covering group for two-dimensional Euclidean group. As was noted repeatedly before, this group has the three parameters u, v and θ. We therefore have to consider a three-dimensional Euclidean space spanned by these parameters. The translation parameters u and v can cover the entire two-dimensional uv plane for both $SE(2)$ and $SW(2)$. However, the rotation parameter θ can extend from $-\pi$ to $+\pi$ in the case of $SE(2)$, while it extends from -2π to $+2\pi$ in $SW(2)$.

As is illustrated in Figure 4.2, all closed curves in $SW(2)$ can be continuously deformed to a point, while this is not always possible for $SE(2)$.

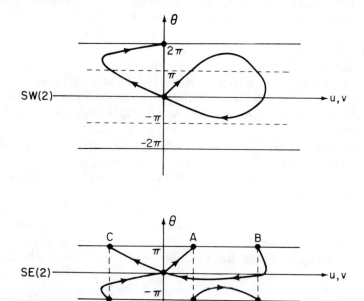

Fig. 4.2. Manifolds for $SW(2)$ and $SE(2)$. Both $SW(2)$ and $SE(2)$ are connected because any two group operators or points in the parameter manifold can be connected by a line lying completely within the manifold. $SW(2)$ is, however, simply connected because any path beginning and ending at the origin can be continuously deformed to a point at the origin. If the path does not touch the boundaries (right, above), it can always be continuously deformed to the origin. If the path does touch the boundary of the manifold, it can still be continuously deformed to the identity (left, above). This is true for $SW(2)$ in which the rotation parameter θ extends from 2π to -2π. In the case of $SE(2)$, if a path in the parameter space cuts the surface of the manifold and returns to the identity, it cannot be continuously deformed to a point. If, however, the path cuts the surface twice, it can be continuously deformed to a point at the origin. Hence $SE(2)$ is not simply connected. $SW(2)$ is the universal covering group for the two-dimensional Euclidean group.

Lie Groups and Lie Algebras

For this reason, $SW(2)$ is the universal covering group for the two-dimensional Euclidean group.

There is a simple manner in which all Lie groups with the same Lie algebra may be obtained from the universal covering group. Let S be a simply-connected Lie group and N be one of its discrete invariant subgroups. Then

$$S_x S_n S_x^{-1} = S_{n'} \tag{4.1}$$

for all S_n, $S_{n'}$ an element of N, and S_x an element of S. Then by Equation (2.2) of Chapter I, we can form the factor or quotient group:

$$G = S/N, \tag{4.2}$$

such that it is a Lie group whose Lie algebra A is isomorphic to the Lie algebra of S. G is multiply connected when N contains more than one element.

In $SU(2)$, the two-by-two identity matrix and its negative are the elements of the group. These two elements form a discrete invariant subgroup. We can therefore form a quotient group as is described in Equation (4.2). This quotient group is $SO(3)$. The story is the same for $SW(2)$ and $SE(2)$.

5. Further Theorems of Lie Groups

The theory of Lie groups is a fascinating subject. It is still a developing subject. In general, representations of compact groups are well understood. Those for simple and semi-simple Lie groups have also been thoroughly studied. However, mathematicians are still working hard on non-compact groups and on groups which are neither simple nor semi-simple. It appears that what we need in the literature are examples, rather than different versions of proofs. We therefore list here without proof some theorems of Lie groups for which examples are discussed in later chapters.

The most crucial point in studying space-time symmetries of elementary particles is that we have to deal with non-compact groups. In addition, the groups are neither simple nor semi-simple. The inhomogeneous Lorentz group or the Poincaré group and some of its subgroups are such groups. The basic advantage of this group is that the space-time symmetry or common sense allows us to carry out explicit calculations even if we do not understand relevant mathematical theorems. Therefore, the explicit construction of its representations useful in physics will provide the examples for studying mathematical theorems. The study of the Poincaré group will provide examples for the following mathematical theorems.

(a) *Theorems on the Dimensionality of Representations*

If a group is compact, it is possible to construct finite-dimensional unitary representations, as in the case of the three-dimensional rotation group.

If a group is non-compact and does not contain Abelian invariant subgroups, its finite-dimensional representations are non-unitary, and its unitary representations are infinite-dimensional. The $SE(2)$ group discussed throughout Chapters I and II is non-compact, and its three-by-three matrix representation given in Equation (I.1.4) is non-unitary. However, this group contains an Abelian invariant subgroup. For this reason, we cannot apply this theorem directly to the case of $SE(2)$ (Weinberg, 1964). We have to resort to explicit calculations!

As we shall see in later chapters, whether a representation is unitary or not depends largely on whether there exists a square-integrable Hilbert space in which the generators of the group act as Hermitian operators. This investigation will require a non-trivial amount of calculation for each problem.

(b) *Cartan's Criterion on Semi-simple Lie Algebras*

The criterion for determining whether a given group contains an Abelian invariant subgroup, called "Cartan's criterion" is discussed in many textbooks (Gilmore, 1974). Cartan's theorem can be formulated in terms of a symmetrical tensor of the second rank which can be constructed from the structure constants:

$$g_{\rho\sigma} = C^{\mu}_{\rho\lambda} C^{\lambda}_{\sigma\mu}. \tag{5.1}$$

Then the necessary and sufficient condition for the group to be semi-simple is

$$\det(g_{\rho\sigma}) \neq 0. \tag{5.2}$$

Indeed, if we compute the above determinant for $SO(3)$, it does not vanish. The determinant vanishes for $SE(2)$ (Problem 5 in Section 6). In Chapter VIII, we shall study this point in more detail when we discuss a contraction of $SO(3)$ to $SE(2)$.

The distinction between groups which have Abelian invariant subgroups and those which do not is important, because Abelian invariant subgroups, though apparently easier to deal with can actually be more troublesome from the point of view of representations. Let us consider again the $SE(2)$ group. As we have seen in Section 2 of Chapter I, the translation subgroup of the $SE(2)$ group is an Abelian invariant subgroup. The three-by-three matrix

Lie Groups and Lie Algebras 45

$T(u, v)$ of Equation (I.2.4) is one of its possible representations. The T matrix has the pleasant Abelian property that

$$T(u_1, v_1) T(u_2, v_2) = T(u_1, v_1) T(u_2, v_2)$$
$$= T(u_1 + u_2, v_1 + v_2). \quad (5.3)$$

However, this matrix can never be brought to a diagonal form through a similarity transformation, while most of the mathematical theorems known to physicists are based on diagonalizable matrices.

6. Exercises and Problems

Exercise 1. Work out explicitly the parameter transformation of Equation (1.3) for the three-parameter group of $SE(2)$.

If we use the notation

$$\beta^1 = \theta', \quad \beta^2 = u', \quad \text{and} \quad \beta^3 = v', \quad (6.1)$$

in addition to α^1, α^2 and α^3 given in Equation (1.28), then

$$\begin{bmatrix} \cos \phi^1(\alpha; \beta) & -\sin \phi^2(\alpha; \beta) & \phi^3(\alpha; \beta) \\ \sin \phi^1(\alpha; \beta) & \cos \phi^2(\alpha; \beta) & \phi^3(\alpha; \beta) \\ 0 & 0 & 1 \end{bmatrix} \quad (6.2)$$

$$= \begin{bmatrix} \cos \theta' & -\sin \theta' & u' \\ \sin \theta' & \cos \theta' & v' \\ 0 & 0 & 1 \end{bmatrix} \begin{bmatrix} \cos \theta & -\sin \theta & u \\ \sin \theta & \cos \theta & v \\ 0 & 0 & 1 \end{bmatrix}.$$

Thus

$$\phi^1 = \theta + \theta',$$
$$\phi^2 = u \cos \theta' - v \sin \theta' + u', \quad (6.3)$$
$$\phi^3 = u \sin \theta' + v \cos \theta' + v'.$$

Exercise 2. Discuss active and passive transformations of the $SE(2)$ group.

Let us define the transformation given in Equation (I.2.5) to be active. This transformation first rotates the coordinate point (x, y) by angle ϕ around the origin. It then translates the rotated point by u and v along the x and y directions respectively.

On the other hand, if we perform the same rotation on the function

$$g(x, y) = (x + iy)^m = r^m e^{im\phi}, \quad (6.4)$$

using X_θ given in Equation (1.31), where

$$r = (x^2 + y^2)^{1/2}, \quad \phi = \tan^{-1}(y/x), \tag{6.5}$$

we get

$$(e^{-i\theta X_\theta}) g(x, y) = r^m e^{im(\phi - \theta)}. \tag{6.6}$$

If we apply the translation operators on the above expression,

$$(e^{-i(uX_u + vX_v)}) (e^{-i\theta X_\theta}) g(x, y) = (x'' + iy'')^m$$
$$= g(x'', y''), \tag{6.7}$$

where

$$x'' = (x - u) \cos \theta + (y - v) \sin \theta,$$
$$y'' = -(x - u) \sin \theta + (y - v) \cos \theta.$$

The above linear transformation can also be written as

$$\begin{bmatrix} x'' \\ y'' \\ 1 \end{bmatrix} = \begin{bmatrix} \cos \theta & \sin \theta & -(u \cos \theta + v \sin \theta) \\ -\sin \theta & \cos \theta & (u \sin \theta - v \cos \theta) \\ 0 & 0 & 1 \end{bmatrix} \begin{bmatrix} x \\ y \\ 1 \end{bmatrix}. \tag{6.8}$$

The matrix in this expression is precisely the inverse of that of the active transformation matrix of Equation (I.2.5).

Problem 1. Referring to Equation (1.3), show that for α, β, γ,

$$\phi[\phi(\alpha; \beta); \gamma] = \phi[\alpha; \phi(\beta; \alpha)], \tag{6.9}$$

independent of x. This is an associative property.

Problem 2. Use the identity property to show that $\alpha_0 = \phi(\alpha; \alpha^{-1})$.

Problem 3. Using the Taylor expansion in $\delta\alpha$, show that Equation (1.14) is indeed identical to Equation (1.15).

Problem 4. It is essential that the dx^i of Equation (1.16) be linearly independent. Show that a necessary and sufficient condition for this is that the α^ρ's are linearly independent.

Problem 5. Calculate Cartan's determinant of Equation (5.2) for $SO(3)$ and for $SE(2)$. Show that it is non-zero for $SO(3)$, while it vanishes for $SE(2)$.

Problem 6. Calculate the generators fo $SO(3)$ and $SO(2, 1)$. Construct the

Lie Groups and Lie Algebras

Casimir operators. What are the subgroups of these groups. Carry out the same procedure for $SO(4)$ and $SO(3, 1)$.

Problem 7. Consider $SL(2, c)$. Because there are four complex elements and one complex (or two real) constraints, this group is generated by six independent generators. Show that the generators can be chosen as

$$S_i = \tfrac{1}{2}\sigma_i, \quad K_i = \tfrac{1}{2}\sigma_i, \quad i = 1, 2, 3, \tag{6.10}$$

where σ_i are the usual Pauli spin matrices. The commutation relations among S_i are well known. Complete the algebraic relations for all six generators of this group. Show that

$$C_1 = \sum_{i=1}^{3}(S_i S_i - K_i K_i)$$

and

$$C_2 = \sum_{i=1}^{3} S_i K_i \tag{6.11}$$

are the Casimir operators.

Problem 8. Show that $SO(2, 1)$ and $SL(2, r)$ are locally isomorphic to each other. $SL(2, r)$ is a subgroup of $SL(2, c)$ consisting of real matrices.

Problem 9. Using Cartan's criterion, show that $SL(2, c)$ does not have Abelian invariant subgroups. However, the subgroup of $SL(2, c)$ consisting of the matrices of the form

$$\begin{bmatrix} \alpha & \beta \\ 0 & 1/\alpha \end{bmatrix} \tag{6.12}$$

contains an invariant subgroup. The invariant subgroup in this case is generated by

$$N_1 = \begin{bmatrix} 0 & 1 \\ 0 & 0 \end{bmatrix}, \quad N_2 = \begin{bmatrix} 0 & -i \\ 0 & 0 \end{bmatrix}. \tag{6.13}$$

The above *set of generators* commute with the generators of the subgroup represented by Equation (6.11). However, this set does not commute with all the generators of the $SL(2, c)$ group. Show therefore that the subgroup generated by the above N_1 and N_2 operators is a subgroup but is not an invariant subgroup of $SL(2, c)$.

Problem 10. Consider the differential operators:

$$J_3 = \frac{1}{2}\left(x\frac{\partial}{\partial x} + y\frac{\partial}{\partial y}\right),$$

$$J_1 = \frac{1}{2}\left(x\frac{\partial}{\partial y} + y\frac{\partial}{\partial x}\right), \tag{6.14}$$

$$J_2 = \frac{1}{2i}\left(x\frac{\partial}{\partial y} - y\frac{\partial}{\partial x}\right).$$

Show that these differential operators satisfy the commutation relations for the three-dimensional rotation group. Construct a vector space in which these operators act like the generators of the rotation group.

Problem 11. Consider two matrices A and B of the same size. They do not necessarily commute with each other. Show that

$$e^{-A} B e^A = B + [B, A] + \frac{1}{2!}[[B, A], A] + \cdots \tag{6.15}$$

Problem 12. Prove then that

$$e^A e^B = e^C, \tag{6.16}$$

where

$$C = A + B + \tfrac{1}{2}[A, B] + \tfrac{1}{12}[A, [A, B]] - \tfrac{1}{12}[B, [B, A]] + \cdots$$

This formula is commonly known as the Baker-Campbell-Hausdorff or BCH formula, and is discussed extensively in the literature (Miller, 1972).

Problem 13. Let us consider the coupled first-order differential equations which can be written in the form:

$$i\frac{d\mathbf{x}}{dt} = A\mathbf{x}, \tag{6.17}$$

where $\mathbf{x}$ is a column vector and A is a square matrix. If A does not depend on t, the solution of the above differential equation can be written as

$$\mathbf{x}(t) = e^{-iAt}\mathbf{x}(0). \tag{6.18}$$

Using this formula, find the solutions of the following differential equations:

$$\frac{d}{dt}x_1 = x_2 + x_3, \quad \frac{d}{dt}x_2 = x_1 + x_3, \quad \frac{d}{dt}x_3 = x_2 + x_3. \tag{6.18}$$

Lie Groups and Lie Algebras

Problem 14. Let us go back to Equation (6.16). If the matrix A depends on time t, then the solution may be written formally as

$$\mathbf{x}(t) = \exp\left(-i \int_{t_0}^{t} A(t')\,dt'\right) x(t_0). \tag{6.20}$$

Give a physical interpretation of this formula for an electron in a time-dependent magnetic field. Study the problem in detail when the magnetic field is the superposition of a strong constant magnetic field along the z axis and a weak sinusoidal field along the x direction. The above expression is one of the fundamental formulas in the interaction representation which eventually leads us to Feynman diagrams (Schweber, 1961). For a discussion of this problem with a strong oscillating magnetic field, see Shirley (1965).

Problem 15. We are quite familiar with the spherical harmonics. They can be obtained either from the Legendre differential equation or from the irreducible tensor products of Cartesian coordinate variables. Let us take spherical harmonics with $\ell = 1$. Then its transformation property is the same as that of the coordinate transformation. We are also familiar with the differential form of the rotation operators applicable to the spherical harmonics. Do these operators generate active or passive transformations?

Problem 16. $SU(4)$ is the group of four-by-four unitary unimodular matrices. Construct the generators of $SU(4)$. How many generators does this group have? Construct these generators from those of two $SU(2)$ groups, each having three generators. Discuss all possible subgroups of $SU(4)$. What is the physical motivation for studying this group? See Wigner (1937). See also Oneda and Takasugi (1976).

Chapter III

Theory of the Poincaré Group

The Poincaré group is the group of inhomogeneous Lorentz transformations, namely Lorentz transformations followed by space-time translations. In order to study this group, we have to understand first the group of Lorentz transformations, the group of translations, *and* how these two groups are combined to form the Poincaré group.

As far as the coordinate transformations are concerned, the study of representations is a matter of constructing four-by-four and five-by-five matrices for the homogeneous and inhomogeneous Lorentz groups respectively. However, we are interested in the symmetry problems associated with quantum states and operators. We are particularly interested in Lorentz-transformation properties of Hilbert spaces and in operators acting on a Hilbert space.

To attack these problems, we use the original method of Wigner based on the little groups which leave the four-momentum of a given free particle invariant (Wigner, 1939). We should realize however that the mathematical techniques available to today's physicists are much richer and far more diverse than those of 1939. We should note in particular the following points.

(a) While the Poincaré group is a continuous group, Wigner did not employ the mathematical techniques available for Lie groups in his 1939 paper. The fact that the generators can serve useful purposes was not recognized until the appearance of the paper of Bargmann and Wigner (1946).

(b) The concept of covering group is widely known by physicists these days. The covering group for the Lorentz group is $SL(2, c)$ and is double-valued. For this reason, we do not have to repeat the lengthy argument Wigner gave on the double-valuedness of the representation.

(c) In his 1939 paper, Wigner was mostly interested in constructing representations of the little groups of the Poincaré group, although the concept of orbit completion was also clearly spelled out. These days, the study of relativistic particles includes calculations of orbit completion. The word "orbit completion," in practical terms, means the calculation of Lorentz boosts.

(d) In 1939, electrons and photons were known to exist. Wigner's 1939

Theory of the Poincaré Group 51

paper makes contact with the physical world through the representation for electrons. We now have a better understanding of photons and neutrinos. In addition, we now have quarks, hadrons and vector mesons. It is therefore necessary to extend Wigner's original form of the representation theory to include explanations for these new particles.

(e) In 1939, there did not exist a covariant formulation of quantum field theory. While Wigner was primarily interested in unitary representations in his original 1939 paper, finite-dimensional non-unitary representations play important roles in quantum field theory. It is therefore of interest to add discussions of non-unitary representations when we study representations of the Poincaré group.

In spite of the development of quantum field theory in which finite-dimensional non-unitary representations are commonly used, Wigner's original proposal to construct unitary representations is still the most fundamental issue. We have not forgotten this original proposal, and shall disucss in detail the method of constructing concrete models in this chapter and continue the discussion in Chapters V and VI. Chapter VII includes a discussion of the unitary representation of the photon polarization vectors.

In Section 1, we study the four-by-four Lorentz transformation matrices and the resulting Lie algebra. In Section 2, we study the orbits and little groups of the Lorentz group, in preparation for constructing representations of the Poincaré group. In Section 3, the little-group decomposition of the Poincaré group is carried out, and the Casimir operator is constructed for each little group.

Section 4 deals with the problem of constructing unitary representations of Lorentz transformations applicable to wave functions. It is shown that the problem reduces to that of finding a Hilbert space consisting of functions square integrable over the spatial and time-like coordinates.

We discuss in Section 5 finite-dimensional representations commonly used in quantum field theory. The little group and the corresponding orbit completion are discussed for each type of field. In Section 6, we discuss space and time reflections applicable to the representations commonly used in physics, with a particular emphasis on the little groups. Section 7 consists of exercises and problems.

1. Group of Lorentz Transformations

We shall use throughout this book the four-vector notation:

$$x^\mu = (x, y, z, t) \quad \text{and} \quad x_\mu = (x, y, z, -t). \tag{1.1}$$

The superscript and subscript will denote the row and column indices respectively. Therefore, x^μ and x_μ are column and row vectors respectively. The conversion from x^μ to x_μ is achieved through the relation:

$$x_\mu = g_{\mu\nu} x^\nu, \tag{1.2}$$

where $g_{11} = g_{22} = g_{33} = -g_{44} = 1$ while all other $g_{\mu\nu}$ are zero. It is often convenient to use the matrix:

$$G = \begin{bmatrix} 1 & 0 & 0 & 0 \\ 0 & 1 & 0 & 0 \\ 0 & 0 & 1 & 0 \\ 0 & 0 & 0 & -1 \end{bmatrix}. \tag{1.3}$$

We shall use the notation x_i to cover the first three components of the four-vector and use x_0 for the fourth component.

$$x^i = x_i = (x, y, z), \quad x^4 = -x_4 = t = x_0. \tag{1.4}$$

The lower case Greek and Roman letters will be used for the four-vector and three-component spatial vectors respectively.

The general (homogeneous) Lorentz group consists of all real four-by-four matrices A satisfying the condition:

$$A G A^T = G, \tag{1.5}$$

where A^T is the transpose of A. The A matrix leaves the quantity $(x^2 + y^2 + z^2 - t^2)$ invariant. This group is therefore often called $O(3, 1)$. If a^μ_ν are the elements of the matrix A, where μ and ν are the row and column indices respectively, the above equation can be written as

$$g_{\mu\nu} a^\mu_\sigma a^\nu_\lambda = g_{\sigma\lambda}. \tag{1.6}$$

This expression contains ten conditions on the sixteen elements of the four-by-four matrix A. Therefore, the Lorentz transformation matrix contains six independent parameters.

Since the determinant of A^T is the same as that of A, the condition of Equation (1.5) leads to

$$\det(A) = \pm 1. \tag{1.7}$$

The Lorentz group with $\det(A) = +1$ is called $SO(3, 1)$. Furthermore, if we take the $\sigma = \lambda = 0$ component of Equation (1.6),

$$|a^4_4|^2 = 1 + \sum_{i=1}^{3} |a^i_4|^2. \tag{1.8}$$

Thus

$$|a^4_4| \geq 1. \tag{1.9}$$

Transformations with positive a^4_4 are called orthochronous Lorentz transformations. The group of orthochronous Lorentz transformations with $\det(A) = 1$ is called the *proper Lorentz group*.

The proper Lorentz group is an invariant subgroup of the general Lorentz group. $O(3, 1)$ can therefore be expanded in terms of the cosets of the

Theory of the Poincaré Group

proper Lorentz group. The cosets in this case form a quotient group isomorphic to the four-element group consisting of the following matrices:

$$I(++) = \begin{bmatrix} 1 & 0 \\ 0 & 1 \end{bmatrix}, \quad I(+-) = \begin{bmatrix} 1 & 0 \\ 0 & -1 \end{bmatrix},$$

$$I(-+) = \begin{bmatrix} -1 & 0 \\ 0 & 1 \end{bmatrix}, \quad I(--) = \begin{bmatrix} -1 & 0 \\ 0 & -1 \end{bmatrix}.$$

(1.10)

These matrices represent the four group of type B discussed in Chapter I. The $I(++)$ component represents the proper Lorentz group, and is the identity. The upper-left element of the above matrix specifies the space-inversion property. If the element is (-1), then the spatial coordinates are to be inverted. The lower-right element is for the time reflection property. If (-1), the time variable is to change its sign. The sign of the determinant of the above matrix is the same as that of the four-by-four Lorentz transformation matrix.

The matrices which perform rotations on the three spatial coordinates form the $SO(3)$-like rotation subgroup of the proper Lorentz group. The coordinate transformation matrix takes the reduced form

$$A_r(\theta, \phi, \psi) = \begin{bmatrix} \begin{array}{c} \text{3-by-3} \\ \text{Rotation} \\ \text{Matrix} \end{array} & \begin{array}{c} 0 \\ 0 \\ 0 \end{array} \\ \hline 0 \quad 0 \quad 0 & 1 \end{bmatrix},$$

(1.11)

where ψ is the amount of rotation, and the angle variables θ and ϕ specify the direction of the rotation axis. As is well known, the rotation group has three independent parameters.

The boost matrix along the z direction takes the form

$$A_b(z, \eta) = \begin{bmatrix} 1 & 0 & 0 & 0 \\ 0 & 1 & 0 & 0 \\ 0 & 0 & \cosh\eta & \sinh\eta \\ 0 & 0 & \sinh\eta & \cosh\eta \end{bmatrix}.$$

(1.12)

A pure boost along the (θ, ϕ) direction is achieved by

$$A_b(\theta, \phi, \eta) = A_r(\theta, \phi, 0) A_b(z, \eta) [A_r(\theta, \phi, 0)]^{-1}$$

(1.13)

A pure boost also has three independent parameters.

While rotation matrices are orthogonal:

$$(A_r)^{-1} = (A_r)^T,$$

(1.14)

it is easy to prove using Equation (1.13) that pure boost matrices are symmetric:

$$A_b = (A_b)^T. \tag{1.15}$$

A multiplication of two A_r matrices results in another A_r. However, it is easy to check that a multiplication of two different A_b matrices does not give another A_b matrix. Instead,

$$A_b A'_b = A_r A''_b. \tag{1.16}$$

The most general transformation matrix for the proper Lorentz group can be decomposed into the form:

$$A = A_r A_b, \tag{1.17}$$

where both A_r and A_b have their respective three independent parameters. The proper Lorentz group is therefore a six-parameter group.

The transformation operators for the proper Lorentz group can be exponentiated as

$$A = \exp\left[-i \sum_{i=1}^{3} (\theta_i L_i + \eta_i K_i)\right], \tag{1.18}$$

where L_i and K_i are the generators of rotations and proper Lorentz boosts respectively. They form six independent generators of the Lorentz group. When applied to the four vector x^μ, namely the column vector (x, y, z, t), the generators take the matrix form:

$$L_1 = \begin{bmatrix} 0 & 0 & 0 & 0 \\ 0 & 0 & -i & 0 \\ 0 & i & 0 & 0 \\ 0 & 0 & 0 & 0 \end{bmatrix}, \quad K_1 = \begin{bmatrix} 0 & 0 & 0 & i \\ 0 & 0 & 0 & 0 \\ 0 & 0 & 0 & 0 \\ i & 0 & 0 & 0 \end{bmatrix},$$

$$L_2 = \begin{bmatrix} 0 & 0 & i & 0 \\ 0 & 0 & 0 & 0 \\ -i & 0 & 0 & 0 \\ 0 & 0 & 0 & 0 \end{bmatrix}, \quad K_2 = \begin{bmatrix} 0 & 0 & 0 & 0 \\ 0 & 0 & 0 & i \\ 0 & 0 & 0 & 0 \\ 0 & i & 0 & 0 \end{bmatrix}, \tag{1.19}$$

$$L_3 = \begin{bmatrix} 0 & -i & 0 & 0 \\ i & 0 & 0 & 0 \\ 0 & 0 & 0 & 0 \\ 0 & 0 & 0 & 0 \end{bmatrix}, \quad K_3 = \begin{bmatrix} 0 & 0 & 0 & 0 \\ 0 & 0 & 0 & 0 \\ 0 & 0 & 0 & i \\ 0 & 0 & i & 0 \end{bmatrix}.$$

In general, the generators of the proper Lorentz group applicable to an arbitrary function of the coordinate variables are

$$L_i = -i\varepsilon_{ijk} x_j(\partial/\partial x_k),$$
$$K_i = -i[t(\partial/\partial x_i) + x_i(\partial/\partial t)]. \quad (1.20)$$

These generators satisfy the following commutation relations:

$$[L_i, L_j] = i\varepsilon_{ijk} L_k,$$
$$[L_i, K_j] = i\varepsilon_{ijk} K_k, \quad (1.21)$$
$$[K_i, K_j] = -i\varepsilon_{ijk} L_k.$$

The first commutator indicates that transformations generated by L_i form a rotation subgroup of the proper Lorentz group. The second and third commutators imply that boosts alone do not form a group, and that a multiplication of two pure boosts will result in a multiplication of a boost and a rotation. The rotation subgroup is not an invariant subgroup of the Lorentz group.

The six generators given in Equation (1.20) can be combined into the covariant form

$$L_{\mu\nu} = -i\{x_\mu(\partial/\partial x^\nu) - x_\nu(\partial/\partial x^\mu)\}, \quad (1.22)$$

where

$$L_i = \varepsilon_{ijk} L_{kj},$$
$$K_i = L_{0i}.$$

From the transformation properties of x^μ and x_μ and those of their derivatives, it is clear that $L_{\mu\nu}$ transforms like a second-rank tensor:

$$L'_{\mu\nu} = a_\mu^\rho a_\nu^\sigma L_{\rho\sigma}, \quad (1.23)$$

under the Lorentz transformation. The commutation relations for $L_{\mu\nu}$ are

$$[L_{\mu\nu}, L_{\rho\sigma}] = i(g_{\mu\rho}L_{\nu\sigma} - g_{\nu\rho}L_{\mu\sigma} + g_{\mu\sigma}L_{\rho\nu} - g_{\nu\sigma}L_{\rho\mu}). \quad (1.24)$$

2. Orbits and Little Groups of the Proper Lorentz Group

According to the definition given in Section 3 of Chapter I, the little group of the proper Lorentz group consists of the subset $A^{(p)}$ of the proper transformation matrices satisfying the condition

$$A^{(p)} p = p, \quad (2.1)$$

where p is a column four-vector p^μ. Since the above relation is a linear form, the little group for p is the same as that for αp, where α is an arbitrary real number.

Again, according to the definition given in Section 4 of Chapter I, the orbit of p is the "surface" in the four-dimensional p space on which the quantity:

$$p^2 = p_\mu p^\mu = -m^2 \tag{2.2}$$

is constant. There is a one-to-one correspondence between this surface and the coset space $A/A^{(p)}$.

For the proper Lorentz group, there are six such orbits in the p space. The first two orbits consist of two time-like surfaces with positive m^2, and with positive and negative values of p_0 respectively. The third and fourth orbits are the forward and backward light cones respectively. The fifth orbit is a space-like surface with negative values of m^2. The sixth orbit is the origin in the p space where $p_1 = p_2 = p_3 = p_0 = 0$. These orbits are illustrated in Figure 2.1.

These orbits enable us to study all possible little groups by studying the little group for one representative four-vector on each orbit. For the first and

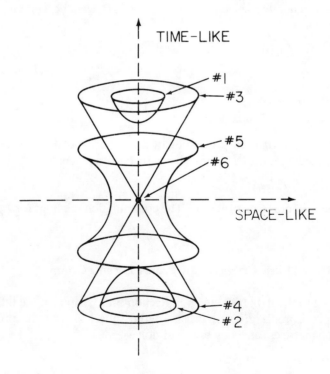

Fig. 2.1. Orbits of the Lorentz group at all possible four-momenta. The four-momentum can be time-like, light-like and space-like, or zero. There are six orbits. (This figure has been adapted from Halpern and Branscomb, 1965).

Theory of the Poincaré Group

second orbits, we can consider the little group as the group of transformations which leave the vector $(0, 0, 0, \pm m)$ invariant. The little groups for the third and fourth orbits can be represented by the matrices which leave $(0, 0, \omega, \pm \omega)$ invariant. The little group representing the fifth orbit leaves invariant the four-vector $(0, 0, p_3, 0)$. The little group for the sixth orbit leaves only the null vector invariant.

The task of finding representations of the little groups is now reduced to solving the linear equations:

$$\begin{bmatrix} a_1^1 & a_2^1 & a_3^1 & a_4^1 \\ a_1^2 & a_2^2 & a_3^2 & a_4^2 \\ a_1^3 & a_2^3 & a_3^3 & a_4^3 \\ a_1^0 & a_2^0 & a_3^0 & a_4^0 \end{bmatrix} \begin{bmatrix} p^1 \\ p^2 \\ p^3 \\ p^4 \end{bmatrix} = \begin{bmatrix} p^1 \\ p^2 \\ p^3 \\ p^4 \end{bmatrix}, \tag{2.3}$$

for each representation p subject to the condition $g_{\sigma\rho} a_\mu^\sigma a_\nu^\rho = g_{\mu\nu}$. Because this matrix equation imposes three additional conditions if the column vector is non-zero, the little groups for the first five orbits have three independent parameters. There are no additional conditions for the sixth orbit. Let us study each little group carefully.

(i) For the first and second orbits, the column vector $(0, 0, 0, \pm m)$ should remain invariant. From the above matrix equation and from the relation given in Equation (1.8), it is clear that $a_i^4 = a_4^i = 0$. This condition can be written explicitly as

$$\begin{bmatrix} a_1^1 & a_2^1 & a_3^1 & 0 \\ a_1^2 & a_2^2 & a_3^2 & 0 \\ a_1^3 & a_2^3 & a_3^3 & 0 \\ 0 & 0 & 0 & 1 \end{bmatrix} \begin{bmatrix} 0 \\ 0 \\ 0 \\ \pm m \end{bmatrix} = \begin{bmatrix} 0 \\ 0 \\ 0 \\ \pm m \end{bmatrix}. \tag{2.4}$$

The little group in this case is therefore the rotation subgroup of the proper Lorentz group.

(ii) For the third and fourth orbits, we have to consider Lorentz transformations which leave the four-vectors $(0, 0, \omega, \pm \omega)$ invariant. Let us study first the third orbit represented by $(0, 0, \omega, \omega)$. It is clear that rotations around the z axis will leave this four-vector invariant. The rotation matrix in this case takes the form

$$R(\phi) = \begin{bmatrix} \cos \phi & -\sin \phi & 0 & 0 \\ \sin \phi & \cos \phi & 0 & 0 \\ 0 & 0 & 1 & 0 \\ 0 & 0 & 0 & 1 \end{bmatrix}, \tag{2.5}$$

In addition, as is shown in Exercise 1 in Section 7, the representative four-vector remains invariant under the transformation (Wigner, 1939 and 1948):

$$T(u, v) = \begin{bmatrix} 1 & 0 & -u & u \\ 0 & 1 & -v & v \\ u & v & 1 - (u^2 + v^2)/2 & (u^2 + v^2)/2 \\ u & v & -(u^2 + v^2)/2 & 1 + (u^2 + v^2)/2 \end{bmatrix}. \quad (2.6)$$

This matrix has the algebraic property:

$$T(u, v) = T(u, 0) \, T(0, v) = T(0, v) \, T(u, 0),$$

$$T(u_2, v_2) \, T(u_1, v_1) = T(u_2 + u_1, v_2 + v_1), \quad (2.7)$$

which is identical to that of the translation subgroup of $SE(2)$ discussed in Chapter I.

The most general form of the three-parameter four-by-four matrix representing this little group can be written as

$$D(u, v, \phi) = T(u, v) \, R(\phi). \quad (2.8)$$

The D matrix has the algebraic property:

$$D(u_2, v_2, \phi_2) \, D(u_1, v_1, \phi_1) = T(u', v') \, R(\phi_2 + \phi_1)), \quad (2.9)$$

where

$$u' = u_2 + u_1 \cos \phi_2 - v_1 \sin \phi_2,$$

$$v' = v_2 + u_1 \sin \phi_2 + v_1 \cos \phi_2. \quad (2.10)$$

This algebraic property is identical to that for the F matrices representing the $SE(2)$ group which were discussed in Chapter I.

The calculations are essentially the same for the fourth orbit represented by $(0, 0, \omega, -\omega)$. The R matrix remains the same. The T matrix in this case is the transpose or the Hermitian conjugate of the T matrix given in Equation (2.6), which represents an Abelian group. Therefore, the algebraic property of the new D matrix is the same as that of the original D matrix for the third orbit given in Equation (2.8). Indeed, the little groups for the third and fourth orbits are isomorphic to $SE(2)$.

(iii) For the fifth orbit, the little group is represented by the three-parameter matrices which leave $(0, 0, q, 0)$ invariant. Here again the rotation around the z axis will satisfy the requirement. Boosts along the x and y axes will also leave the representative four vector invariant. These boosts together with the rotation around the z axis form a three-dimensional Lorentz group operating on two spatial dimensions and one time-like dimension. This group is often called $SO(2, 1)$.

(iv) The sixth orbit consists of only the origin. The six-parameter matrices of the proper Lorentz group will leave this point invariant.

Let us summarize this section by writing down the generators of the little groups and the commutation relations. For the first and second orbits consisting of time-like surfaces in the four-vector spaces, the little group is generated by L_1, L_2 and L_3, and their commutation relations are given in Equation (1.21).

For the third orbit consisting of the forward light cone, the little group is generated by L_3 given in Equation (1.19) and by N_1 and N_2 where

$$N_1 = \begin{bmatrix} 0 & 0 & -i & i \\ 0 & 0 & 0 & 0 \\ i & 0 & 0 & 0 \\ i & 0 & 0 & 0 \end{bmatrix}, \quad N_2 = \begin{bmatrix} 0 & 0 & 0 & 0 \\ 0 & 0 & -i & i \\ 0 & i & 0 & 0 \\ 0 & i & 0 & 0 \end{bmatrix}. \quad (2.11)$$

These matrices are the infinitesimal limits of the $T(u, 0)$ and $T(0, v)$ matrices given in Equations (2.6) and (2.7). From the matrices given in Equation (1.19), it is clear that

$$N_1 = K_1 - L_2, \quad \text{and} \quad N_2 = K_2 + L_1. \quad (2.12)$$

These operators have the commutation relations:

$$[N_1, N_2] = 0,$$
$$[N_1, L_3] = -iN_2, \quad (2.13)$$
$$[N_2, L_3] = iN_1.$$

These commutation relations are identical to those for the $SE(2)$ group. If the N operators in the above equations are replaced by their respective Hermitian conjugates, the commutation relations remain invariant. The little group for the fourth orbit is generated by L_3 and the Hermitian conjugates of N_1 and N_2.

The little group for the fifth orbit consisting of space-like surfaces is generated by L_3, K_1 and K_2.

$$[K_1, K_2] = -iL_3,$$
$$[K_1, L_3] = -iK_2, \quad (2.14)$$
$$[K_2, L_3] = iK_1.$$

As was pointed out before, the little group for the sixth orbit consisting only of the origin is the proper Lorentz group generated by the matrices given in Equation (1.19).

These little groups will be discussed in more detail in later chapters in connection with representations of the Poincaré group. In the meantime, we should keep in mind the fact that the local isomorphism does not necessarily mean that groups having the same Lie algebra are identical. Clearly $SU(2)$ is not identical to $SO(3)$. The little group for massless particles has the same

Lie algebra as that of $SE(2)$, but they are quite different groups. We shall use the word "like" to specify the local isomorphism. For instance, $SU(2)$ is an "$O(3)$-like" subgroup of $SL(2, c)$. The little group for massless particles is an "$E(2)$-like" subgroup of the Lorentz group.

3. Representations of the Poincaré Group

The study of the Poincaré group starts with the group of inhomogeneous Lorentz transformations on the four-dimensional Minkowski space in the form

$$x'^\mu = a^\mu_\nu x^\nu + b^\nu. \tag{3.1}$$

The matrix performing the above linear transformation is not difficult to construct. Let us consider the five-by-five matrices of the form

$$A = \begin{bmatrix} & & & & 0 \\ & \text{4-by-4} & & & 0 \\ & \text{Lorentz Trans.} & & & 0 \\ & \text{Matrix} & & & 0 \\ \hline 0 & 0 & 0 & 0 & 1 \end{bmatrix}, \quad B = \begin{bmatrix} 1 & 0 & 0 & 0 & b^1 \\ 0 & 1 & 0 & 0 & b^2 \\ 0 & 0 & 1 & 0 & b^3 \\ 0 & 0 & 0 & 1 & b^0 \\ 0 & 0 & 0 & 0 & 1 \end{bmatrix}. \tag{3.2}$$

applicable to the five-element column vector $(x, y, z, t, 1)$. Then the inhomogeneous Lorentz transformation of Equation (3.1) can be written as

$$x' = B(b) A(a) x, \tag{3.3}$$

where x, x', a, and b are abbreviations of their respective quantities with indices. This matrix is indeed the representation of the inhomogeneous Lorentz transformation of the space-time coordinate.

The A matrix has the same properties as the four-by-four Lorentz transformation matrices which we discussed in Sections 1 and 2. The B matrix representing translations in the four-dimensional Minkowski space is Abelian and is an invariant subgroup of the inhomogeneous Lorentz group:

$$B(b') B(b) = B(b' + b),$$
$$A(a) B(b) = B(aba^{-1}) A(a). \tag{3.4}$$

The group BA therefore has the algebraic property:

$$[B(b') A(a')] [B(b) A(a)] = B(b' + a'ba'^{-1}) A(a'a). \tag{3.5}$$

Indeed, the Poincaré group is a semidirect product of the Lorentz group $A(a)$ and the translation group $B(b)$.

Because B is an invariant subgroup, matrices of the form AB, instead of BA, are equally adequate for representing the group. This form of representation is convenient when we study space-time symmetries of a relativistic particle with a given four-momentum p, whose state vector has the momentum dependence in the form

$$\exp(\pm ip \cdot x), \tag{3.6}$$

where

$$p \cdot x = p_\mu x^\mu.$$

From this expression and also from the five-by-five matrix form for B, it is clear that the translation is generated by

$$P_\mu = -i(\partial/\partial x^\mu). \tag{3.7}$$

From the discussions given in Sections 1 and 2, we can write down the most general form for the generators of the Lorentz transformation:

$$M_{\mu\nu} = L_{\mu\nu} + S_{\mu\nu}, \tag{3.8}$$

where the $L_{\mu\nu}$ part is applicable to coordinate variables, while $S_{\mu\nu}$ is applicable to all other variables which are not affected by $L_{\mu\nu}$. In the case of massive Dirac particles, the S operators are to generate Lorentz transformations on the Dirac spinor. In the case of photons, we have to construct $S_{\mu\nu}$ applicable to the polarization vector. In the case of particles with a finite size and with internal motion of constituent particles, the S opertaor has to generate Lorentz transformations on the internal coordinate variables.

With this in mind, we can write the commutation relations among the generators of the Poincaré group as

$$\begin{aligned}[P_\mu, P_\nu] &= 0, \\ [M_{\mu\nu}, P_\rho] &= i(g_{\mu\rho}P_\nu - g_{\nu\rho}P_\mu), \\ [M_{\mu\nu}, M_{\rho\sigma}] &= i(g_{\mu\rho}M_{\nu\sigma} - g_{\nu\rho}M_{\mu\sigma} + g_{\mu\sigma}M_{\rho\nu} - g_{\nu\sigma}M_{\rho\mu}).\end{aligned} \tag{3.9}$$

The fact that P_μ does not commute with $M_{\mu\nu}$ is a clear indication that the Poincaré group is not a direct product of the Lorentz and translation groups.

In constructing irreducible representations of the group, it is convenient to construct, from the generators of the group, the maximal set of operators which commute with all the generators. These are the Casimir operators discussed in Section 3 of Chapter II. In the case of the Poincaré group, we can show after a lengthy but straightforward calculation (Problem 5 in Section 7) that the Casimir operators are

$$\begin{aligned}P^2 &= P_\mu P^\mu, \\ W^2 &= W_\mu W^\mu,\end{aligned} \tag{3.10}$$

where

$$W_\mu = \tfrac{1}{2} \varepsilon_{\mu\nu\alpha\beta} P^\nu M^{\alpha\beta}. \tag{3.11}$$

W_μ transforms like a four-vector but is constrained to satisfy

$$P_\mu W^\mu = 0. \tag{3.12}$$

The four-vector W therefore has only three independent components. Furthermore, it commutes with P_μ:

$$[P_\mu, W_\nu] = 0. \tag{3.13}$$

When the system is a free-particle state, the operator P^2 determines the (mass)^2 of the particle. Because of Equation (3.13), P_μ in Equations (3.10) and (3.11) can be replaced by its eigenvalue p_μ. We are then led to the suspicion that three independent components of W_μ are the generators of the little group at p^μ. Let us examine this point for each orbit.

(i) For the first and second orbits where $p^\mu = (0, 0, 0, \pm m)$, $W_0 = 0$, and

$$W_i = \pm (m/2)\, \varepsilon_{ijk} M_{jk}, \tag{3.14}$$

where m is the particle mass. Clearly three W_i's generate an $SO(3)$-like little group with its spin oriented in an arbitrary direction. The L_{ij} part of M_{ij} has no effect on the system, and therefore can be dropped. The S_{ij} part represents the internal angular momentum.

We are already quite familiar with the spin 1/2 electron case where the internal angular momentum comes from the spinor representation of the rotation group. For the case of massive vector mesons, represented by a four-vector A^μ with the constraint $p_\mu A^\mu = 0$, the S^{ij} matrices should be those of the $SO(3)$ subgroup of the proper Lorentz group. They are four-by-four matrices. If the particle is a relativistic extended hadron with internal quark coordinate y_i, then the S_{ij} operators take the form

$$S_{ij} = -i[y_i(\partial/\partial y_j) - y_j(\partial/\partial y_i)]. \tag{3.15}$$

The internal space-time symmetry of the extended hadron will be studied in detail in Chapters V, VI, XI and XII.

(ii) The third and fourth orbits are specified by the four-vectors $p^\mu = (0, 0, \omega, \pm \omega)$. Here again the L_{ij} part drops out, and

$$\begin{aligned} W_1 &= \omega(S^{02} - S^{23}), \\ W_2 &= \omega(S^{01} - S^{31}), \\ W_3 &= W_0 = \omega S^{12}. \end{aligned} \tag{3.16}$$

W_0 in this case is redundant. It is clear that the first three W operators are proportional to the generators of the $SE(2)$-like little groups discussed in Section 2.

The massless particle most familiar to us is of course the photon, which is commonly represented by its four-vector potential A^μ. The photon is known to have spin 1 parallel or antiparallel to its momentum. A^μ depends on gauge parameters which do not produce any measurable effect. Can the symmetry of the little group explain all these? The matrices of the little group generated

Theory of the Poincaré Group

by W_i operators applicable to A^μ are the four-by-four matrices $D(u, v, \phi)$ given in Equation (2.8).

There are also interesting stories about neutrinos which are spin 1/2 massless particles. The basic difference between photons and neutrinos is that neutrinos are polarized whereas the photon spin can be both parallel and antiparallel. Does the symmetry of the little group play any role in the polarization of neutrinos? This problem, together with other massless particles including photons, will be discussed in Chapter VII.

(iii) For the fifth orbit specified by the four-vector $p^\mu = (0, 0, q, 0)$, the W operators are

$$W_1 = q\,S_{02}, \quad W_2 = q\,S_{10}, \quad W_3 = 0, \quad W_0 = q\,S_{12}. \tag{3.17}$$

The three non-vanishing generators are proportional to $S_3 = S_{12}$, $K_1 = S_{02}$, and $K_2 = S_{02}$. These generators satisfy the commutation relations for the $SO(2, 1)$ group given in Equation (2.14).

(iv) The sixth orbit is specified by a four-momentum whose components vanish. The W operators do not exist in this case. Since this null four-vector is invariant under Lorentz transformations, there is no need for orbit completion once the little group is known.

The little groups are summarized in the table given in the Introduction. They are subgroups of $O(3, 1)$ and $SL(2, c)$. For the case of time-like and light-like p^μ, their representations describe internal space-time symmetries of observed free particles. Although we do not see immediate physical applications if the little groups for the space-like and point-like orbits, they are of mathematical interest and should be included in the discussion of the Poincaré group.

The $O(3)$-like little group for massive particles is compact and can be studied with the techniques developed for the three-dimensional rotation group. The $O(2, 1)$-like little group is non-compact and its representations are either unitary and infinite-dimensional or non-unitary and finite-dimensional. The same is true for representations of $O(3, 1)$. The $E(2)$-like little group for massless particles is expected to have the same algebraic property as $E(2)$. As was discussed in Chapter II, the $E(2)$ group is non-compact, but it has an Abelian invariant subgroup. We shall see how this affects its representations in Chapter VII.

Let us summarize the method of constructing representations of the Poincaré group. The inhomogeneous Lorentz group is a ten-parameter Lie group and is a semi-direct product of the four-parameter space-time translation group and the six-parameter group of Lorentz transformations. We are interested here only in those representations which are relevant to physics. The fundamental contribution Wigner made in 1939 was to observe that the little group which leaves the four-momentum of a given relativistic particle describes the internal space-time symmetry. In the case of electrons, for instance, the $O(3)$-like little group describes the spin angular momenta of electrons.

Indeed, Wigner's little group serves as the starting point for a more ambitious program for studying representations of the Lorentz group. After constructing representations of the little group for the four-momentum of a given particle, we have to study their Lorentz transformation properties as the four-momentum traces its orbit. This process of orbit completion is necessarily the study of a non-compact group, whose finite-dimensional representations are non-unitary, and whose unitary representations are infinite-dimensional.

The little group is definitely a subgroup of the Lorentz group. Indeed, the above-mentioned orbit completion leads to the idea of constructing representations of a given group starting from those of its subgroups. The representation constructed in this manner is called the *induced representation*.

When we construct representations of the Poincaré group by taking a semi-direct product of the space-time translation group and the homogeneous Lorentz group, it is an induced representation. In Chapter I, it was noted that the representations of $E(2)$ can be constructed from those of $O(2)$ and the two-dimensional translation group. This construction is also an induced representation. From the mathematical point of view, the study of the Poincaré group consists of constructing induced representations. The mathematics of induced representation is a very rich subject (Mackey, 1968, 1978). However, in this book, we shall restrict ourselves to the examples which explain the physical world.

4. Lorentz Transformations of Wave Functions

In Sections 2 and 3, we have been preoccupied with the little groups of the Poincaré group whose transformations leave the four-momentum of the free particle invariant. As we noted in Section 2, we still have to *complete the orbit* of each little group. In order to discuss fully the relativistic space-time symmetry, we have to study the effect of Lorentz transformations which change the four-momentum.

We already know that the Lorentz scalars such as P^2 and W^2 should remain invariant under Lorentz transformations. Since, however, we are dealing with quantum mechanics, we have to study the transformation properties of the Hilbert space and of the operators applicable to the Hilbert space. Indeed, this was the ultimate purpose of Wigner's original paper (Wigner, 1939). The 1946 paper of Bargmann and Wigner contains the following paragraphs.

In quantum mechanics, we deal with the wave functions, ψ, satisfying wave equations. They describe the possible states of a quantum mechanical system and form a linear vector space V which is either finite or infinite dimensional, and on which a positive definite inner product (ϕ, ψ) is defined for any two wave functions ϕ and ψ (i.e., they form a Hilbert space). The inner product usually involves a summation over the spin indices, as well as an integration over the whole configuration or momentum space.

If the wave function in question refers to a free particle and satisfies relativistic equations, there exists a correspondence between the wave functions describing the same state in different frames. The transformations considered here form the group of all inhomogeneous Lorentz transformations. Let ψ_a and $\psi_{a'}$ be the wave functions of the same state in two different Lorentz frames a and a' respectively. Then $\psi_{a'} = U(A)\psi_a$, where $U(A)$ is a linear unitary operator which depends on Lorentz transformation A leading from a to a'.

Bargmann and Wigner continue:

By proper normalization, U is determined by A up to a factor ± 1. Moreover, the operator U form a single- or double-valued representation of the inhomogeneous Lorentz group, i.e., for a succession of two Lorentz transformations A_1, and A_2, we have

$$U(A_1 A_2) = \pm U(A_1) U(A_2). \tag{4.1}$$

Since all Lorentz frames are equivalent for the description of our system, it follows that, together with ψ, $U(A)\psi$ is also a possible state viewed from the original Lorentz frame ℓ. Thus, the vector space V contains, with every ψ, all transforms $U(A)\psi$, where A is any Lorentz transformation.

A modern interpretation of the double-valuedness of Equation (4.1) is that $SL(2, c)$ is the universal covering group for the group of Lorentz transformations. In his original paper (1939), Wigner was primarily interested in unitary representations $U(A)$ with positive definite metric:

$$\psi_{a'}^n = \sum_k u_k^n(a', a) \psi_a^k, \tag{4.2}$$

with

$$\sum_k |u_k^n(a', a)|^2 = 1,$$

where the indices n and k specify quantum states. The summation is expected to extend from zero to infinity for unitary representations. Indeed, the construction of representations satisfying the above requirement is one of the primary purposes of this book.

In order that the representation be unitary, the generators of Lorentz transformations have to be Hermitian. We are familiar with the case in which the rotation operators are Hermitian. However, attempts to construct finite-dimensional matrices lead to non-unitary representations. The last hope therefore is in the *differential form of the generators* given in Equation (1.20). Here again there is no problem with the generators of rotations. In order that the boost operators be Hermitian, the operators K_i should satisfy the condition

$$(\phi, K_i \psi) = (K_i \phi, \psi). \tag{4.3}$$

K_i contains a derivative with respect to time. For this reason, the wave functions to which K_i is applicable have to be square integrable in the time variable as well as the spatial variables.

It is not common in quantum mechanics to construct functions square-integrable in the time variable, because there is no Hilbert space associated with this variable. Therefore, we are tempted to consider functions which do not depend on time. In this case, the time derivative in K_i disappears, and K_i is still Hermitian. The resulting unitary representations have been studied in detail by Naimark (1957). However, functions which do not depend on time in one Lorentz frame are not independent of time in other frames. It is therefore essential that we understand the physics of this time dependence before making any attempt to construct representations.

Another complication in constructing the above-mentioned unitary representations is that we are discussing here a free particle with a definite four-momentum. If this is the case, is there any non-trivial space-time distribution? Are there space-time variables which are other than those conjugate to the components of the four-momentum?

By definition, the problem of finding unitary representations of the Lorentz transformation is that of constructing a square-integrable Hilbert space in both space and time variables. We shall discuss in Chapter V a concrete model which satisfies this criterion. This problem is much deeper than a simple mathematical construction. The issue is how to construct a localized probability distribution which can be Lorentz-transformed. We shall discuss in Chapter VI the physical interpretation of the mathematical construction given in Chapter V. The phenomenological aspects of this formalism will be discussed in Chapters XI and XII.

Let us return to the question of dimensionality of representations. As was noted before, the summation given in Equation (4.2) is an infinite sum indicating that unitary representations of non-compact group are infinite-dimensional. If the dimension is to be finite, as in the case of most of the particles in quantum field theory, the representation is non-unitary. The question then is whether this lack of unitarity is the final word for the representations we use in field theory.

As we discussed in Section 4 of Chapter I, the word unitarity is associated with the positive definite metric. As was also pointed out there, the concept of pseudo-unitarity or pseudo-orthogonality plays an important role in both mathematics and physics. In the pseudo-unitary representation, the metric is not positive definite, and some components of the vector space have negative metric.

The four-by-four Lorentz transformation matrix applicable to four-vectors constitute a pseudo-orthogonal representation. The four-by-four transformation matrix applicable to Dirac spinors belong to a pseudo-unitary representation. The Lorentz transformation matrix applicable to the photon four-potential is of course a pseudo-orthogonal matrix. However, in the case of photon four-potentials, it is possible to eliminate the time-like component imposing a gauge condition in a Lorentz-invariant manner. In this case, the transformation matrix becomes unitary. This photon problem will be discussed in Chapters VII and VIII. We shall study in Section 6 of the present chapter some of the non-unitary representations which commonly appear in the existing literature.

In his original paper (1939), Wigner was interested in constructing unitary representations. Theories based on finite-dimensional non-unitary representations are also consistent with Wigner's original spirit, if physical interpretations can be given to the pseudo-unitary representation in question.

5. Lorentz Transformations of Free Fields

In quantum field theory, most of the particles and fields have finite components and are Lorentz-transformed by finite-dimensional non-unitary matrices, with the metric which is not positive definite. We explained in Section 4 why the discussions of these representations do not destroy Wigner's original purpose of constructing unitary representations.

The particles most familiar to us through our experience with quantum field theory are scalar or pseudoscalar mesons, photons, vector mesons, electrons and neutrinos. The Lorentz transformation properties of these fields based on finite-dimensional matrices have been discussed extensively in the literature. We are interested here in the little groups for these fields.

Let us discuss first the free scalar field which satisfies the homogeneous Klein-Gordon equation. The solution of the Klein-Gordon equation takes the form

$$\phi_p(x) = e^{\pm ip \cdot x}. \tag{5.1}$$

This solution is in an eigenstate of the four-momentum. If we boost this scalar particle in such a way that it will have the four-momentum p', then p' is obtained through an active Lorentz transformation:

$$p'^\mu = a^\mu_\nu p^\nu. \tag{5.2}$$

The transformed solution takes the form

$$\phi_{p'}(x) = e^{\pm ip' \cdot x}. \tag{5.3}$$

Because the exponent is a Lorentz scalar, we can make arbitrary Lorentz transformations on p and x simultaneously. If we make a transformation which is inverse to that of Equation (5.2), then the above expression can be written as

$$\phi_{p'}(x) = e^{\pm ip \cdot x'}, \tag{5.4}$$

where

$$x^\mu = a^\mu_\nu x'^\nu.$$

The transformation of x to x' is the inverse of that of the transformation for p in Equation (5.2). This is a passive transformation. Indeed, we can achieve the same purpose by making an active transformation on p or by making a passive transformation on the space-time coordinate. The scalar particles have no internal space-time structure subject to transformations of the little group.

Let us next consider electromagnetic fields. In spite of the well-known inconvenience associated with gauge degrees of freedom, the four-potential form of electromagnetic fields is still the most useful representation in all branches of physics, especially in quantum field theory. For free photons, the solution of the homogeneous wave equation becomes

$$A^\mu(x) = A^\mu e^{\pm ip \cdot x}. \qquad (5.5)$$

The coordinate dependence is only through the exponential form. The Lorentz-transformation property of this exponential form is the same as that for the scalar field. As for the four-vector A^μ, we have to apply an active Lorentz transformation on it.

The complication in using the four-vector form is that not all the components are independent. There are in fact only two independent components. For instance, if the momentum of the photon is in the z direction, we can let the third and fourth components vanish:

$$A^\mu = (A_1, A_2, 0, 0), \qquad (5.6)$$

or equivalently impose the conditions:

$$p_\mu A^\mu = 0 \quad \text{and} \quad \mathbf{p} \cdot \mathbf{A} = 0. \qquad (5.7)$$

The first of the above conditions is Lorentz-invariant, but the second is not. For this reason, the Lorentz transformation of the photon four-vector is more complicated than that for the momentum four-vector.

However, we should note that the four-by-four matrix of the little group applicable to the photon case is given in Equation (2.6). Although this matrix does not change the four-momentum, it changes A^μ. Thus, there is a reason to expect that the little group matrix will play a role in making the conditions of Equation (5.7) covariant. This problem will be discussed in Chapters VII and VIII.

The four-vector representation of the free massive vector particle can also be written as Equation (5.5), with the subsidiary condition:

$$p_\mu A^\mu = 0. \qquad (5.8)$$

Unlike the massless case, there are three independent components. If the particle is at rest, then the fourth component of A^μ vanishes. The little group in this case is generated by L_1, L_2 and L_3 given in Equation (1.19). If the particle gains a velocity along the z direction, we have to perform an active Lorentz boost on the A^μ four-vector using the boost matrix

$$B(\eta) = e^{-i\eta K_3} = \begin{bmatrix} 1 & 0 & 0 & 0 \\ 0 & 1 & 0 & 0 \\ 0 & 0 & \cosh\eta & \sinh\eta \\ 0 & 0 & \sinh\eta & \cosh\eta \end{bmatrix}. \tag{5.9}$$

The little group for this moving vector meson is generated by L'_1, L'_2, and L'_3, where

$$L'_i = B(\eta) L_i [B(\eta)]^{-1}. \tag{5.10}$$

These generators also satisfy the commutation relations for the three-dimensional rotation group:

$$[L'_i, L'_j] = i\varepsilon_{ijk} L'_k. \tag{5.11}$$

Free spin-1/2 Dirac particle wave functions satisfy the Dirac equation:

$$\left(i\gamma^\mu \frac{\partial}{\partial x^\mu} + m\right)\psi(x) = 0. \tag{5.12}$$

This equation has four linearly independent solutions. The Dirac matrices satisfy the algebraic relations:

$$\gamma_\mu \gamma_\nu + \gamma_\nu \gamma_\mu = -2g_{\mu\nu}. \tag{5.13}$$

The smallest matrices satisfying the above commutation relations are four-by-four. We can write them as

$$\gamma_i = \begin{bmatrix} 0 & \sigma_i \\ -\sigma_i & 0 \end{bmatrix}, \tag{5.14}$$

$$\gamma_0 = \begin{bmatrix} 0 & 1 \\ 1 & 0 \end{bmatrix}, \quad \gamma_5 = \begin{bmatrix} 1 & 0 \\ 0 & -1 \end{bmatrix}. \tag{5.15}$$

This form of the Dirac matrices is called the Weyl representation, which we shall use throughout this book.

In terms of these γ matrices, we can construct the generators of Lorentz transformations:

$$S_{\mu\nu} = (1/4i)[\gamma_\mu, \gamma_\nu]. \tag{5.16}$$

In the Weyl representation, this quantity can be written out as

$$S_i = \frac{1}{2} \varepsilon_{ijk} S_{jk} = \frac{1}{2} \begin{bmatrix} \sigma_i & 0 \\ 0 & \sigma_i \end{bmatrix}, \qquad (5.17)$$

$$K_i = S_{0i} = \frac{1}{2} \begin{bmatrix} i\sigma_i & 0 \\ 0 & -i\sigma_i \end{bmatrix}. \qquad (5.18)$$

The form of S_i in the above expression is in block diagonal form. The boost generators K_i are also in block diagonal form. However, the first and second blocks of the boost generators have the opposite sign to each other. In order to see whether this is acceptable, let us go back to $SL(2, c)$ discussed in Section 6 (Problem 7) of Chapter II. The generators of $SL(2, c)$ satisfy the same Lie algebra as that for $O(3, 1)$ given in Equation (1.21). These commutation relations of Equation (1.21) remain unchanged when the boost operators change their sign. For this reason, the above rotation and boost operators satisfy the same commutation relations as those for $O(3, 1)$ or $SL(2, c)$. Furthermore, they can be combined into the covariant form

$$[S_{\mu\nu}, S_{\rho\sigma}] = i(g_{\mu\rho} S_{\nu\sigma} - g_{\nu\rho} S_{\mu\sigma} - g_{\mu\sigma} S_{\rho\nu} - g_{\nu\sigma} S_{\rho\mu}). \qquad (5.19)$$

The little group for the Dirac particle is generated by three S_i operators of Equation (5.17) when the particle is at rest. If the particle gains a velocity along the z direction, the generators of the little group become

$$S'_i = B(\eta) S_i [B(\eta)]^{-1}, \qquad (5.20)$$

where $B(\eta)$ is the boost matrix and takes the form

$$B(\eta) = \begin{bmatrix} e^{\eta/2} & 0 & 0 & 0 \\ 0 & e^{-\eta/2} & 0 & 0 \\ 0 & 0 & e^{-\eta/2} & 0 \\ 0 & 0 & 0 & e^{\eta/2} \end{bmatrix}. \qquad (5.21)$$

If the mass of the Dirac particle vanishes, the little group is generated by S_3 in Equation (5.17), and N_1 and N_2:

$$\begin{aligned} N_1 &= K_1 + S_2, \\ N_2 &= K_2 - S_1, \end{aligned} \qquad (5.22)$$

where $K_{1,2}$ and $S_{1,2}$ are four-by-four matrices given in Equations (5.17) and (5.18). However, in this case, the Dirac equation becomes separated, and the upper two components of the Dirac spinor becomes completely decoupled from the lower components. Therefore, we can choose either the upper component or lower component, by choosing the eigenvalue of β_5 to be $+1$ or -1 respectively. Accordingly, the matrices representing the little group become two-by-two.

We shall continue the discussion of the symmetry of the Dirac equation in Chapter IV.

6. Discrete Symmetry Operations

We have so far studied representations of the Poincaré group and its little groups without playing much attention to the fact that there are also dicrete symmetries associated with this group. As was noted in Section 1, if the space and time reflections are taken into account, the Lorentz group has four branches. In addition, we have to note the fact that wave functions and operators in quantum mechanics are complex, and the results of calculations to be compared with experiment remain unchanged if all mathematical symbols in the formalism are replaced by their complex conjugates. Thus the study of discrete symmetry should include space reflection, time reflection, and complex conjugation.

The primary purpose of studying these discrete symmetry operations is to understand properties of elementary particles under space inversion (P, parity) charge conjugation (C), and time reversal (T). We expect that the momentum will reverse its direction under P and T, and that the angular momentum will change its sign under C and T.

We shall examine these operations using the generators of Lorentz transformations. The above-mentioned discrete symmetry operations will in general change the signs of the generators. The rotation operators correspond to physically measurable dynamical quantities. Thus if they change their sign, the direction of the angular momentum is reversed. However, there is no measurable quantity associated with the boost operators. Furthermore, the change in sign of the boost generators does not necessarily mean that the direction of motion is changed.

In order to examine this point more effectively, let us rewrite the commutation relations for the J_i and K_i operators:

$$[J_i, J_j] = i\varepsilon_{ijk} J_k,$$
$$[J_i, K_j] = i\varepsilon_{ijk} K_k, \qquad (6.1)$$
$$[K_i, K_j] = -i\varepsilon_{ijk} J_k.$$

It is necessary to change J_i to $-J_i$ when J_i changes its sign. However, the commutation relations remain invariant when the boost generators change their sign. Thus the change in sign of K_i alone does not lead to any definite physical conclusion.

In order to study the effect of the discrete symmetry operation on the boost generator, we have to see the final form of the transformation matrix. If a massive particle at rest is boosted along one direction, it will move along that direction. If it is boosted along the opposite direction, it will move in the opposite direction.

Let us examine first the effect of the discrete symmetry operations on the representations based on the differential forms of the generators given in Equation (1.20). It is quite clear that all the generators will change sign under complex conjugation. However, complex conjugation is not the same as Hermitian conjugation. The angular momentum operators remain invariant under space and time reflections, while the boost operators change their sign under space reflection or time reflection. These are summarized in Table 6.1.

TABLE 6.1
Properties of the differential forms of J_i and K_i under discrete symmetry operations.

	Space reflection	Time reflection	Complex conjugation	Hermitian conjugation
J_i	J_i	J_i	$-J_i$	J_i
K_i	$-K_i$	$-K_i$	$-K_i$	K_i (unitary) $-K_i$ (non-unitary)

Because the generators are pure imaginary, the transformation matrices are real and remain invariant under complex conjugation. Thus the parity operation is achieved by space-reflection. Charge conjugation is simply complex conjugation. Time reversal is time reflection followed by complex conjugation.

The above-mentioned symmetry operations are applicable to bosons with integer spin in which the differential forms of the generators are used, including photons, vector mesons, and gravitons. Pseudo-scalar mesons have an intrinsic parity.

In addition, we would expect that these symmetry operations will be applicable to the unitary representation which we will discuss in Chapter V. As is seen in Table 6.1, the Hermitian conjugation in the unitary representation is different from that for the non-unitary representation.

Let us next discuss massive Dirac particles. If we use for the γ matrices the expressions given in Equation (5.17), space inversion or P is achieved through

$$(\gamma_0) S_i (\gamma_0)^{-1} = S_i,$$
$$(\gamma_0) K_i (\gamma_0)^{-1} = -K_i, \tag{6.2}$$

Theory of the Poincaré Group

followed by space reflection. The parity operation will leave the direction of the spin invariant, but will reverse the direction of the momentum.

Charge conjugation or *C* requires sign changes in both the rotation and boost generators. This is done through:

$$(\gamma_2) S_i^* (\gamma_2)^{-1} = -S_i,$$
$$(\gamma_2) K_i^* (\gamma_2)^{-1} = -K_i. \tag{6.3}$$

This will lead to the spin flip and change in the energy state.

The time reversal or *T* operation requires a change in sign of S_i:

$$(\gamma_1 \gamma_3) S_i^* (\gamma_1 \gamma_3)^{-1} = -S_i,$$
$$(\gamma_1 \gamma_3) K_i^* (\gamma_1 \gamma_3)^{-1} = K_i. \tag{6.4}$$

This will lead to the reversal in the direction of the angular momentum as well as in the direction of the momentum.

It is now evident that, under the successive operations of *PCT*, each generator is brought back to itself, or *PCT* is the identity operation. We often call this effect the *PCT* theorem applicable to massive fermions. These effects are summarized in Table 6.2. In Table 6.2, we used the following forms for the rotation and boost matrices respectively:

$$R(\mathbf{n}, \alpha) = \exp(-i\alpha \mathbf{n} \cdot \mathbf{S}),$$

and

$$B(\mathbf{n}, \eta) = \exp(-i\eta \mathbf{n} \cdot \mathbf{K}), \tag{6.5}$$

and took into account the fact that complex conjugation is required for *C* and *T*.

It is easy to study the effect of *P*, *C* and *T* on little groups for massive particles. For massive Dirac particles, the three-parameter group is generated by S_i which corresponds to the spin of the particle. The direction of the spin remains unchanged under parity operation. However, it changes sign under time-reversal and charge conjugation.

TABLE 6.2
Effects of *P*, *C* and *T* on the generators of Lorentz transformation.

	P	C	T	PCT
S_i	S_i	$-S_i$	$-S_i$	S_i
$R(\mathbf{n}, \alpha)$	R	R	R	R
K_i	$-K_i$	$-K_i$	K_i	K_i
$B(\mathbf{n}, \eta)$	B^{-1}	B	B^{-1}	B

For massless Dirac particles, there is no Lorentz frame in which the particle is at rest, and the generators of the little group are linear combinations of S_i and K_i. For this reason, the generators of the little group undergo more changes than a simple change in sign. This will lead to a non-trivial result, which we shall discuss in Chapter VII.

Likewise, the generators of the little group for photons will undergo non-trivial transformations under discrete symmetry operations. However, while neutrinos share some properties with electrons, photons share properties with massive vector mesons.

Unlike photons, there is a Lorentz frame in which the vector meson is at rest. We can study P, C, and T operations in this Lorentz frame. It is important to realize that these operations are not Lorentz-invariant. For example, $I(+,+)$ and $I(-,-)$ matrices in Equation (1.10) are Lorentz invariant, but $I(-,+)$ and $I(+,-)$ are not. If we boost the $I(+,-)$ matrix along the z direction,

$$B_z(\eta) \, I(+,-) \, B_z(-\eta)$$

$$= \begin{bmatrix} \cosh \eta & \sinh \eta \\ \sinh \eta & \cosh \eta \end{bmatrix} \begin{bmatrix} 1 & 0 \\ 0 & -1 \end{bmatrix} \begin{bmatrix} \cosh \eta & -\sinh \eta \\ -\sinh \eta & \cosh \eta \end{bmatrix}$$

$$= \begin{bmatrix} \cosh (2\eta) & -\sinh (2\eta) \\ \sinh (2\eta) & -\cosh (2\eta) \end{bmatrix}. \tag{6.6}$$

The above form is quite different from the rest-frame expression for $I(+,-)$.

Finally, for unitary representations in which both the rotation and boost generators are Hermitian, we will have to deal with the differential forms for these operators. It is clear from Table 6.1 and 6.2 that P is simply a space reflection in this case. C is complex conjugation, and T is complex conjugation followed by time reflection. Since all the generators are imaginary, the transformation matrices are real and are unaffected by complex conjugation.

7. Exercises and Problems

Exercise 1. The translation-like matrix of the $E(2)$-like little group for massless particles is given in Equation (2.6). Show that this matrix is a product of one rotation matrix and two boost matrices.

In order to solve this problem, we have to construct non-trivial four-by-four transformation matrices which leave the four-momentum invariant. If a massless particle moves along the z direction, we can consider a Lorentz boost of the momentum along the x axis, followed by a rotation around the y axis which will bring us back to the z axis. A boost along the z axis will bring the momentum to its original condition. In matrix notation, these operations can be written as a product of three matrices:

$$T_x(u) = \begin{bmatrix} 1 & 0 & 0 & 0 \\ 0 & 1 & 0 & 0 \\ 0 & 0 & (2-u^2)/2w & -u^2/2w \\ 0 & 0 & -u^2/2w & (2-u^2)/2w \end{bmatrix} \times$$

$$\times \begin{bmatrix} w & 0 & -u & 0 \\ 0 & 1 & 0 & 0 \\ u & 0 & w & 0 \\ 0 & 0 & 0 & 1 \end{bmatrix} \begin{bmatrix} 1/w & 0 & 0 & u/w \\ 0 & 1 & 0 & 0 \\ 0 & 0 & 1 & 0 \\ u/w & 0 & 0 & u/w \end{bmatrix}, \quad (7.1)$$

where

$$w = (1 - u^2)^{1/2}. \quad (7.2)$$

In the above expression, the first matrix is a boost along the z direction, the second a rotation around the y axis, and the third a boost along the x direction. The multiplication of the above three matrices is straightforward, and

$$T_x(u) = \begin{bmatrix} 1 & 0 & -u & u \\ 0 & 1 & 0 & 0 \\ u & 0 & 1 - u^2/2 & u^2/2 \\ u & 0 & -u^2/2 & 1 + u^2/2 \end{bmatrix}. \quad (7.3)$$

The above procedure is physically meaningful only when u^2 is smaller than one. If $|u|$ is greater than one, w in Equation (7.2) becomes pure imaginary. However, the above $T_x(u)$ has the property:

$$T_x(u_1 + u_2) = T_x(u_1) T_x(u_2). \quad (7.4)$$

We can thus divide an arbitrarily large u into u/n, where n is an integer large enough to make $|u/n|$ smaller than one, and repeat the same process n times.

By carrying out a similar procedure on the yz plane, we can get another four-by-four T matrix:

$$T_y(v) = \begin{bmatrix} 1 & 0 & 0 & 0 \\ 0 & 1 & -v & v \\ 0 & v & 1 - v^2/2 & v^2/2 \\ 0 & v & -v^2/2 & 1 + v^2/2 \end{bmatrix}. \quad (7.5)$$

Then it is straightforward to show that T_x and T_y matrices commute. Thus the matrices:

$$T(u, v) = T_y(v) \, T_x(u) \tag{7.6}$$

form a two-parameter Abelian subgroup of the little group.

Exercise 2. Consider a Lorentz boost along the x direction followed by another along the y direction. Calculate the resulting Lorentz transformation matrix. Does the resulting matrix represent a pure boost or a boost times a rotation?

Since the z direction is not affected by these transformations, we can use the vector (x, y, t), and three-by-three matrices to perform boosts. Let us start with the boost matrix along the x direction:

$$B_x(\xi) = \begin{bmatrix} \cosh \xi & 0 & \sinh \xi \\ 0 & 1 & 0 \\ \sinh \xi & 0 & \cosh \xi \end{bmatrix}. \tag{7.7}$$

We can obtain a boost matrix along the ϕ direction by rotating the above matrix:

$$B_\phi(\lambda) = R(\phi) \, B_x(\lambda) \, [R(\phi)]^{-1}, \tag{7.8}$$

where

$$R(\phi) = \begin{bmatrix} \cos \phi & -\sin \phi & 0 \\ \sin \phi & \cos \phi & 0 \\ 0 & 0 & 1 \end{bmatrix}. \tag{7.9}$$

Since the inverse of the rotation matrix is its transpose, boost matrices are symmetric.

Let us next consider the boost matrix along the y direction:

$$B_y(\eta) = \begin{bmatrix} 1 & 0 & 0 \\ 0 & \cosh \eta & \sinh \eta \\ 0 & \sinh \eta & \cosh \eta \end{bmatrix}. \tag{7.10}$$

Then the boost along the x direction followed by another boost along the y direction can be represented by the three-by-three matrix $B_y(\eta) \, B_x(\xi)$. However, this matrix is not symmetric. For this reason, the resulting matrix cannot be a pure boost matrix. On the other hand, we can still consider the form:

$$B_y(\eta) \, B_x(\xi) = B_\phi(\lambda) \, R(\alpha). \tag{7.11}$$

Theory of the Poincaré Group

The computation of the above matrix equation is straightforward. There are three parameters to determine. There are however nine matrix elements to compare. The result is

$$\cosh \lambda = (\cosh \xi) \cosh \eta,$$
$$\tan \phi = (\coth \xi) \sinh \eta, \qquad (7.12)$$
$$\tan \frac{\alpha}{2} = \tanh \frac{\xi}{2} \tanh \frac{\eta}{2}.$$

When the amount of boost is the same for both x and y directions: $\xi = \eta$,

$$\cosh \lambda = (\cosh \xi)^2, \quad \tan \phi = \cosh \xi, \quad \tan \frac{\alpha}{2} = \left(\tanh \frac{\xi}{2}\right)^2. \qquad (7.13)$$

This means that, when ξ is very small,

$$\lambda = \sqrt{2}\,\xi, \quad \phi = 45°, \quad \text{and} \quad \alpha = \xi^2/2.$$

This result is consistent with what we expect from nonrelativistic kinematics. On the other hand, when ξ is very large,

$$\lambda = 2\xi, \quad \phi = 90°, \quad \alpha = 90°. \qquad (7.14)$$

This result is somewhat unexpected. However, we should be able to see this effect in connection with massless particles.

A more interesting problem woud be two successive Lorentz boosts along arbitrary directions. We can handle this problem using the mathematical technique given here. However, it is not an efficient method to work with nine different equations to determine three parameters. We shall discuss this problem in greater detail in Chapter IX.

Problem 1. Consider two four-vectors V and W and their "inner" product

$$(V, W) = (V_\mu)^* W^\mu. \qquad (7.15)$$

Show that, if they are orthogonal to each other: $(V, W) = 0$, and if one of them is time-like, then the other is space-like. See Section 4B of Wigner's paper (1939).

Problem 2. The four-by-four Lorentz transformation matrix has four eigenvalues and four eigenvectors. Show that, if λ is an eigenvalue, $1/\lambda$, λ^* and $1/\lambda^*$ are also eigenvalues. Show that the eigenvectors V_1 and V_2 corresponding respectively to λ_1 and λ_2 are orthogonal if $\lambda_1^* \lambda_2 \neq 1$. See Section 4B of Wigner's paper (1939).

Problem 3. Show that the complex eigenvalue of the Lorentz transformation matrix has unit modulus, and that the real eigenvalue is positive. See Section 4B of Wigner's paper.

Problem 4. Wigner's paper (1939) contains a proof that the group of Lorentz transformations is simple. Use Cartan's criterion to show that this group is a simple group.

Problem 5. Show that Casimir operators for the homogeneous Lorentz group are

$$C_1 = \tfrac{1}{2} M^{\mu\nu} M_{\mu\nu}, \quad \text{and} \quad C_2 = \tfrac{1}{4} \varepsilon_{\mu\nu\alpha\beta} M^{\mu\nu} M^{\alpha\beta}, \tag{7.6}$$

and show that the operators given in Equation (3.10) are the Casimir operators for the Poincaré group. Show also that C_1 and C_2 for the homogeneous Lorentz group cannot serve as the Casimir operators for the inhomogeneous Lorentz group.

Problem 6. Solve the matrix equation in Equation (2.3) starting from the most general form of the real four-by-four matrix with sixteen elements. Use the condition that it be a Lorentz transformation matrix to reduce the number of independent parameters from sixteen to six. For the little groups for non-vanishing four-momenta, solve the matrix equation to reduce further the number of independent parameters from six to three (Halpern and Branscomb, 1965; Halpern, 1968).

Problem 7. Write down the Dirac equation for an electron in an external electromagnetic field. Using this equation, show that the operations on the Dirac matrices in Equations (6.2), (6.3) and (6.4) are parity, charge conjugation and time reversal respectively.

Chapter IV

Theory of Spinors

One easy way for physicists to understand group theory is in terms of coordinate transformations. Indeed, as we did in Chapter III, the study of the Lorentz group naturally starts with the group of coordinate transformation matrices operating on four-component Minkowskian vectors. The question then is whether those four-by-four matrices are the smallest matrices having the algebraic properties of the proper Lorentz group (Cartan, 1966). The answer to this question is "No".

The three-dimensional rotation group is locally isomorphic to $SU(2)$ generated by three Hermitian traceless two-by-two matrices. Thus it is quite natural for us to look for six two-by-two matrices which satisfy the same set of commutation relations as that for the Lorentz group. Indeed, there are six traceless complex two-by-two matrices satisfying this requirement. They are called the generators of $SL(2, c)$. We have already begun discussion of this group in connection with the Dirac equation in Chapter III. The purpose of this chapter is to study the general properties of the $SL(2, c)$ group.

There are several advantages in studying this subject. First, there are particles in nature whose space-time symmetry is that of $SL(2, c)$. Second, $SL(2, c)$ is a covering group for the Lorentz group, and therefore more fundamental from a mathematical point of view. Third, since every value of spin can be regarded as a multiple of 1/2, the theory of spinors together with spin addition mechanisms will constitute a theory of angular momenta. Fourth, because we are dealing with two-by-two matrices, calculation of Lorentz transformations is much easier than when dealing with four-by-four matrices. Finally, in $SL(2, c)$, the boost generators are simply $\pm i$ times the rotation generators. For this reason, we can discuss this group with three generators, instead of six, by complexifying the group parameters.

In Section 1, we discuss $SL(2, c)$ as the covering group of the group of Lorentz transformations. We study in Section 2 subgroups of $SL(2, c)$, particularly those which correspond to the little groups discussed in Chapter III. $SU(2)$ is the $SO(3)$-like subgroup of $SL(2, c)$. Since this group is discussed extensively in the literature, we give in Section 3 a summary of what is already known to most readers.

The generators of $SL(2, c)$ satisfy the same set of commutation relations as those for $SO(3, 1)$. We are of course interested in knowing more than this local property. Section 4 deals with the problem of constructing explicitly the four-by-four Lorentz transformation matrix from the two-by-two matrices of $SL(2, c)$, and with the problem of constructing four vectors from $SL(2, c)$ spinors. In physics, the representation of $SL(2, c)$ appears most often in the form of the Dirac equation. For this reason, in Section 5, we discuss in detail the symmetry of the Dirac equation. In Section 6, we present some of the items not discussed in Sections 1–5 in the form of exercises and problems.

1. SL(2, c) as the Covering Group of the Lorentz Group

Let us consider the group generated by the following six two-by-two matrices:

$$S_i = (1/2)\sigma_i, \quad \text{and} \quad K_i = (i/2)\sigma_i, \tag{1.1}$$

where σ_i are the Pauli spin matrices. The above operators take the following explicit forms.

$$S_1 = \begin{bmatrix} 0 & 1/2 \\ 1/2 & 0 \end{bmatrix}, \quad K_1 = \begin{bmatrix} 0 & i/2 \\ i/2 & 0 \end{bmatrix},$$

$$S_2 = \begin{bmatrix} 0 & -i/2 \\ i/2 & 0 \end{bmatrix}, \quad K_2 = \begin{bmatrix} 0 & 1/2 \\ -1/2 & 0 \end{bmatrix}, \tag{1.2}$$

$$S_3 = \begin{bmatrix} 1/2 & 0 \\ 0 & -1/2 \end{bmatrix}, \quad K_3 = \begin{bmatrix} i/2 & 0 \\ 0 & -i/2 \end{bmatrix}.$$

These matrices satisfy the commutation relations:

$$[S_i, S_j] = i\varepsilon_{ijk} S_k,$$
$$[S_i, K_j] = i\varepsilon_{ijk} K_k, \tag{1.3}$$

and

$$[K_i, K_j] = -i\varepsilon_{ijk} S_k.$$

These relations are identical to those for the proper Lorentz group.

There is therefore every reason to expect that the group of two-by-two matrices of the form

$$W = \exp\left[-i \sum_{i=1}^{3} (\theta_i S_i + \eta_i K_i)\right] \tag{1.4}$$

Theory of Spinors

will have the same algebraic properties as the proper Lorentz group. This group is often called $SL(2, c)$. Because the generators are traceless, the determinant of W is one. $SL(2, c)$ is therefore a unimodular group. Since the K_i matrices are not Hermitian, $SL(2, c)$ is not a unitary group.

We note in Equation (1.3) that the Lie algebra for the generators is invariant under the sign change of the boost operators K_i. In the case of $O(3, 1)$ which was discussed in Chapter III, the sign of the boost generators K_i is unambiguously defined in terms of the space and time variables. However, in the present case of $SL(2, c)$, we have to consider both signs. Since the sign change can be performed easily, we shall choose the positive sign given in Equation (1.2) unless otherwise required.

Let us look at the expression of Equation (1.4). Since K_i is just i times S_i, W of Equation (1.4) can be written as

$$W = \exp\left[-\frac{i}{2}(\zeta_1 \sigma_1 + \zeta_2 \sigma_2 + \zeta_3 \sigma_3)\right], \tag{1.5}$$

with

$$\zeta_i = \theta_i + i\eta_i.$$

This means that, when needed, we can study $SL(2, c)$ with three generators and three complex parameters.

In order to see the correspondence with the matrices of Lorentz transformations, let us note first that the three-parameter subgroup generated by S_i is already familiar to us from the rotation group. This is the $SU(2)$ group homomorphic to $SO(3)$. We know how the matrices of the $SU(2)$ group change the direction of the spin operators. With this in mind, let us consider the two-by-two matrix

$$X = Ix_0 + \sigma_i x_i$$

$$= \begin{bmatrix} t+z & x-iy \\ x+iy & t-z \end{bmatrix}. \tag{1.6}$$

where I is the two-by-two unit matrix. Then

$$\det(X) = t^2 - x^2 - y^2 - z^2. \tag{1.7}$$

Therefore unimodular transformations on the above matrix are Lorentz transformations (Wigner, 1939).

Let us consider the transformation

$$X' = WXW^\dagger. \tag{1.8}$$

Since W and $W^\dagger$ are unimodular, $\det(X)$ is preserved. If W is restricted to the $SU(2)$ subgroup W_r generated by the three S_i matrices, then W_r is unitary and does not affect the t component in Equation (1.6). The effect of a W_r

transformation on the $\sigma_i x_i$ is a rotation in the three-dimensional (x, y, z) space. The rotation by ψ around the (θ, ϕ) direction is represented by

$$W_r(\theta, \phi, \psi) = \exp\left[-i\frac{\psi}{2}\mathbf{n}\cdot\boldsymbol{\sigma}\right] = I\cos\frac{\psi}{2} - i\mathbf{n}\cdot\boldsymbol{\sigma}\sin\frac{\psi}{2}, \quad (1.9)$$

where $\mathbf{n}$ is the unit vector along the (θ, ϕ) direction.

Let us next consider the effect of $W_b(z, \eta) = \exp[-i\eta K_3]$. The explicit form of this matrix is

$$W_b(z, \eta) = \begin{bmatrix} \exp(\eta/2) & 0 \\ 0 & \exp(-\eta/2) \end{bmatrix}, \quad (1.10)$$

The effect of this matrix on X is clearly a boost along the z axis:

$$z' = z\cosh\eta + t\sinh\eta,$$
$$t' = z\sinh\eta + t\cosh\eta. \quad (1.11)$$

The boost along the (θ, ϕ) direction is achieved through

$$W_b(\theta, \phi) = W_r(\theta, \phi, 0)\, W_b(z, \eta)\, [W_r(\theta, \phi, 0)]^\dagger. \quad (1.12)$$

As in the case of Equation (1.17) of Chapter III for the proper Lorentz group, the most general form for the W matrix is

$$W = W_r W_b. \quad (1.13)$$

Once we write down the above expression for two-by-two matrices containing six-parameters, we are led to the question of whether there is a one-to-one correspondence between a given matrtix of $SL(2, c)$ and a four-by-four matrix in the proper Lorentz group. The answer to this question is "No". If we require that the W matrix be the two-by-two unit matrix when all the parameters are zero, the rotation parameter ψ is allowed to vary from 0 to 4π for the W matrices while the corresponding parameter for the four-by-four A matrices for the proper Lorentz group takes values between 0 and 2π. For this reason, for each A, there are two W matrices, namely W and $-W$ which produce the same effect on the transformation defined in Equation (1.9). $SL(2, c)$ is simply connected and is therefore the universal covering group for the proper Lorentz group.

2. Subgroups of SL(2, c)

As we noted in Chapter I, $SL(2, c)$ consists of non-singular two-by-two matrices of the form:

$$W = \begin{bmatrix} a & b \\ c & d \end{bmatrix}, \quad \text{with } (ad - bc) = 1, \quad (2.1)$$

where a, b, c and d are complex numbers. This matrix is applicable to spinors of the form

$$x = \begin{bmatrix} u \\ v \end{bmatrix}, \qquad (2.2)$$

Like the proper Lorentz group, $SL(2, c)$ is a simple group, i.e., it has no invariant subgroups. We are interested in those subgroups which are locally isomorphic to the little groups of the Lorentz group discussed in Chapter III, and we call for simplicity those subgroups the little groups of $SL(2, c)$.

Among the subgroups of $SL(2, c)$, we would expect that the $SO(3)$-like little group corresponding to the time-like orbit is generated by S_1, S_2, and S_3. The $SE(2)$-like little groups for the forward and backward light-like orbits are generated by S_3, N_1 and N_2, where

$$N_1 = K_1 - S_2, \quad N_2 = K_2 + S_1, \quad S_3, \qquad (2.3)$$

and their Hermitian conjugates respectively. The $SO(2, 1)$-like little group for space-like orbit is generated by S_3, K_1 and K_2. The $SO(3, 1)$-like little group corresponding to the point-like orbit confined to the origin is $SL(2, c)$ itself.

In obtaining these little groups, the easiest and quickest method would be to use the above-mentioned generators. However, in order to study the properties of $SL(2, c)$ more effectively, we use a somewhat inefficient way of obtaining subgroups by imposing additional conditions on the parameters a, b, c, and d.

First, we can put the restriction that the above matrix be unitary. Then this subgroup is $SU(2)$ generated by three S_i matrices satisfying the commutation relations for the three-dimensional rotation group. In this case, the elements of the matrix of Equation (2.1) should satisfy

$$|a|^2 + |b|^2 = 1, \quad |c|^2 + |d|^2 = 1, \quad a^*b + c^*d = 0, \qquad (2.4)$$

and consequently the two-by-two matrix of Equation (2.1) becomes

$$\begin{bmatrix} a & b \\ -b^* & a^* \end{bmatrix}, \qquad (2.5)$$

with

$$aa^* + bb^* = 1. \qquad (2.6)$$

This matrix has three independent real parameters. When applied to the spinor of Equation (2.6), this group preserves the norm

$$|u|^2 + |v|^2. \qquad (2.7)$$

$SU(2)$ is indeed a unitary group. $SU(2)$ is a simple group and has no invariant subgroups. It has a one-parameter unitary subgroup generated by S_3

of Equation (1.2). Since this group plays the pivotal role in the rotation and Lorentz groups, we shall discuss its representations in detail in Section 3.

Second, the matrices with real elements form a subgroup. This subgroup is called $SL(2, r)$. This group is equivalent to the group of matrices of the form:

$$M = \begin{bmatrix} a & b \\ b^* & a^* \end{bmatrix}, \quad \text{with } (aa^* - bb^*) = 1, \tag{2.8}$$

This group is generated by S_3, K_1 and K_2, and is locally isomorphic to $SO(2, 1)$. The above matrix satisfies the condition:

$$M^\dagger J M = J, \tag{2.9}$$

where

$$J = \begin{bmatrix} 1 & 0 \\ 0 & -1 \end{bmatrix},$$

and therefore the transformation of the spinor of Equation (2.2) preserves the quantity:

$$|u|^2 - |v|^2. \tag{2.10}$$

For this reason, this subgroup is sometimes called $SU(1, 1)$. The group of two-by-two real unimodular matrices satisfying the condition of Equation (2.9) is sometimes known as the real symplectic group of dimension 2 or $Sp(2, r)$ (Weyl, 1946). This group which is called $SL(2, r)$, $SU(1, 1)$ or $Sp(2, r)$ is also simple. It has two one-parameter subgroups. One is the unitary group generated by S_3, and the other is a non-unitary group generated by K_1.

Third, let us see whether $SL(2, c)$ has a subgroup consisting of triangular matrices of the form

$$F = \begin{bmatrix} a & b \\ 0 & 1/a \end{bmatrix}. \tag{2.11}$$

The determinant of this matrix is 1. If a and b are allowed to be complex, there are four parameters. In fact this subgroup is generated by S_3, K_3, N_1 and N_2, where

$$N_1^{(+)} = \begin{bmatrix} 0 & i \\ 0 & 0 \end{bmatrix}, \quad N_2^{(+)} = \begin{bmatrix} 0 & 1 \\ 0 & 0 \end{bmatrix}, \tag{2.12}$$

and the expressions for S_3 and K_3 are given in Equation (1.2). These four operators satisfy the commutation relations (Problem 16 in Section 6):

$$[S_3, N_1] = iN_2, \quad [S_3, N_2] = -iN_1, \quad [N_1, N_2] = 0, \tag{2.13}$$

$$[K_3, N_1] = -N_2, \quad [K_3, N_2] = N_1, \quad [K_3, S_3] = 0. \tag{2.14}$$

Theory of Spinors

Among the above six commutation relations, the first three form a closed algebraic relation. This means that matrices of the form

$$D(b_1, b_2, \phi) = \begin{bmatrix} e^{-i\phi/2} & b_1 - ib_2 \\ 0 & e^{i\phi/2} \end{bmatrix} \quad (2.15)$$

represent a three-parameter subgroup generated by S_3, N_1 and N_2. As we did in Chapter II, we call this group $SW(2)$.

The fundamental algebraic relations for the generators of $SL(2, c)$ given in Equation (1.3) remain invariant under the sign change in K_i. Under this sign change, the N operators become their Hermitian conjugates. Thus the resulting matrices are of the lower triangular form. The algebraic properties of the lower triangular matrices are the same as those for the upper triangular matrices.

Let us go back to the upper triangular matrix of Equation (2.15). We can decompose this matrix as

$$D(b_1, b_2, \phi) = T(b_1, b_2) R(\phi). \quad (2.16)$$

where

$$T(b_1, b_2) = \begin{bmatrix} 1 & b_1 - ib_2 \\ 0 & 1 \end{bmatrix},$$

$$R(\phi) = \begin{bmatrix} e^{-i\phi/2} & 0 \\ 0 & e^{i\phi/2} \end{bmatrix}, \quad (2.17)$$

where b_1 and b_2 are real parameters. $T(b_1, b_2)$ can further be decomposed as

$$T(b_1, b_2) = T(b_1, 0) T(0, b_2) = T(0, b_2) T(b_1, 0). \quad (2.18)$$

Furthermore, it can be shown that $D(b_1, b_2, \phi)$ is a semi-direct product of $R(\phi)$ and $T(b_1, b_2)$ as in the case of the $SE(2)$ group discussed in Chapters I and III.

Another interesting property of the N matrices is that they cannot be brought to a diagonal form by any similarity transformation. They however have the property that

$$(N_1)^2 = (N_2)^2 = 0. \quad (2.19)$$

Thus the exponentiation takes the following simple form:

$$T(b_1, b_2) = \exp(-i[b_1 N_1 + b_2 N_2]) = \begin{bmatrix} 1 & b_1 - ib_2 \\ 0 & 1 \end{bmatrix},$$

As in the case of $SE(2)$ group, $T(b_1, b_2)$ is an invariant subgroup of $SW(2)$. Since $SW(2)$ a subgroup of $SL(2, c)$, $T(b_1, b_2)$ is also a subgroup of

$SL(2, c)$, but is not an invariant subgroup of $SL(2, c)$. $SL(2, c)$ does not have any invariant subgroup.

3. SU(2)

The $SU(2)$ group has been studied exhaustively in the literature in connection with the rotation group, and it is unnecessary to give another full-fledged treatment of this group. We mention in this section only those features which will be useful in our later discussions of $SL(2, c)$, starting from what we already know.

Let us discuss first its connection with $SO(3)$. It is possible to construct three-by-three rotation matrices from the elements of the two-by-two matrix given in Equation (2.5) (Problem 6 in Section 6). The rotation matrix applicable to the three-dimensional space of (x, y, z) takes the form

$$\begin{bmatrix} \text{Re}(a^2 - b^2) & -\text{Im}(a^2 + b^2) & -2\text{Re}(ab) \\ \text{Im}(a^2 - b^2) & \text{Re}(a^2 + b^2) & -2\text{Im}(ab) \\ 2\text{Re}(aa^*) & -2\text{Im}(aa^*) & aa^* - bb^* \end{bmatrix}. \qquad (3.1)$$

The elements of the above three-by-three matrix are quadratic and homogeneous in the elements of the two-by-two matrix of Equation (2.1). Therefore, to a given three-by-three matrix of $SO(3)$, there correspond two two-by-two matrices. One is the expression given in Equation (3.1), and the other is the negative of Equation (2.5). This is of course the manifestation of the fact that the correspondence between $SU(2)$ and $SO(3)$ is two-to-one.

The two-by-two matrix of Equation (2.1) can be constructed from the generators of $SU(2)$, namely S_1, S_2, and S_3. Then it is convenient to use the spinor notation α and β, where

$$\alpha = \begin{bmatrix} 1 \\ 0 \end{bmatrix}, \quad \beta = \begin{bmatrix} 0 \\ 1 \end{bmatrix}, \qquad (3.2)$$

which are the eigenstates of S_3 with the eigenvalues $\pm 1/2$ respectively. The total angular momentum for these spinors is $1/2$. The application of the two-by-two matrix of Equation (2.1) results in a rotation of the spinor. For instance, the matrix

$$\exp(-i\theta S_y) = \begin{bmatrix} \cos\dfrac{\theta}{2} & -\sin\dfrac{\theta}{2} \\ \sin\dfrac{\theta}{2} & \cos\dfrac{\theta}{2} \end{bmatrix} \qquad (3.3)$$

rotates the spin in the z direction to the direction which makes the angle θ in the xz plane (Rose, 1957; Edmonds, 1957). The corresponding matrix in the $SO(3)$ group applicable to the Cartesian vector (x, y, z) is

$$\begin{bmatrix} \cos\theta & 0 & -\sin\theta \\ 0 & 1 & 0 \\ \sin\theta & 0 & \cos\theta \end{bmatrix}, \tag{3.4}$$

where we have used the sign convention of the passive transformation, since we will be dealing with functions of the coordinates rather than coordinates themselves.

It is often more convenient to use the spherical vector:

$$\begin{bmatrix} -(x+iy)/\sqrt{2} \\ z \\ (x-iy)/\sqrt{2} \end{bmatrix} = \begin{bmatrix} -1/\sqrt{2} & -i/\sqrt{2} & 0 \\ 0 & 0 & 1 \\ 1/\sqrt{2} & -i/\sqrt{2} & 0 \end{bmatrix} \begin{bmatrix} x \\ y \\ z \end{bmatrix}, \tag{3.5}$$

which is obtained from the Cartesian form by a unitary transformation. The rotation matrix applicable to this spherical vector is

$$\begin{bmatrix} \dfrac{1+\cos\theta}{2} & (1/\sqrt{2})\sin\theta & \dfrac{1-\cos\theta}{2} \\ -(1/\sqrt{2})\sin\theta & \cos\theta & (1/\sqrt{2})\sin\theta \\ \dfrac{1-\cos\theta}{2} & -(1/\sqrt{2})\sin\theta & \dfrac{1+\cos\theta}{2} \end{bmatrix}, \tag{3.6}$$

which is obtained from Equation (3.4) through the conjugate transformation with the unitary matrix in Equation (3.5):

$$\begin{bmatrix} -1/\sqrt{2} & -i/\sqrt{2} & 0 \\ 0 & 0 & 1 \\ 1/\sqrt{2} & -i/\sqrt{2} & 0 \end{bmatrix} \begin{bmatrix} \cos\theta & 0 & -\sin\theta \\ 0 & 1 & 0 \\ \sin\theta & 0 & \cos\theta \end{bmatrix} \begin{bmatrix} -1/\sqrt{2} & 0 & 1/\sqrt{2} \\ i/\sqrt{2} & 0 & i/\sqrt{2} \\ 0 & 1 & 0 \end{bmatrix}. \tag{3.7}$$

It is possible to add the angular momenta of two spin 1/2 particles by taking the direct product of their spinors. For particles 1 and 2, the sums of the spin operators are

$$J_i = S_i^{(1)} + S_i^{(2)}. \tag{3.8}$$

If we add spins of two spin-1/2 particles, the total spin is either zero or 1. If the total angular momentum is zero, the spin wave function is

$$|j=0, m=0\rangle = (\alpha_1\beta_2 - \beta_1\alpha_2)/\sqrt{2}, \tag{3.9}$$

where the subscripts 1 and 2 are for particles 1 and 2 respectively. The rotation matrix applicable to this one-dimensional representation is 1. On the other hand, if the total angular momentum is 1, then the spin wave functions take the form

$$|1,1\rangle = \alpha_1 \alpha_2,$$
$$|1,0\rangle = (\alpha_1 \beta_2 + \beta_1 \alpha_2)/\sqrt{2}, \qquad (3.10)$$
$$|1,-1\rangle = \beta_1 \beta_2.$$

The rotation matrix applicable to this set of wave functions is identical to that for the spherical vector of Equation (3.5) (Problem 7 in Section 6).

We started above with two spin 1/2 particles. If we take a direct product of two wave functions, it can be decomposed into one symmetric and one antisymmetric combination. The symmetric wave functions form a representation space for $j = 1$, while j for the antisymmetric combination is 0. Because this symmetry property is invariant under rotations, this method is extensively used in constructing representations of all finite groups.

Let us next see how this symmetry method works for the three-particle case. J_i is now a sum of all three spin operators.

$$J_i = S_i^{(1)} + S_i^{(2)} + S_i^{(3)}. \qquad (3.11)$$

The resulting angular momentum is either 3/2 or 1/2. Because each particle has two possible states, there are eight wave functions for this three particle system (Problem 8 in Section 6). They are, for $j = 3/2$,

$$|\tfrac{3}{2}, \tfrac{3}{2}\rangle = \alpha_1 \alpha_2 \alpha_3,$$
$$|\tfrac{3}{2}, \tfrac{1}{2}\rangle = (1/\sqrt{3})(\alpha_1 \alpha_2 \beta_3 + \alpha_1 \beta_2 \beta_3 + \beta_1 \alpha_2 \alpha_3),$$
$$|\tfrac{3}{2}, -\tfrac{1}{2}\rangle = (1/\sqrt{3})(\alpha_1 \beta_2 \beta_3 + \beta_1 \alpha_2 \beta_3 + \beta_1 \beta_2 \alpha_3), \qquad (3.12)$$
$$|\tfrac{3}{2}, -\tfrac{3}{2}\rangle = \beta_1 \beta_2 \beta_3.$$

These wave functions are totally symmetric. There cannot be a totally antisymmetric wave function because each spinor has only two possible spin states. We can next construct two wave functions which are symmetric in the first two indices:

$$|\tfrac{1}{2}, \tfrac{1}{2}\rangle = (1/\sqrt{6})(\alpha_1 \alpha_2 \beta_3 + \alpha_1 \beta_2 \alpha_3 - 2\beta_1 \alpha_2 \alpha_3),$$
$$|\tfrac{1}{2}, -\tfrac{1}{2}\rangle = (1/\sqrt{6})(\beta_1 \beta_2 \alpha_3 + \beta_1 \alpha_2 \beta_3 - 2\alpha_1 \beta_2 \beta_3). \qquad (3.13)$$

There are two additional wave functions antisymmetric in the first two indices.

$$|\tfrac{1}{2}, \tfrac{1}{2}\rangle = (1/\sqrt{2})(\alpha_1 \alpha_2 \beta_3 - \alpha_1 \beta_2 \alpha_3),$$
$$|\tfrac{1}{2}, -\tfrac{1}{2}\rangle = (1/\sqrt{2})(\beta_1 \beta_2 \alpha_3 - \beta_1 \alpha_2 \beta_3). \qquad (3.14)$$

All four of these wave functions should be orthogonal to the totally

Theory of Spinors

symmetric wave functions. We shall discuss the symmetry properties of the three-particle wave functions in detail in Chapter XI. What is important is that, if we rotate each of the spinors in the wave functions, the resulting transformation is exactly like that of the one spinor with total spin 1/2 (Problem 9 in Section 6).

The rotation matrix applicable to the above spin 1/2 wave functions is the same as the one for α and β. The rotation matrix applicable to the $j = 3/2$ wave functions are four-by-four. How can we construct this rotation matrix?

For a given value of j which is either an integer or half integer, we start with the diagonal $(2j+1)$-by-$(2j+1)$ matrix J_z:

$$J_z = \begin{bmatrix} j & 0 & 0 & 0 & 0 \\ 0 & j-1 & 0 & 0 & 0 \\ 0 & 0 & * & 0 & 0 \\ 0 & 0 & 0 & -j+1 & 0 \\ 0 & 0 & 0 & 0 & -j \end{bmatrix}. \tag{3.15}$$

The J_x and J_y can be constructed from the commutation relations:

$$[J_i, J_j] = i\varepsilon_{ijk} J_k. \tag{3.16}$$

In practice, we take the combinations:

$$J_\pm = J_x \pm i J_y, \tag{3.17}$$

and write commutation relations as

$$[J_\pm, J_z] = \pm J_z, \quad [J_+, J_-] = iJ_z. \tag{3.18}$$

We then construct the matrices using the relations

$$\begin{aligned} J_z |j, m\rangle &= m |j, m\rangle, \\ J_\pm |j, m\rangle &= [(j \mp m)(j \pm m + 1)]^{1/2} |j, m \pm 1\rangle. \end{aligned} \tag{3.19}$$

We often couple a spin 1/2 wave function with a wave function of an arbitrary integer angular momentum ℓ, which is represented by the spherical harmonics $Y_\ell^m(\theta, \phi)$. The resulting total angular momentum is either $j = \ell + 1/2$ or $j = \ell - 1/2$. The wave functions in this case are

$$\Phi_{j=\ell+1/2}^m = \begin{bmatrix} \sqrt{(\ell + m + 1/2)/(2\ell + 1)} \; Y_\ell^{m-1/2}(\theta, \phi) \\ \sqrt{(\ell - m + 1/2)/(2\ell + 1)} \; Y_\ell^{m+1/2}(\theta, \phi) \end{bmatrix},$$

$$\Phi_{j=\ell-1/2}^m = \begin{bmatrix} -\sqrt{(\ell - m + 1/2)/(2\ell + 1)} \; Y_\ell^{m-1/2}(\theta, \phi) \\ \sqrt{(\ell - m + 1/2)/(2\ell + 1)} \; Y_\ell^{m+1/2}(\theta, \phi) \end{bmatrix}. \tag{3.20}$$

The point of the above discussion is that there are many different ways to add angular momenta. However, they all result in irreducible representations or invariant vector space s with total angular momentum j. The rotation matrix applicable to a given j is $(2j + 1)$-by-$(2j + 1)$. For instance, the rotation matrix applicable to one spin 1/2 particle is the same as the one for the spin-1/2 state constructed from three spin-1/2 particles. For this reason, it is sufficient to study the rotation matrix which is applicable to the maximum value of j obtainable from a given number of spin-1/2 particles. If, for instance, we are interested in constructing representations for $j = 5/2$, it is sufficient to calculate the rotation matrix for five spin 1/2 particles whose spins are added to give the maximum value of the total angular momentum.

In order that the spinors give the maximum value of the total angular momentum, they should form a totally symmetric combination. We are thus interested in finding all irreducible representations of $SU(2)$ by constructing the representations of totally symmetric products. Starting from u and v and transforming according to Equation (3.3), we consider the set of products (Wigner, 1959; Hamermesh, 1962):

$$f_m = \frac{u^{j+m} v^{j-m}}{\sqrt{(j+m)!(j-m)!}} \qquad (3.21)$$

with

$$m = -j, -j+1, \ldots, j-1, j,$$

where j is integral or half-integral. This form satisfies the relation

$$\sum_m |f_m|^2 = \sum_m \frac{|u^{j+m} v^{j-m}|^2}{(j+m)!(j-m)!} = (|u|^2 + |v|^2)^j/(2j)!.$$

This sum is independent of m, and will be invariant under transformations which preserve $(|u|^2 + |v|^2)$. We are therefore led to consider whether the above form will serve as the basis vector for representations of $SU(2)$.

With this point in mind, let us consider the operators:

$$J_3 = \frac{1}{2}\left(u\frac{\partial}{\partial u} - v\frac{\partial}{\partial v}\right),$$

$$J_+ = u\frac{\partial}{\partial v}, \qquad (3.22)$$

$$J_- = v\frac{\partial}{\partial u}.$$

Theory of Spinors

These operators satisfy the commutation relations given in Equation (3.18). Consequently, these operators satisfy the relations of Equation (3.19):

$$J_3 f_j^m = m f_j^m,$$

$$J_\pm f_j^m = \sqrt{(j \mp m)(j \pm m + 1)} \, f_j^{m \pm 1}. \tag{3.23}$$

The form of Equation (3.21) forms the basis for a $(2j+1)$-dimensional representation of $SU(2)$.

Next, let us consider the transformation of the spinor of Equation (2.2) by the $SU(2)$ matrix given in Equation (2.5). For a fixed j, the homogeneous polynomials f_m are transformed among themselves by the linear transformation:

$$u' = au + bv,$$

$$v' = -b^*u + a^*v. \tag{3.24}$$

Thus

$$R(a, b) f_m = \sum_{m'} D^j_{mm'}(a, b) f_{m'}, \tag{3.25}$$

where

$$D^j_{mm'}(a, b) = \sum_\mu \frac{\sqrt{(j+m)!(j-m)!(j+m')!(j-m')!}}{k!(j+m-k)!(j-m'-k)!(m'-m+k)!}$$

$$\times a^{j+m-k} (a^*)^{j-m'-k} b^k (-b^*)^{m'-m+k}.$$

The above formula gives a complete calculation of all irreducible representations of the rotation group. For example, for $j = 1/2$, we have $f_{1/2} = u$ and $f_{-1/2} = v$. We can calculate each element in the two-by-two matrix which rotates the spinor around the y axis by θ using the above formula. The result of this calculation will confirm the matrix given in Equation (3.3). For $j = 1$, we have $f_1 = u^2/\sqrt{2}$, $f_0 = uv$, $f_{-1} = v^2/\sqrt{2}$ for the three-dimensional representation. Here there are nine matrix elements for the three-by-three matrix. We can check this calculation for the rotation around the y axis and compare with the expression given in Equation (3.6).

4. SL(2, c) Spinors and Four-Vectors

Let us go back to the commutation relations of Equation (1.3) for the generators of $SL(2, c)$. While it is not possible to change the sign of S_i without changing the commutation relations, the boost operators K_i can take two different signs. For $S_i = \frac{1}{2}\sigma_i$, the boost generators K_i can be either iS_i or

$-iS_i$. We shall continue to use α and β for the spinor space in which $K_i = iS_i$. However, we shall use $\dot\alpha$ and $\dot\beta$ for the spinor space in which the boost generators are $\dot K_i = -iS_i$. Under rotations generated by S_i, the spinors in the undotted and dotted spaces behave in the same manner as is described in Section 3. The boost operators applicable to the dotted space are the inverse of those applicable to the undotted space.

We often wonder why the Dirac matrices are four-by-four instead of two-by-two. The reason is very simple, as was discussed in Chapter III and will be studied further in Section 5 of the present chapter, the Dirac wave function is a direct sum of both undotted and dotted spinors.

The next question is whether two spin 1/2 states can be combined to give a spin-1 state. This possibility within the framework of $SU(2)$ and $SO(3)$ is widely discussed in quantum mechanics textbooks. This case is also mentioned in Section 3. In the $SU(2)$ case, the symmetric product gives a vector and the antisymmetric combination becomes a scalar. The question then is whether we can extend this procedure to the $SL(2, c)$ regime to include Lorentz boosts.

Let us start with the spinor combinations for massive particles at rest, and consider two spin 1/2 particles. Let us assume that particle 1 and particle 2 are boosted by $K_i = \pm \frac{1}{2}i\sigma_i$ respectively. Then this spinor combination $\alpha_1 \dot\alpha_2$ is invariant under boosts along the z direction, and is like $-(x + iy)$ under rotations. Thus we are led to write:

$$\alpha_1 \dot\alpha_2 = (-1, -i, 0, 0). \tag{4.1}$$

Likewise, for other combinations, we can consider

$$\begin{aligned} \beta_1 \dot\beta_2 &= (1, -i, 0, 0), \\ \alpha_1 \dot\beta_2 &= (0, 0, 1, 1), \\ \beta_1 \dot\alpha_2 &= (0, 0, 1, -1). \end{aligned} \tag{4.2}$$

Indeed, $\beta_1 \dot\beta_2$ is like $(x - iy)$, and $\frac{1}{2}(\alpha_1 \dot\beta_2 + \beta_1 \dot\alpha_2)$ and $\frac{1}{2}(\alpha_1 \dot\beta_2 - \alpha_2 \dot\beta_1)$ are like z and t under rotations in which dotted and undotted spinors are transformed in the same manner.

Under the boost along the z direction, $\alpha_1 \dot\alpha_2$ and $\beta_1 \dot\beta_2$ remain invariant, $\frac{1}{2}(\alpha_1 \dot\beta_2 \pm \beta_1 \dot\alpha_2)$ behave like z and t respectively. As for the x direction, the boost operator is

$$B_x(\eta) = B_{1x}^{(+)}(\eta) B_{2x}^{(-)}(\eta), \tag{4.3}$$

where $B_{1x}^{(+)}(\eta)$ and $B_{2x}^{(-)}(\eta)$ are the boost operators applicable to particle 1 and 2 respectively. The boost operators applicable to the first and second states are

$$B_x^{(\pm)}(\eta) = \begin{bmatrix} \cosh(\eta/2) & \pm \sinh(\eta/2) \\ \pm \sinh(\eta/2) & \cosh(\eta/2) \end{bmatrix}, \tag{4.4}$$

Theory of Spinors 93

respectively. The effect of this boost on $\alpha_1 \dot{\alpha}_2$ is

$$B_x(\eta)(-\alpha_1\dot{\alpha}_2) = -\alpha_1\dot{\alpha}_2 \cosh^2(\eta/2) + \beta_1\dot{\beta}_2 \sinh^2(\eta/2) +$$
$$+ \tfrac{1}{2}(\alpha_1\dot{\beta}_2 - \beta_1\dot{\alpha}_2)\sinh\eta$$
$$= (\cosh\eta, i, 0, \sinh\eta). \qquad (4.5)$$

This result is identical to that of boosting $-(x + iy)$. We can carry out similar calculations for $B_x(\eta)\beta_1\dot{\beta}_2$, $B_x(\eta)\alpha_1\dot{\beta}_2$, and $B_x(\eta)\beta_1\dot{\alpha}_2$ (Problem 15 in Section 6). As for the boost along the y direction, we know how to rotate this system of spinors around the z axis by 90°.

The above procedure of constructing four-vectors from $SL(2, c)$ spinors will be useful when we study gauge transformations for massless particles in Chapters VII and VIII. We are then led to the question of what happens to the combination of two undotted spinors or two dotted spinors. This question will be discussed in Chapter X.

In the meantime, our immediate interest is in constructing the real four-by-four Lorentz transformation matrix from the complex elements of a, b, c, d of the two-by-two matrix given in Equation (2.1) representing $SL(2, c)$. The problem of converting the two-by-two $SU(2)$ matrix of Equation (2.5) to the three-by-three rotation matrix of Equation (3.1) is widely discussed in textbooks. The question here is whether it is possible to use the same procedure in the $SL(2, c)$ case.

If we go back to Equation (1.8), the calculation is straightforward. It is a matter of solving the linear equations resulting from Equation (1.8) using the explicit form of Equation (2.1):

$$\begin{bmatrix} t' + z' & x' - iy' \\ x' + iy' & t' - z' \end{bmatrix}$$

$$= \begin{bmatrix} a & b \\ c & d \end{bmatrix} \begin{bmatrix} t + z & x - iy \\ x + iy & t - z \end{bmatrix} \begin{bmatrix} a^* & c^* \\ b^* & d^* \end{bmatrix}. \qquad (4.6)$$

The result of the above calculation is the following four-by-four Lorentz transformation matrix (Problem 10 in Section 6):

$$\begin{bmatrix} \text{Re}(ad^* + bc^*) & -\text{Im}(ad^* - bd^*) & \text{Re}(ac^* - bd^*) & \text{Re}(ac^* + bd^*) \\ \text{Im}(ad^* + bc^*) & \text{Re}(ad^* - bd^*) & \text{Im}(ac^* - bd^*) & \text{Im}(ac^* + bd^*) \\ \text{Re}(ab^* - cd^*) & -\text{Im}(ab^* - cd^*) & \tfrac{1}{2}(aa^* - bb^* - cc^* + dd^*) & \tfrac{1}{2}(aa^* + bb^* - cc^* - dd^*) \\ \text{Re}(ab^* + cd^*) & -\text{Im}(ab^* + cd^*) & \tfrac{1}{2}(aa^* - bb^* + cc^* + dd^*) & \tfrac{1}{2}(aa^* + bb^* + cc^* + dd^*) \end{bmatrix}. \qquad (4.7)$$

This procedure is the same as taking the direct product of the spinors α

and β to which the transformation matrix W of Equation (1.4) is applicable and those for the complex conjugate of W. If we use the notation α^* and β^* as the spinors to which W^* is applicable, the elements of the matrix of Equation (4.6) can be identified as (Novozhilov, 1975)

$$\alpha_1 \alpha_2^* = (t+z), \quad \alpha_1 \beta_2^* = (x-iy),$$
$$\beta_1 \alpha_2^* = (x+iy), \quad \beta_1 \beta_2^* = (t-z). \tag{4.8}$$

The expressions are quite different from combinations given in Equations (4.1) and (4.2) for undotted and dotted representations. The completion of $SL(2, c)$ through complex conjugation is mathematically convenient (Bargmann, 1947; Naimark, 1957), especially when we use the conformal form

$$z' = \frac{az+b}{cz+d}, \tag{4.9}$$

and the complex conjugate of this expression. However, the dotted representation is more convenient when we deal with the Dirac spinors.

Let us see how these two representations are related to each other. If we write W of Equation (1.4) as

$$W = \exp\left[-\frac{i}{2} \sum_i (\theta_i \sigma_i + i\eta_i \sigma_i)\right], \tag{4.10}$$

then

$$\dot{W} = \exp\left[-\frac{i}{2} \sum_i (\theta_i \sigma_i - i\eta_i \sigma_i)\right], \tag{4.11}$$

and

$$W^* = \exp\left[-\frac{i}{2} \sum_i (-\theta_i \sigma_i^* + i\eta_i \sigma_i^*)\right].$$

On the other hand, in the standard notation for the Pauli spin matrices,

$$\sigma_2 \sigma_i^* \sigma_2 = -\sigma_i. \tag{4.12}$$

Therefore,

$$\dot{W} = \sigma_2 W^* \sigma_2, \tag{4.13}$$

and

$$\dot{\alpha} = \sigma_2 \alpha^* \quad \text{and} \quad \dot{\beta} = \sigma_2 \beta^*, \tag{4.14}$$

up to a constant factor of unit modulus. We choose the convention:

$$\alpha^* = \dot{\beta} \quad \text{and} \quad \beta^* = -\dot{\alpha}. \tag{4.15}$$

Theory of Spinors

In order to make a connection with Equation (4.2), we have to make a further adjustment to accommodate the usual passive transformation applicable to the spherical harmonics by reversing the sign of y.

5. Symmetries of the Dirac Equation

We discussed the Dirac equation briefly in Chapter III as an illustrative example of the representation of the Poincaré group for free particles with internal angular momentum. We noted there that if we use the Weyl representation of the Dirac matrices, the generators of the rotations and boosts take the form (Problem 1 in Section 6).

$$S_i = \begin{bmatrix} (1/2)\sigma_i & 0 \\ 0 & (1/2)\sigma_i \end{bmatrix}, \tag{5.1}$$

$$K_i = \begin{bmatrix} (i/2)\sigma_i & 0 \\ 0 & (-i/2)\sigma_i \end{bmatrix}. \tag{5.2}$$

It was noted in Section 1 that the boost operators can take two different signs in the $SL(2, c)$ regime. Indeed, the significance of the Dirac equation is that it represents a *direct sum* of the representations with the two different signs of the boost operators.

It is also very important to note that $\gamma_5 = i\gamma^1\gamma^2\gamma^3\gamma^0$ takes the form

$$\gamma_5 = \begin{bmatrix} 1 & 0 \\ 0 & -1 \end{bmatrix}. \tag{5.3}$$

This means that γ_5 commutes with both the rotation and boost generators, and remains invariant under Lorentz transformations. Furthermore, the eigenvalues of this matrix determine the sign of the boost generators.

With this preparation, let us consider the Dirac equation for a free particle with four-momentum p_μ. For a given sign of p_0, there are two different solutions corresponding to two different signs of the energy:

$$\psi^{(\pm)} = U(\pm p) \, e^{\pm ip \cdot x}. \tag{5.4}$$

In terms of $U(p)$, the Dirac equation becomes

$$(\gamma \cdot p - M) \, U(p) = 0, \tag{5.5}$$

where the explicit forms for the γ matrices are given in Section 5 of Chapter III. M is the mass of the particle. We can consider also the equation in which four-vector p_μ is replaced by $-p_\mu$. This is one way to deal with both signs of the energy, and this sign convention is used frequently in quantum field theory.

However, when we discuss symmetry problems, it is more convenient to use a different convention. This different but equivalent way is to rotate around the y axis the above-mentioned negative-energy solution by 180°. Then the sign of the momentum is the same as that of the positive energy solution, while the sign of the energy is opposite to that of the positive-energy solution. The positive-energy solution takes the form

$$\psi^{(+)} = U(\mathbf{p}) \, e^{i(\mathbf{p}\cdot\mathbf{x} - Et)}, \qquad (5.6)$$

and the negative energy solution is

$$\psi^{(-)} = V(\mathbf{p}) \, e^{i(\mathbf{p}\cdot\mathbf{x} + Et)}, \qquad (5.7)$$

where

$$E = +[\mathbf{p}^2 + M^2]^{1/2}.$$

Therefore, we can obtain the negative-energy solution for the positive-energy solution by simply reversing the sign of E. If we write the Dirac spinor $U(\mathbf{p})$ as

$$U(\mathbf{p}) = \begin{bmatrix} x_+ \\ x_- \end{bmatrix}. \qquad (5.8)$$

Then the above spinor satisfies the matrix equation

$$\begin{bmatrix} -M & E + \boldsymbol{\sigma}\cdot\mathbf{p} \\ E - \boldsymbol{\sigma}\cdot\mathbf{p} & -M \end{bmatrix} \begin{bmatrix} x_+ \\ x_- \end{bmatrix} = 0. \qquad (5.9)$$

The solution of this equation is

$$x_- = \frac{E - \boldsymbol{\sigma}\cdot\mathbf{p}}{M} x_+, \quad \text{or} \quad x_+ = \frac{E + \boldsymbol{\sigma}\cdot\mathbf{p}}{M} x_-. \qquad (5.10)$$

Let us assume without loss of generality that the momentum of the particle is along the z direction. The spin along the direction of the momentum is called the *helicity* (Wigner, 1939; Jacob and Wick, 1959). If the spin is parallel (anti-parallel) to the momentum, the helicity is said to be positive (negative). It can be shown that the energy operator commutes with the helicity operator S_3 (Problem 3 in Section 6). Since $\boldsymbol{\sigma}\cdot\mathbf{p}$ acting on the helicity state does not change the helicity, we can let $x_\pm$ be proportional to either α (positive helicity) or β negative helicity. Since there are two possible helicity states for a given sign of E, there are four linearly independent solutions.

For positive p_0 or $p_0 = E$, the positive helicity spinor takes the form

$$U_+(\mathbf{p}) = \begin{bmatrix} \left(\dfrac{E+P}{E-P}\right)^{1/4} \alpha \\ \left(\dfrac{E-P}{E+P}\right)^{1/4} \dot{\alpha} \end{bmatrix}, \qquad (5.11)$$

where $P = p_z$, and we used the identity:

$$E - M = \frac{M^2}{E + M}, \tag{5.12}$$

in deriving the above expression from Equation (5.9). The negative helicity state is

$$U_-(\mathbf{p}) = \begin{bmatrix} \left(\dfrac{E - P}{E + P}\right)^{1/4} \beta \\ \left(\dfrac{E + P}{E - P}\right)^{1/4} \beta \end{bmatrix}. \tag{5.13}$$

For negative p_0 or $p_0 = -E$,

$$V_+(\mathbf{p}) = \begin{bmatrix} -\left(\dfrac{E - P}{E + P}\right)^{1/4} \alpha \\ \left(\dfrac{E + M}{E - M}\right)^{1/4} \dot{\alpha} \end{bmatrix}, \tag{5.14}$$

for positive helicity. The negative helicity spinor is

$$V_-(\mathbf{p}) = \begin{bmatrix} \left(\dfrac{E + P}{E - P}\right)^{1/4} \beta \\ -\left(\dfrac{E + P}{E - P}\right)^{1/4} \beta \end{bmatrix}. \tag{5.15}$$

In order to study the symmetry of the little groups, let us start with the spinors for the particle at rest:

$$U_+(0) = \begin{bmatrix} \alpha \\ \dot{\alpha} \end{bmatrix}, \quad U_-(0) = \begin{bmatrix} \beta \\ \beta \end{bmatrix}, \tag{5.16}$$

$$V_+(0) = \begin{bmatrix} -\alpha \\ \dot{\alpha} \end{bmatrix}, \quad V_-(0) = \begin{bmatrix} \beta \\ -\beta \end{bmatrix}. \tag{5.17}$$

For a massive particle, the little group is $SU(2)$ applicable to both upper and lower components of the above spinors. Indeed, the above spinors form the representation space for rotations generated by the spin operators given in Equation (5.1).

Let us boost along the z axis the above spinors by applying the boost operators generated by K_3 of Equation (5.2):

$$B_z(\eta) = \begin{bmatrix} B^{(+)}(\eta) & 0 \\ 0 & B^{(-)}(\eta) \end{bmatrix}, \tag{5.18}$$

where

$$B_z^{(\pm)}(\xi) = \exp\left(\pm \frac{\xi}{2}\sigma_z\right) = \begin{bmatrix} e^{\pm\xi/2} & 0 \\ 0 & e^{\mp\xi/2} \end{bmatrix}, \qquad (5.19)$$

with

$$\xi = \sinh^{-1}\left(\frac{P}{E}\right) \quad \text{or} \quad \exp\left(\frac{\xi}{2}\right) = \left(\frac{E+P}{E-P}\right)^{1/4}.$$

The result of this calculation is

$$B_z(\eta)\, U_\pm(0) = U_\pm(P), \quad B_z(\eta)\, V_\pm(0) = V_\pm(P). \qquad (5.20)$$

We have thus obtained the solutions of the Dirac equation purely from the symmetry considerations. We constructed first the representation space of the $O(3)$-like little group, and then completed the orbit by boosting the system. Thus the solutions of the free-particle Dirac equation, which does not contain any dynamical effects, constitute a manifestation of the space-time symmetry properties.

For the moving Dirac particle, the S_i matrices given in Equation (5.1) are no longer the generators of the little group. The generators are

$$S_i' = B_z(\xi)\, S_i\, B_z^{-1}(\xi), \qquad (5.21)$$

where $B_z(\xi)$ is given in Equation (5.19). These new generators are not Hermitian but have the same algebraic property as that of S_i.

From the spinor expression given in Equations (5.11), (5.13), (5.14) and (5.15), it is likely that we can obtain the spinors for massless particles by taking the large-momentum/zero-mass limit of those spinors. However, the limiting process is far more complicated than this, because the little group for massless particles is not like $O(3)$ but is like $E(2)$. This problem is discussed in detail in Chapters VII and VIII. In the meantime, we shall study the Dirac equation for massless particles as an independent entity.

The Dirac equation for a massless particle can be written as:

$$\begin{bmatrix} \sigma\cdot\mathbf{p} & 0 \\ 0 & -\sigma\cdot\mathbf{p} \end{bmatrix} \begin{bmatrix} x_1 \\ x_2 \end{bmatrix} = p_0 \begin{bmatrix} x_1 \\ x_2 \end{bmatrix}. \qquad (5.22)$$

There are two uncoupled equations in the above expressions. The basic difference between the above equation and the equation for a massive particle in Equation (5.5) is that the above form is invariant under the simultaneous sign changes in p_0 and $\mathbf{p}$. We thus expect an additional symmetry.

This additional symmetry comes from the fact that γ_5 commutes with the Hamiltonian operator, and can therefore be simultaneously diagonalized with the energy. The eigenvalue of this matrix is $+1$ or -1, as is specified in

Equation (5.3). In the real world, this symmetry appears as the polarization of neutrinos. What implication does this $\gamma_5 = \pm 1$ symmetry have in terms of the little groups?

Let us assume again that the massless particle moves along the z direction, and the particle has a definite helicity. If the energy has the same sign as the momentum, the solution of the above matrix equation is

$$U_+(\mathbf{p}) = \begin{bmatrix} \alpha \\ 0 \end{bmatrix} \quad \text{or} \quad U_-(\mathbf{p}) = \begin{bmatrix} 0 \\ \beta \end{bmatrix}. \tag{5.23}$$

If the sign of the energy is opposite to that of the momentum,

$$V_+(\mathbf{p}) = \begin{bmatrix} 0 \\ \dot{\alpha} \end{bmatrix} \quad \text{or} \quad V_-(\mathbf{p}) = \begin{bmatrix} \beta \\ 0 \end{bmatrix}. \tag{5.24}$$

If we rotate the above wave functions by 180° around the y axis, they become identical to those of Equation (5.22). Thus, unlike the case of massive particles, there are only two independent solutions, and we can restrict ourselves to the solutions of Equation (5.22).

Within the system of spinors given in Equation (5.22), we have two ways to construct the neutrino wave function. The first way is to choose

$$\psi_+(x) = U_+(\mathbf{p})\, e^{i\omega(z-t)}, \tag{5.25}$$

and the second choice is

$$\psi_-(x) = U_-(\mathbf{p})\, e^{i\omega(z-t)}. \tag{5.26}$$

The eigenvalues of γ_5 for the first and second choices are $+1$ and -1 respectively. The helicity for the first and the second wave functions are positive and negative respectively.

We can obtain the anti-particle wave functions by performing charge conjugation on each of the above wave functions. The charge conjugation of Equation (5.25) is

$$(\psi_+(x))^C = U_-(\mathbf{p})\, e^{-i\omega(z-t)}. \tag{5.27}$$

The charge conjugation for the second choice is

$$(\psi_-(x))^C = U_+(\mathbf{p})\, e^{-i\omega(z-t)}. \tag{5.28}$$

The charge conjugation changes the helicity in both cases, while γ_5 remains invariant.

It is clear from the above analysis that if $U_+(\mathbf{p})$ is the spinor for the particle (anti-particle), then $U_-(\mathbf{p})$ is the spinor for the anti-particle (particle). It is thus sufficient to study only the spinors in Equations (5.25) and (5.26). If $\gamma_5 = +1$, then the particle has positive helicity, while the helicity of the anti-particle is negative. If $\gamma_5 = -1$, then it is in the other way around. In the world of neutrinos, their polarization characteristic shows consistently $\gamma_5 = -1$ (Goldhaber *et al.*, 1958).

In order to study the problem of what fundamental space-time symmetry causes the separation of the $\gamma_5 = 1$ state from that of $\gamma_5 = -1$, let us go back to the generators of Lorentz transformations applicable to the four-component spinors given in Equation (5.22). The boost operator applicable to the lower component is the inverse of that for the upper component.

The little group in this case is generated by S_3, N_1 and N_2, which can be written explicitly as

$$S_3 = \frac{1}{2}\begin{bmatrix} \sigma_3 & 0 \\ 0 & \sigma_3 \end{bmatrix},$$

$$N_1 = K_1 + S_2 = \begin{bmatrix} i\sigma_- & 0 \\ 0 & i\sigma_+ \end{bmatrix}, \quad (5.29)$$

$$N_2 = K_2 - S_1 = -\begin{bmatrix} \sigma_- & 0 \\ 0 & \sigma_+ \end{bmatrix},$$

where

$$\sigma_+ = \begin{bmatrix} 0 & 1 \\ 0 & 0 \end{bmatrix}, \quad \text{and} \quad \sigma_- = \begin{bmatrix} 0 & 0 \\ 1 & 0 \end{bmatrix}.$$

These matrices satisfy the commutation relations for the generators of the $SE(2)$ group.

Since we demand that the wave function be diagonal in S_3, there is no need to construct the rotation matrix it generates. The four-by-four transformation matrices generated by N_1 and N_2 take the form

$$D(u, v) = \exp[-i(uN_1 + vN_2)]$$
$$= I - iuN_1 - ivN_2.$$
$$= \begin{bmatrix} I + (u + iv)\sigma_- & 0 \\ 0 & I + (u + iv)\sigma_+ \end{bmatrix}. \quad (5.30)$$

The spinor solutions U_+ and U_- given in Equation (5.23) are invariant under the operation of the above matrix. Therefore, the $\gamma_5 = \pm 1$ symmetry can be translated into the invariance under the transformation of the $E(2)$-like little group.

We noted before in the case of massive particles that the Dirac spinors can be constructed from the symmetry considerations alone starting from the representation space of the little group. The question then is whether this can be done in the massless case. More specifically, let us see whether we can get the result equivalent to $\gamma_5 = \pm 1$ by symmetry property of the little group. We can start with the wave functions diagonal in the S_3 operator:

$$\psi_\pm(x) = W_\pm \exp[\pm i\omega(z-t)], \qquad (5.31)$$

where

$$W_\pm = \begin{bmatrix} A & x_\pm \\ B & x_\pm \end{bmatrix}.$$

Because all six generators of Lorentz transformations are in block diagonal form, each of the upper and lower components of the above four-spinor forms a representation basis for the $SL(2, c)$ group. The above wave function satisfies the second-order D'Alembertian equation, but is not a solution of the Dirac equation. If we demand that the wave function satisfy the Dirac equation for massless particles, the result is the separation of $\gamma_5 = \pm 1$. The question is whether we can obtain the same result from the symmetry of the little group without using the Dirac equation.

The four-by-four D matix of Equation (5.30) consists of the two-by-two $\sigma_\pm$ matrices. When these step-up and step-down operators are applied to $x_\pm$ contained in the wave function of Equation (5.31), they produce:

$$\sigma_+ x_+ = 0, \quad \sigma_+ x_- = x_-,$$
$$\sigma_- x_+ = x_-, \quad \sigma_- x_- = 0. \qquad (5.32)$$

Thus if we insist on the invariance under $D(u, v)$, w has to be

$$W_-^{(+)} = \begin{bmatrix} \alpha \\ 0 \end{bmatrix} \quad \text{or} \quad W_+^{(+)} = \begin{bmatrix} 0 \\ \beta \end{bmatrix}. \qquad (5.33)$$

Indeed, we can therefore obtain $\gamma_5 = \pm 1$ from the condition that the world be invariant under the $D(u, v)$ transformation.

It can be said further that the polarization of neutrinos is a consequence of the invariance under this D transformation (Han et al., 1982c). We know that this transformation can be interpreted in terms of the basic space-time transformations. What is then its physical significance? What interpretation does this kind of transformation have in the case of photons? We shall discuss these questions in Chapters VII and VIII.

6. Exercises and Problems

Exercise 1. Let us go back to the commutation relations of Equation (1.3). They are invariant under Hermitian conjugation, but not under complex conjugation. Show how the complex conjugated generators of $SL(2, c)$ satisfy the original commutation relations.

From the form of the Pauli spin matrices,

$$(S_1)^* = S_1, \quad (S_2)^* = -S_2, \quad (S_3)^* = S_3. \tag{6.1}$$

Thus the commutation relations

$$[(S_i)^*, (S_j)^*] = i\varepsilon_{ijk} (S_k)^*, \tag{6.2}$$

are satisfied. As for the boost operators,

$$(K_1)^* = -K_1, \quad (K_2)^* = K_2, \quad (K_3)^* = -K_3. \tag{6.3}$$

From Equations (6.1) and (6.3), we can construct the commutation relations:

$$[(S_i)^*, (K_j)^*] = i\varepsilon_{ijk} (K_k)^*,$$
$$[(K_i)^*, (K_j)^*] = -i\varepsilon_{ijk} (S_k)^*. \tag{6.4}$$

This property of $SL(2, c)$ is not shared by $SO(3, 1)$.

Let us check some of the transformation properties of the spinor combinations for this complex conjugate representation (Naimark, 1957; Novozhilov, 1975). Rotations around x and z axes are in the same direction as in the original representation, but the rotation around the y axis is in the opposite direction. Boosts along the x and z directions are in the opposite directions, but the boost along the y axes is in the same direction.

With this point in mind, let us check some of the transformation properties of the spinor combinations given in Equation (4.8). $\alpha_1 \alpha_2^*$ and $\beta_1 \beta_2^*$ should be invariant under rotations around the z axis. If we rotate around the x or y axis, the combination $\frac{1}{2}(\alpha_1 \alpha_2^* + \beta_1 \beta_2^*)$ remains invariant. Therefore, this combination can be identified as t, and $\frac{1}{2}(\alpha_1 \alpha_2^* - \beta_1 \beta_2^*)$ as z. Under rotations around the z axis, $\alpha_1 \beta_2^*$ and $\beta_1 \alpha_2^*$ behave like $(x - iy)$ and $(x + iy)$ respectively.

Under boosts along the z direction, $\alpha_1 \beta_2^*$ and $\beta_1 \alpha_2^*$ remain invariant just like $(x \pm iy)$. If we boost the system along the z axis with the parameter ξ, then

$$\alpha_1 \alpha_2^* \rightarrow e^\xi \alpha_1 \alpha_2^*, \quad \text{and} \quad \beta_1 \beta_2^* \rightarrow e^{-\xi} \beta_1 \beta_2^*, \tag{6.5}$$

just like $(t + z)$ and $(t - z)$ respectively. $\alpha_1 \beta_2^*$ and $\beta_1 \alpha_2^*$ remain invariant under this boost. If we boost the spinors along the x direction, the combination $(\alpha_1 \alpha_2^* - \beta_1 \beta_2^*)$ which corresponds to z remains invariant. The combination $(1/2i)(\beta_1 \alpha_2^* - \alpha_1 \beta_2^*)$ which corresponds to y also remains invariant. Under the same boost, the combinations $\frac{1}{2}(\alpha_1 \alpha_2^* + \beta_1 \beta_2^*)$ and $\frac{1}{2}(\alpha_1 \beta_2^* + \beta_1 \alpha_2^*)$, which behave like t and x respectively, transform like t and x respectively.

Exercise 2. Show that the most general form of the $SL(2, c)$ matrix given in Equation (2.1) can be written as a multiplication of a unimodular diagonal matrix, an upper and lower triangular matrices with unit diagonal elements (Miller, 1968).

We can write

$$W = \begin{bmatrix} a & b \\ c & d \end{bmatrix} = \begin{bmatrix} 1 & b' \\ 0 & 1 \end{bmatrix} \begin{bmatrix} 1 & 0 \\ c' & 0 \end{bmatrix} \begin{bmatrix} e^{a'/2} & 0 \\ 0 & e^{-a'/2} \end{bmatrix}, \qquad (6.6)$$

which can be written as

$$[\exp(b'\sigma_+)][\exp(c'\sigma_-)][\exp(a'\sigma_3)], \qquad (6.7)$$

where the parameters b', c' and a' are complex numbers. We have thus reduced the number of parameters into three complex numbers, using the fact that the determinant of the matrix is one. This is quite consistent with the form given in Equation (1.5).

The result of the matrix multiplications in Equation (6.6) is

$$W = \begin{bmatrix} (1 + b'c')e^{a'/2} & b'e^{-a'/2} \\ -c'e^{a'/2} & e^{-a'/2} \end{bmatrix}. \qquad (6.8)$$

We can now identify $d = \exp(-a'/2)$, $b = b'd$, $c = -c'/d$. a is determined from the condition that the determinant of W is one.

Exercise 3. In Equation (4.9), and also in Section 4 of Chapter I, we noted that $SL(2, c)$ is homomorphic to the conformal transformation

$$z' = \frac{az + b}{cz + d}. \qquad (6.9)$$

Since the above expression is invariant under the sign change of a, b, c and d, the correspondence is two-to-one. When $a = d = 1$ and $b = c = 0$, this transformation is an identity transformation. When $a = d = 1$, and b and c are small, we can define an infinitesimal transformation. Is it possible to formulate a Lie group and its Lie algebra for the function $f(z)$?

Let us go back to the expression of W given in Equation (1.5). If the group parameters are allowed to be complex, it is sufficient to use only three generators for $SL(2, c)$. It is convenient to choose three rotation generators: J_1, J_2 and J_3, satisfying the commutation relations:

$$[J_3, J_\pm] = \pm J_\pm, \quad [J_+, J_-] = 2J_+. \qquad (6.10)$$

where

$$J_\pm = J_x \pm iJ_y.$$

We are quite familiar with the differential forms of the J operators depending on three variables. In Equation (3.22), we introduced those depending on two variables. This time, we have to introduce the J operators depending on only one variable. The one-variable forms are available in the literature (Bargmann, 1947; Miller, 1968), and they are:

$$J_3 = i\left(-n + z\frac{d}{dz}\right), \quad J_+ = i\left(2nz + z^2\frac{d}{dz}\right), \quad J_- = -i\frac{d}{dz}. \quad (6.11)$$

These operators satisfy the commutation relations of Equation (6.10). For functions of z, we should use the operator:

$$T = (\exp(-ia'J_3))(\exp(-ic'J_-))(\exp(-ib'J_+)). \quad (6.12)$$

in which the relation between the active and passive transformations is used (see Exercise 2 in Section 6 of Chapter II). The result of this operation on $f(z)$ is (Miller, 1968)

$$Tf(z) = (bz + d)^{2n} f\left(\frac{az + b}{bz + d}\right). \quad (6.13)$$

In addition to the change in the z variable according to Equation (6.9), there is a multiplication factor $(bz + d)^{2n}$. For this reason, this type of representation is called a *multiplier representation* (Bargmann, 1947).

Problem 1. In Equations (5.14) and (5.15) of Chapter III, we introduced the Weyl representation of the Dirac matrices. However, in addition, there is another representation commonly used in the literature, in which γ_0 and γ_5 become interchanged:

$$\gamma_0 = \begin{bmatrix} 1 & 0 \\ 0 & -1 \end{bmatrix}, \quad \gamma_5 = \begin{bmatrix} 0 & 1 \\ 1 & 0 \end{bmatrix}, \quad (6.14)$$

but the matrices γ_i remain unchanged. This is called the Dirac representation. Show that the Dirac representation is unitarily equivalent to the Weyl representation (Bjorken and Drell, 1964). What form do the generators of rotations and boosts take in the Dirac representation?

Problem 2. A more conventional way of solving the Dirac equation is to solve the eigenvalue equation for the Hamiltonian in the Dirac representation. The Hamitonian in the Weyl representation takes the form:

$$H = \begin{bmatrix} \boldsymbol{\sigma} \cdot \mathbf{p} & M \\ M & -\boldsymbol{\sigma} \cdot \mathbf{p} \end{bmatrix}. \quad (6.15)$$

Solve the eigenvalue equation in the Weyl representation, and compare the result with the expressions given in Section 5. Show that the helicity operator commutes with the Hamiltonian.

Problem 3. Show that the helicity and γ_5 can be diagonalized simultaneously, and show that, in this case, the Hamiltonian is not diagonal. Show that, if the helicity operator commutes with the Hamiltonian, they can be simultaneously diagonalized. Show in this case that γ_5 is not diagonal. The Hamiltonian is diagonal in the Dirac representation, but is not in the Weyl representation.

Problem 4. The Dirac representation of the Dirac equation is commonly used for solving the hydrogen atom problem in the non-relativistic limit. Use the Weyl representation to do the same.

Problem 5. The Hamiltonian is not diagonal in the Dirac or Weyl representation of the Dirac equation. Is it possible to obtain a representation in which the Hamiltonian is diagonal by making a unitary transformation of the Dirac or Weyl representation? The answer to this question is "yes," and the transformation in question is called the Foldy Wouthuysen transformation (Foldy and Wouthuysen, 1950). Solve the hydrogen atom problem in the Foldy-Wouthuysen representation. For recent applications of the Foldy-Wouthuysen transformation, see Streater and Wightman (1964), Bogoliubov *et al.* (1975), Barut and Raczka (1977), and Ali (1981).

Problem 6. Construct the three-by-three rotation matrix from the two-by-two $SU(2)$ matrix. This problem is discussed in many textbooks (Rose, 1957; Edmonds, 1957).

Problem 7. Let us go to Equation (3.10). Starting from the rotation matrix for each spinor, show that the rotation matrix for the states $|1, 1\rangle, |1, 0\rangle$ and $|1, -1\rangle$ is the same as the rotation matrix for $Y_1^m(\theta, \phi)$.

Problem 8. Calculate the rotation matrix for the degenerate states of $j = 3/2$ given in Equation (3.12). Compare the result with the matrix which can be computed from the general Equation (3.25).

Problem 9. Show that the rotation matrices for the states given in Equation (3.13) and in Equation (3.14) are identical and are the same as the rotation matrix for the spinors α and β.

Problem 10. Starting from the most general form of the two-by-two matrix representing $SL(2, c)$ given in Equation (2.1) with six independent real parameters, calculate the four-by-four Lorentz transformation matrix in terms of a, b, c and d given in Equation (4.7). Use the equation given in Equation (4.6) (Naimark, 1957 and 1964; Smirnov, 1961).

Problem 11. The four-by-four matrix of Equation (4.7) is a complicated expression. Check whether this expression becomes that of each little group. Check in particular whether the expression reduces to that of Equation (3.1) for the $SU(2)$ subgroup. Check also whether it becomes the $T(u, v)$ matrix of Equation (2.6) in Chapter III when the $SL(2, c)$ transformation takes the form:

$$\begin{bmatrix} 1 & u - iv \\ 0 & 1 \end{bmatrix}. \tag{6.16}$$

Problem 12. Wigner's paper (1939) contains a proof that the group of proper Lorentz transformations is simple. Use Cartan's criterion to show that $SL(2, c)$ is a simple group. Show also that $SU(2)$ and $SU(1, 1)$ are simple groups, but $SW(2)$ [$SE(2)$-like subgroup of $SL(2, c)$] is not.

Problem 13. Let us consider the conformal representation:

$$z' = \frac{az + b}{cz + d}. \tag{6.17}$$

For the $SE(2)$-like little group, $c = 0$, and the modulus of a is unity. The transformation is only a linear transformation. Discuss the property of transformations for $SU(2, c)$ and $SU(1, 1)$ in the complex z plane.

Problem 14. There are two equivalent ways of representing the $SO(2, 1)$-like subgroup of $SL(2, c)$. One is $SU(1, 1)$, and the other is $SL(2, r)$. Translate this equivalence into the language of the four-by-four matrix.

Problem 15. Calculate the boost matrices along the x direction for the spinor combinations given in Equation (4.2).

Problem 16. The operators S_3, N_1 and N_2 of Equation (2.13) satisfy the Lie algebra of the $SE(2)$ group. If we add K_3, the four generators satisfy the closed algebra of Equations (2.13) and (2.14). What are the transformations generated by these operators? See Janner and Janssen (1972).

Chapter V

Covariant Harmonic Oscillator Formalism

Because wave functions play a central role in nonrelativistic quantum mechanics, one method of combining quantum mechanics and special relativity takes the form of efforts to construct relativistic wave functions with an appropriate probability interpretation. The story of relativistic wave functions begins with Schrödinger's original attempt to formulate his wave mechanics using the relativistic wave equation which is known today as the Klein-Gordon equation (Dirac, 1972). Unlike the Schrödinger wave function, the solutions of the Klein-Gordon equation have the well-known negative-energy problem. The Klein-Gordon wave function also has difficulties with a probability interpretation. The negative-energy problem was later "solved" by the second quantization procedure, but the difficulty with a probability interpretation still persists.

We can next mention Dirac's equation for electrons. This equation has been strikingly successful in explaining the properties of the electron in the static limit. However, its relativistic feature is not different from that of the Klein-Gordon equation.

In spite of continued efforts to construct relativistic wave functions, this approach was overshadowed by the successes of quantum field theory in quantum electrodynamics. It is quite fair to say that we are still living under this shadow. One promising approach which emerged from the field theory framework was the Bethe-Salpeter equation. However, the physical meaning of the Bethe-Salpeter wave functions has not yet been clearly defined (Wick, 1954).

Because of its mathematical simplicity, the harmonic oscillator has served as the first concrete solution to many new physical theories. It plays a key role in the developing stages of nonrelativistic quantum mechanics, statistical mechanics, theory of specific heat, molecular theory, quantum field theory, theory of superconductivity, theory of coherent light, and many others. It is, therefore, quite natural to expect that the first non-trivial relativistic wave function would be a relativistic harmonic oscillator wave function.

As early as in 1945, Dirac suggested the use of normalizable relativistic oscillator wave functions for studying relativistic Fock space needed in quantum electrodynamics. In connection with relativistic particles with internal space-time structure, Yukawa attempted to construct relativistic oscillator wave functions in 1953. Yukawa observed that an attempt to solve a relativistic oscillator wave equation in general leads to infinite-component wave functions, and that finite-component wave functions may be chosen if a subsidiary condition involving the four-momentum of the particle is considered. This proposal of Yukawa was further developed by Markov (1956), Takabayasi (1964 and 1979), Sogami (1969), Ishida (1971).

The effectiveness of Yukawa's oscillator wave function in the relativistic quark model was first demonstrated by Fujimura *et al.* (1970) who showed that the Yukawa wave function leads to the correct high-energy asymptotic behavior of the nucleon form factor. The harmonic oscillator wave function was also rediscovered by Feynman *et al.* (1971) who advocated the use of relativistic oscillators instead of Feynman diagrams for studying hadronic structures and interactions. The paper by Feynman *et al.* contains all the problems expected from relativistic wave equations, and the authors of this paper did not make any attempt to hide those problems.

The basic problem facing any relativistic harmonic oscillator equation is the negative-energy spectrum due to time-like excitations. It had once been widely believed that any attempt to obtain finite-component wave functions by eliminating time-like excitations would lead to a violation of probability conservation. This belief did not turn out to be true. It is now possible to construct harmonic oscillator wave functions without time-like wave functions which form the vector spaces for unitary irreducible representations of the Poincaré group.

In Section 1, we formulate the problem by writing down the relativistically invariant differential equation which leads to the covariant harmonic oscillator formalism. In Section 2, we study solutions of the oscillator differential equation which are normalizable in the four-dimensional x, y, z, t space. In Section 3, representations of the Poincaré group for massive hadrons are constructed from the normalizable harmonic oscillator wave functions. It is shown that they form the basis for unitary irreducible representations of the Poincaré group, as well as that for the $O(3)$-like little group for massive particles.

In Section 4, Lorentz transformation properties of the harmonic oscillator wave functions are studied. The linear unitary representation of Lorentz transformation is provided for harmonic oscillator wave functions. In Section 5, we study the relativistic oscillators using the language of the four-dimensional Euclidean coordinate system. We then study in Section 6 how Lorentz transformations are different from rotations in the $O(4)$ coordinate system, and investigate the nature of the Wick rotation commonly used in physics. Section 7 consists of exercises and problems.

1. Covariant Harmonic Oscillator Differential Equations

Developing a new physical theory usually requires a new set of mathematical formulas. There are in general two different approaches to this problem. According to Eddington,* we have to understand all the physical principles before writing down the first mathematical formula. According to Dirac,* however, it is more profitable to construct plausible mathematical devices which can describe quantitatively the real world, and then add physical interpretations to the mathematical formalism. Both special relativity and quantum mechanics were developed in Dirac's way, and most of the new physical models these days are developed in this way.

With this point in mind, let us consider the differential equation of Feynman *et al.* for a hadron consisting of two quarks bound together by a harmonic oscillator potential of unit strength:

$$\left\{ -2\left[\left(\frac{\partial}{\partial x_a^\mu}\right)^2 + \left(\frac{\partial}{\partial x_b^\mu}\right)^2 \right] + \left(\frac{1}{16}\right)(x_a^\mu - x_b^\mu)^2 + m_0^2 \right\} \phi(x_a, x_b) = 0, \quad (1.1)$$

where x_a and x_b are space-time coordinates for the first and second quarks respectively. This partial differential equation has many different solutions depending on the choice of variables and boundary conditions.

In order to simplify the above differential equation, let us introduce new coordinate variables:

$$X = (x_a + x_b)/2,$$
$$x = (x_a - x_b)/2\sqrt{2}. \quad (1.2)$$

The four-vector X specifies where the hadron is located in space-time, while the variable x measures the space-time separation between the quarks. In terms of these variables, Equation (1.1) can be written as

$$\left(\frac{\partial^2}{\partial X_\mu^2} - m_0^2 + \frac{1}{2}\left[\frac{\partial^2}{\partial x_\mu^2} + x_\mu^2 \right] \right) \phi(X, x) = 0. \quad (1.3)$$

This equation is separable in the X and x variables. Thus

$$\phi(X, x) = f(X)\, \psi(x), \quad (1.4)$$

and $f(X)$ and $\psi(x)$ satisfy the following differential equations respectively:

$$\left(\frac{\partial^2}{\partial X_\mu^2} - m_0^2 - (\lambda + 1) \right) f(X) = 0, \quad (1.5)$$

---

* For an interesting discussion on these two contrasting views, see P. A. M. Dirac, *Quantum Electrodynamics,* Comm. Dublin Inst. Adv. Studies, Ser. A, No. 1 (1943).

$$\frac{1}{2}\left(-\frac{\partial^2}{\partial x_\mu^2} + x_\mu^2\right)\psi(x) = (\lambda + 1)\,\psi(x). \tag{1.6}$$

Equation (1.5) is a Klein-Gordon equation, and its solution takes the form

$$f(X) = \exp[\pm ip_\mu X^\mu], \tag{1.7}$$

with

$$-P^2 = -P_\mu P^\mu = M^2 = m_0^2 + (\lambda + 1),$$

where M and P are the mass and four-momentum of the hadron respectively. The eigenvalue λ is determined from the solution of Equation (1.6). We are using the same notation for the operator and eigenvalue for the hadronic four-momentum. This should not cause any confusion since we are dealing only with free hadronic states with a definite four-momentum.

As for the four-momenta of the quarks p_a and p_b, we can combine them into the total four-momentum and momentum-energy separation between the quarks:

$$\begin{aligned} P &= p_a + p_b, \\ q &= \sqrt{2}(p_a - p_b). \end{aligned} \tag{1.8}$$

P is the hadronic four-momentum conjugate to X. The internal momentum-energy separation q is conjugate to x provided that there exist wave functions which can be Fourier-transformed. If the momentum-energy wave functions can be obtained from the Fourier transformation of the space-time wave function, the differential equation in the q space is identical to the harmonic oscillator equation for the x space given in Equation (1.6).

We now have a set of equations to study. The first equation is whether the above set of equations generates a mathematics which has enough aesthetic values. The second question is whether this mathematics can serve as a device which can describe the real world. The third question is whether the mathematical device we use is consistent with the existing rules of quantum mechanics and special relativity. We shall study the first question in this chapter and discuss the second question in Chapters XI and XII. We shall examine the third question in Chapter VI.

2. Normalizable Solutions of the Relativistic Oscillator Equation

We are quite familiar with the fact that the three-dimensional oscillator equation is separable in several coordinate systems. Indeed, the four-dimensional differential equation of Equation (1.6) is separable in at least thirty-four different coordinate systems (Kalnins and Miller, 1977). However, we are only interested in those solutions which are useful for constructing physically relevant representations of the Poincaré group.

There are solutions which are normalizable and which are not. In this section, we discuss one of the solutions in four-dimensional space-time. This solution will form a basis for unitary irreducible representations of the Poincaré group. There are also interesting solutions which are not normalizable. We shall discuss these solutions in Chapters VII and X.

Since we are quite familiar with the three-dimensional harmonic oscillator equation from nonrelativistic quantum mechanics, we are naturally led to consider the separation of the space and time variables and write the four-dimensional harmonic oscillator equation of Equation (1.6) as

$$\left(-\nabla^2 + \frac{\partial^2}{\partial t^2} + [(\mathbf{x})^2 - t^2]\right)\psi(x) = (\lambda + 1)\psi(x). \tag{2.1}$$

However, the $x\,t$ system is not the only coordinate system in which the differential equation takes the above form.

If the hadron moves along the Z direction which is also the z direction, then the hadronic factor $f(X)$ of Equation (1.7) is Lorentz-transformed in the same manner as the scalar particles are transformed, as is discussed in Section 5 of Chapter III. The Lorentz transformation of the internal coordinates from the laboratory frame to the hadronic rest frame takes the form

$$x' = x, \quad y' = y,$$
$$z' = (z - \beta t)/(1 - \beta^2)^{1/2}, \tag{2.2}$$
$$t' = (t - \beta z)/(1 - \beta^2)^{1/2},$$

where β is the velocity of the hadron moving along the z direction. The primed quantities are the coordinate variables in the hadronic rest frame. In terms of the primed variables, the oscillator differential equation is

$$\left(-\nabla'^2 + \frac{\partial^2}{\partial t'^2} + [(\mathbf{x}')^2 - t'^2]\right)\psi(x) = (\lambda + 1)\psi(x). \tag{2.3}$$

This form is identical to that of Equation (2.1), due to the fact that the oscillator differential equation is Lorentz-invariant (Kim and Noz, 1973).

Among many possible solutions of the above differential equation, let us consider the form

$$\psi_\beta(x) = \left(\frac{1}{\pi}\right)\left(\frac{1}{2}\right)^{(a+b+n+k)/2}\left(\frac{1}{a!b!n!k!}\right)^{1/2} H_a(x')\,H_b(y')\,H_n(z')\,H_k(t') \times$$

$$\times \exp\left[-\frac{1}{2}(x'^2 + y'^2 + z'^2 + t'^2)\right], \tag{2.4}$$

where a, b, n, and k are integers, and $H_a(x')$, $H_b(y')$ ... are the Hermite polynomials. This wave function is normalizable, but the eigenvalue takes the values

$$\lambda = (a + b + n - k). \tag{2.5}$$

Thus for a given finite value of λ, there are infinitely many possible combinations of a, b, n and k. The most general solution of the oscillator differential equation is infinitely degenerate (Yukawa, 1953).

Because the wave functions are normalizable, all the generators of the Lorentz transformations given in Equation (1.20) of Chapter III are Hermitian operators. The Lorentz transformation applicable to this function space is therefore a *unitary* transformation. Indeed, we can write any function of the coordinate variables x, y, z and t as a linear combination of the above solutions. In particular, a solution of the oscillator equation with a given set of quantum numbers in the hadronic rest frame can be written as a linear sum of infinitely many solutions in the hadronic rest frame as we shall see in Section 4.

It is very difficult, if not impossible, to give physical interpretations to infinite-component wave functions. For this reason, it is quite natural to seek a finite set from the infinite number of wave functions at least in one Lorentz frame. The simplest way to obtain such a finite set of wave functions is to invoke the restriction that there be no time-like oscillations in the Lorentz frame in which the hadron is at rest and that the integer k in Equations (2.4) and (2.5) be zero. In doing so, we are led to the following two fundamental questions:

(a). Is it possible to give physical interpretations to the wave functions belonging to the resulting finite set?
(b). Is it still possible to maintain Lorentz covariance with this condition?

We shall study the first question in Chapter VI. Let us examine question (b) closely here.

When the hadron moves along the z axis, the $k = 0$ condition is equivalent to

$$\left(t' + \frac{\partial}{\partial t'} \right) \psi_\beta(x) = 0. \tag{2.6}$$

The most general form of the above condition is

$$p_\mu \left(x^\mu - \frac{\partial}{\partial x_\mu} \right) \psi_\beta(x) = 0. \tag{2.7}$$

Thus the $k = 0$ condition is covariant. Once this condition is set, we can write the wave function belonging to this finite set as

$$\psi_\beta(x) = (1/\pi) [1/(2^a 2^b 2^n a! b! n!)]^{1/2} H_a(x') H_b(y') H_n(z') \times$$
$$\times \exp[-\tfrac{1}{2}(x'^2 + y'^2 + z'^2 + t'^2)]. \qquad (2.8)$$

Except for the Gaussian factor in the t' variable, the above expression is the wave function for the three-dimensional isotropic harmonic oscillator. This means that we can use the spherical coordinate system for the x', y' and z' variables. We shall see in Section 3 how these ideas form the basis for constructing representations of the Poincaré group.

Since the above oscillator wave functions are separable in the Cartesian coordinate system, and since the transverse coordinate variables are not affected by the boost along the z direction, we can omit the factors depending on the x and y variables when studying their Lorentz transformation properties. The most general form of the wave function given in Equation (2.4) becomes

$$\psi_\beta^{n,k}(z', t') = [1/(\pi 2^n 2^k n! k!)]^{1/2} \times$$
$$\times H_n(z') H_k(t') \exp[-\tfrac{1}{2}(z'^2 + t'^2)], \qquad (2.9)$$

with

$$\lambda = (n - k).$$

The wave functions satisfying the subsidary condition of Equation (2.7) take the simple form

$$\psi_\beta^n(z, t) = [1/(\pi\, 2^2 n!)]^{1/2} H_n(z') \exp[-(1/2)(z'^2 + t'^2)], \qquad (2.10)$$

with

$$\lambda = n.$$

As we shall see in Chapters VI and XII, this normalizable wave function without excitations along the t' axis describes the internal space-time structure of the hadron moving along the z direction with the velocity parameter β. If $\beta = 0$, then the wave function becomes

$$\psi_0^n(x, t) = [1/(\pi\, 2^n n!)]^{1/2} H_n(z) \exp[-(1/2)(z^2 + t^2)], \qquad (2.11)$$

Thus

$$\psi_\beta^n(z, t) = \psi_0^n(z', t').$$

We have therefore obtained the Lorentz-boosted wave function by making a passive coordinate transformation on the z and t coordinate variables.

Let us next study the orthogonality relations of the wave functions. Since the volume element is Lorentz-invariant:

$$dz\, dt = dz'\, dt', \tag{2.12}$$

there is no difficulty in understanding the orthogonality relation:

$$\int \psi_\beta^{n'}(z, t)\, \psi_\beta^n(z, t)\, dz\, dt = \int \psi_0^{n'}(z, t)\, \psi_0^n(z, t) = \delta_{n'n}. \tag{2.13}$$

However, a more interesting problem is the inner product of two wave functions belonging to different Lorentz frames. As is shown in Exercise 4 in Section 7, the inner product becomes

$$\int \psi_0^{n'}(z, t)\, \psi_\beta^n(z, t)\, dz\, dt = \delta_{n'n}[1 - \beta^2]^{(n+1)/2}. \tag{2.14}$$

The remarkable fact is that the orthogonality in the quantum number n is still preserved, because of the Lorentz invariance of the harmonic oscillator

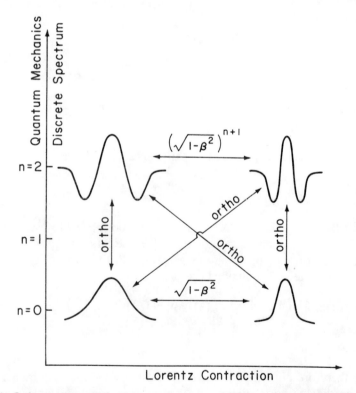

Fig. 2.1. Orthogonality and Lorentz contraction properties of the normalizable harmonic oscillator wave functions. The ground-state wave function is contracted like a rigid rod. The n-th excited state is contracted like a product of $(n+1)$ rigid rods (Kim and Noz, 1978b).

differential equation. The oscillator equation does not depend on the velocity parameter β.

As for the factor $[1 - \beta^2]^{(n+1)/2}$ in Equation (2.14), we note first that, when the oscillator is in the ground state, it becomes like a Lorentz contraction of a rigid rod by $[1 - \beta^2]^{1/2}$. Excited-state wave functions are obtained from the ground state wave function through repeated applications of the step operator:

$$|n, \beta\rangle = \sqrt{1/n!} \left(z' - \frac{\partial}{\partial z'} \right)^n |0, \beta\rangle. \tag{2.15}$$

The transformation property of each step-up operator is like that of z. Therefore, if the ground-state wave function is like a rigid rod along the z direction, the n-th excited state should behave like a multiplication of $(n+1)$ rigid rods (Kim and Noz, 1978b). This contraction property is summarized in Figure 2.1.

3. Irreducible Unitary Representations of the Poincaré Group

As is explained in Chapter III, the Poincaré group consists of space-time translations and Lorentz transformations. Let us go back to the quark coordinates x_a and x_b in Equation (1.2), and consider performing Poincaré transformations on the quarks. The same Lorentz transformation matrix is applicable to x_a, x_b, x as well as X. However, under the space-time translation which changes x_a and x_b to $x_a + a$ and $x_b + b$ respectively,

$$X \to X + a, \tag{3.1}$$

$$x \to x.$$

The quark separation coordinate x is not affected by translations. For this reason, the generators of translations for this system are

$$P_\mu = -i \frac{\partial}{\partial X^\mu}, \tag{3.2}$$

while the generators of Lorentz transformations are

$$M_{\mu\nu} = L^*_{\mu\nu} + L_{\mu\nu}, \tag{3.3}$$

where

$$L^*_{\mu\nu} = -i \left(X_\mu \frac{\partial}{\partial X^\nu} - X_\nu \frac{\partial}{\partial X^\mu} \right),$$

$$L_{\mu\nu} = -i \left(x_\mu \frac{\partial}{\partial x^\nu} - x_\nu \frac{\partial}{\partial x^\mu} \right).$$

It is straightforward to check that the ten generators defined in Equations (3.2) and (3.3) satisfy the commutation relations of the Poincaré group given in Equation (3.9) of Chapter III. We are interested in constructing normalizable wave functions which are diagonal in the Casimir operators P^2 and W^2:

$$P^2 = -\left(\frac{\partial}{\partial X^\mu}\right)^2,$$

$$= \frac{1}{2}\left(-\frac{\partial^2}{\partial x_\mu^2} + x_\mu^2\right) + m_0^2, \tag{3.4}$$

$$W^2 = M^2 (\mathbf{L}')^2. \tag{3.5}$$

where

$$L'_i = -i\varepsilon_{ijk} x'_j \frac{\partial}{\partial x'_k}.$$

The eigenvalue of P^2 is $M^2 = m_0^2 + (\lambda + 1)$, and that for W^2 is $M^2 \ell(\ell + 1)$. M is the hadronic mass, and ℓ is the total intrinsic angular momentum of the hadron due to internal motion of the spinless quarks (Kim et al., 1979d).

In addition, we can choose the solutions to be diagonal in the component of the intrinsic angular momentum along the direction of the motion. This component is often called the helicity. If the hadron moves along the Z direction, the helicity operator is L_3.

Because the spatial part of the harmonic oscillator equation in Equation (2.3) is separable also in the spherical coordinate system, we can write its solution using spherical variables in the hadronic rest frame space spanned by x' y' and z'. The most general form of the solution is

$$\psi_{\beta\lambda\ell}^{k,m}(x) = R_\mu^\ell(r') Y_\ell^m(\theta', \phi') [1/(\sqrt{\pi}\, 2^k k!)]^{1/2} H_k(t')\, e^{-t'^2/2}, \tag{3.6}$$

where

$$r' = [x'^2 + y'^2 + z'^2]^{1/2},$$

$$\cos \theta' = z'/r',$$

$$\tan \phi' = y'/x',$$

and

$$\lambda = 2\mu + \ell - k. \tag{3.7}$$

$R_\mu^\ell(r')$ is the normalized radial wave function for the three-dimensional harmonic oscillator:

$$R_\mu^\ell(r) = (2(\mu!)/[\Gamma\,(\mu + \ell + 3/2)]^3)^{1/2}\, r^\ell\, L_\mu^{\ell+1/2}(r^2)\, e^{-r^2/2}, \tag{3.8}$$

where $L_\mu^{\ell+1/2}(r^2)$ is the associated Laguerre function (Arfken, 1985). The above radial wave function satisfies the orthonormality condition (Morse and Feshbach, 1953):

$$\int_0^\infty r^2 \, R_\mu^\ell(r) \, R_\nu^\ell(r) \, dr = \delta_{\mu\nu}. \tag{3.9}$$

The spherical form given in Equation (3.6) can of course be expressed as a linear combination of the wave functions in the Cartesian coordinate system given in Equation (2.8).

The wave function of Equation (3.6) is diagonal in the Casimir operators of Equations (3.4) and (3.5), as well as in L_3. It indeed forms a vector space for the $O(3)$-like little group (Han et al., 1982b and 1983b). However, the system is infinitely degenerate due to excitations along the t' axis. As we did in Section 2, we can suppress the time-like oscillation by imposing the subsidiary condition of Equation (2.7), or by restricting k to be zero in Equation (3.7). The solution then takes the form

$$\psi_{\beta\lambda\ell}^m(x) = R_\mu^\ell(r') \, Y_\ell^m(\theta', \phi') \, [(1/\pi)^{1/4} \exp(-t'^2/2)], \tag{3.10}$$

with

$$\lambda = 2\mu + \ell.$$

Thus for a given λ, there are only a finite number of solutions. The above spherical form can be expressed as a linear combination of the solutions without time-like excitations in the Cartesian coordinate system given in Equation (2.8).

We can now write the solution of the differential equation of Equation (1.1) as

$$\phi(X, x) = e^{\pm iP \cdot X} \, \psi_{\beta\lambda\ell}^m(x). \tag{3.11}$$

This wave function describes a free hadron with a definite four-momentum having an internal space-time structure which can be described by an irreducible unitary representation of the Poincaré group. The representation is unitary because the portion of the wave function depending on the internal variable x is square-integrable, and all the generators of Lorentz transformations are Hermitian operators. We shall study in Section 4 how these wave functions are Lorentz transformed.

4. Transformation Properties of Harmonic Oscillator Wave Functions

If the hadronic velocity is zero, then its rest frame coincides with the laboratory frame. The wave function then is

$$\psi_0(x) = R_\mu^\ell(r) \, Y_\ell^m(\theta, \phi) \, [(1/\pi)^{1/4} \exp(-t^2/2)]. \tag{4.1}$$

The simplest way to obtain the wave function for the moving hadron is to replace the r, θ and ϕ variables in the above expression by their primed counterparts. This produces Equation (3.6). However, we are interested in obtaining the wave function for a moving hadron as a linear combination of the wave functions for the rest frame. If we apply the boost operator to the wave function for the hadron at rest,

$$\psi_{\beta\lambda}^{\ell m}(x) = [e^{-i\eta K_3}] \, \psi_{0\lambda}^{\ell m}(x), \tag{4.2}$$

where K_3 is the boost generator along the z axis, its form is

$$K_3 = -i\left(z\frac{\partial}{\partial t} + t\frac{\partial}{\partial z}\right), \tag{4.3}$$

and η is related to velocity parameter β by

$$\sinh \eta = \beta/(1 - \beta^2)^{1/2}.$$

Both the rest-frame and moving-frame wave functions have the same set of eigenvalues for the Casimir operators P^2 and W^2 of the Poincaré group.

These eigenstate wave functions are linear combinations of the Cartesian forms in their respective coordinate systems. If the hadron moves along the z direction, the x and y variables remain invariant. Therefore, we use the wave function of Equation (2.10) with $\beta = 0$:

$$\psi_0^{n,0}(z, t) = [1/(\pi 2^n n!)]^{1/2} \, H_n(z) \exp[-(1/2)(z^2 + t^2)]. \tag{4.4}$$

The subscript 0 indicates that there are no time-like excitations: $k = 0$. We are now led to consider the transformation

$$\psi_\beta^{n,0}(z, t) = [\exp(-i\eta K_3)] \, \psi_0^{n,0}(z, t)$$
$$= \psi_0^{n,0}(z', t'), \tag{4.5}$$

and ask what the boost operator $\exp(-i\eta K_3)$ does to $\psi_0^{n,0}(z, t)$.

This boost operator of course changes z and t to z' and t' respectively as is indicated above. However, we are interested in whether the transformation can take the linear form

$$\psi_\beta^{n,0}(z, t) = \sum_{n',k'} A_{n'k'}^{n,0}(\beta) \, \psi_0^{n',k'}(z, t). \tag{4.6}$$

Because the oscillator differential equation is Lorentz invariant, the eigenvalue λ of Equation (2.9) remains invariant, and only the terms which satisfy the condition

$$n = (n' - k') \tag{4.7}$$

make non-zero contributions in the sum. Thus the above expression can be simplified to

$$\psi_\beta^{n,0}(z, t) = \sum_{k=0}^{\infty} A_k^n(\beta) \, \psi_0^{n+k,k}(z, t). \tag{4.8}$$

This is indeed a *linear unitary representation of the Lorentz group*. The representation is infinite-dimensional because the sum over k is extended from zero to infinity (Kim, et al., 1979b).

The remaining problem is to determine the coefficient $A_k^n(\beta)$. Using the orthogonality relation, we can write

$$A_k^n(\beta) = \int dz \, dt \, \psi_0^{n+k,k}(z, t) \, \psi_\beta^{n,0}(z, t)$$

$$= \frac{1}{\pi} \left(\frac{1}{2}\right)^n \left(\frac{1}{2}\right)^{1/2} \left(\frac{1}{n!(n+k)!}\right)^{1/2}$$

$$\times \int dz \, dt \, H_{n+k}(z) \, H_k(t) \, H_n(z')$$

$$\times \exp\left(-\frac{1}{2}(z^2 + z'^2 + t^2 + t'^2)\right). \tag{4.9}$$

In this integral, the Hermitian polynomials and the Gaussian form are mixed with the kinematics of Lorentz transformation. However, if we use the generating function for the Hermite polynomial, the evaluation of the integral is straightforward [Exercise 4 in Section 7], and the result is

$$A_k^n(\beta) = (1 - \beta^2)^{(1+n)/2} \, \beta^k \left(\frac{(n+k)!}{n!k!}\right)^{1/2}, \tag{4.10}$$

Thus the linear expansion given in Equation (4.6) can be written as

$$\psi_\beta^{n,0}(z, t) = [1/(2^n \pi)]^{1/2} (1 - \beta^2)^{(n+1)/2} (\exp[-(z^2 + t^2)/2])$$

$$\times \sum_{k=0}^{\infty} \left(\frac{1}{2}\right)^k \left(\frac{\beta}{2}\right)^k H_{n+k}(z) \, H_k(t). \tag{4.11}$$

As was indicated with respect to Equation (2.11), this linear transformation has to be unitary. Let us check this by calculating the sum

$$S = \sum_{k=0}^{\infty} |A_k^n(\beta)|^2. \tag{4.12}$$

According to Equation (4.10), this sum is

$$S = [1 - \beta^2]^{(n+1)} \sum_{k=0}^{\infty} \frac{(n+k)!}{n!k!} (\beta^2)^k. \tag{4.13}$$

On the other hand, the binomial expansion of $[1 - \beta^2]^{-(n+1)}$ takes the form

$$[1 - \beta^2]^{-(n+1)} = \sum_{k=0}^{\infty} \frac{(n+k)!}{n!k!} \beta^{2k}. \tag{4.14}$$

Therefore the sum S is equal to one. The linear transformation of Equation (4.8) is indeed a unitary transformation.

We have discussed above the transformation of the physical wave function with no time-like excitations. In general, for a given value of λ the infinite-by-infinite transformation matrix can be written as

$$\begin{bmatrix} \psi_\beta^{n,0} \\ \psi_\beta^{n+1,1} \\ \psi_\beta^{n+2,2} \\ \cdots \\ \cdots \\ \cdots \end{bmatrix} = \begin{bmatrix} b_{00} & b_{01} & b_{02} & \cdots & \cdots & \cdots \\ b_{10} & b_{11} & b_{12} & \cdots & \cdots & \cdots \\ b_{20} & b_{21} & b_{22} & \cdots & \cdots & \cdots \\ \cdots & \cdots & \cdots & \cdots & \cdots & \cdots \\ \cdots & \cdots & \cdots & \cdots & \cdots & \cdots \\ \cdots & \cdots & \cdots & \cdots & \cdots & \cdots \end{bmatrix} \begin{bmatrix} \psi_0^{n,0} \\ \psi_0^{n+1,1} \\ \psi_0^{n+2,2} \\ \cdots \\ \cdots \\ \cdots \end{bmatrix}. \tag{4.15}$$

This is the most general form of the unitary irreducible representation of the Lorentz group applicable to the covariant harmonic oscillator wave functions.

The expansion coefficients we calculated from Equation (4.8) to Equation (4.10) constitute the first row of above matrix. The remaining coefficients can also be calculated (Exercise 5 in Section 7). When β is zero, the above matrix becomes an infinite-by-infinite unitary matrix. From the orthogonality of the wave functions:

$$(\psi_\beta^{n+k',k'}, \psi_\beta^{n+k,k}) = \delta_{k'k}, \tag{4.16}$$

it is clear that this matrix has to be unitary and orthogonal:

$$\sum_{k=0}^{\infty} b_{ik} b_{jk} = \delta_{ij}. \tag{4.17}$$

Equations (4.12)–(4.14) constitute an explicit calculation of

$$\sum_k |b_{0k}|^2 = 1. \tag{4.18}$$

We have so far carried out the study of the unitarity of the representation using the solutions in the Cartesian coordinate system. Although the representations diagonal in the Casimir operators of the Poincaré group can be obtained as linear combinations of these solutions, it is still of interest to see how the transformation is achieved directly in terms of the solutions which are eigenstates of the Casimir operators. For this purpose let us note first that these solutions are constructed in terms of the spherical coordinate variables for the three-dimensional (x, y, z) space and treat t separately. If the hadron is at rest,

$$\psi_{0\lambda\ell}^{k,m}(x) = R_\mu^\ell(r)\, Y_\ell^m(\theta, \phi)\, [1/(\sqrt{\pi}\, 2^k k!)]^{1/2}\, H_k(t)\, e^{-t^2/2}. \tag{4.19}$$

Thus we have to write the generators of Lorentz transformations in terms of these variables. The three rotation generators can be written as (Arfken, 1985)

$$\begin{aligned}L_3 &= -i\frac{\partial}{\partial\phi}, \\ L_\pm &= L_1 \pm L_2 \\ &= \pm e^{\pm i\phi}\left(\frac{\partial}{\partial\theta} \pm i\cot\theta\,\frac{\partial}{\partial\phi}\right).\end{aligned} \tag{4.20}$$

It is not difficult to calculate the three boost generators. They take the form

$$\begin{aligned}iK_3 &= \cos\theta\left(r\frac{\partial}{\partial t} + t\frac{\partial}{\partial r}\right) - \frac{t}{r}\sin\theta\,\frac{\partial}{\partial\theta}, \\ iK_\pm &= K_1 \pm iK_2 \\ &= e^{\pm i\phi}\left(r\frac{\partial}{\partial t} + t\sin\theta\,\frac{\partial}{\partial r} - \frac{t}{r}\cos\theta\,\frac{\partial}{\partial\theta} \pm \frac{t}{r\sin\theta}\,\frac{\partial}{\partial\phi}\right).\end{aligned} \tag{4.21}$$

The rotation generators affect only the spherical harmonics in the wave function of Equation (4.19). Thus

$$\begin{aligned}L_3\, \psi_{0\lambda\ell}^{k,m} &= m\, \psi_{0\lambda\ell}^{k,m}, \\ L_\pm\, \psi_{0\lambda\ell}^{k,m} &= \sqrt{(\ell \mp m)(\ell \pm m + 1)}\, \psi_{0\lambda\ell}^{k,m\pm 1}.\end{aligned} \tag{4.22}$$

The above relations mean that rotations do not change the quantum numbers λ, ℓ and k. They only change m. Equation (4.22) indeed corresponds to the fact that the little group for massive hadrons is like $SO(3)$.

On the other hand, if we apply the boost generators, we end up with somewhat complicated formulas:

$$iK_3 \, \psi_{\lambda\ell}^{km} = \left[\frac{(\ell + m + 1)(\ell - m + 1)}{(2\ell + 1)(2\ell + 3)} \right]^{1/2} Y_{\ell+1}^m(\theta, \phi) \, Q_{-\ell} \, F_{\lambda\ell}^k(r, t)$$

$$+ \left[\frac{(\ell + 1)(\ell - m)}{(2\ell + 1)(2\ell - 1)} \right]^{1/2} Y_{\ell-1}^m(\theta, \phi) \, Q_{\ell+1} \, F_{\lambda\ell}^k(r, t),$$

(4.23)

$$iK_{\pm} \, \psi_{\lambda\ell}^{km} = \left[\frac{(\ell \pm m + 1)(\ell \pm m + 2)}{(2\ell + 1)(2\ell + 3)} \right]^{1/2} Y_{\ell+1}^{m\pm 1}(\theta, \phi) \, Q_{\pm\ell} \, F_{\lambda\ell}^k(r, t)$$

$$+ \left[\frac{(\ell \mp m)(\ell \mp m - 1)}{(2\ell + 1)(2\ell - 1)} \right]^{1/2} Y_{\ell-1}^{m\pm 1}(\theta, \phi) \, Q_{\pm(\ell+1)} \, F_{\lambda\ell}^k(r, t).$$

where

$$Q_\ell = \left(t \frac{\partial}{\partial r} + r \frac{\partial}{\partial t} + \ell \frac{t}{r} \right),$$

$$F_{\lambda\ell}^k(r, t) = R_\mu^\ell(r) \, [1/(\sqrt{\pi} \, 2^k \, k!)]^{1/2} \, H_k(t) \exp(-t^2/2).$$

K_3 does not change the value of m, while K_+ and K_- change m by $+1$ and -1 respectively. In addition, unlike the rotation operators, the boost generators change λ, ℓ and k. This is a manifestation of the fact that the unitary representation is infinite-dimensional as is indicated in Equation (4.8) (Problem 1 in Section 7).

It is possible to finish the calculation by explicitly carrying out the differentiations contained in the Q_ℓ operators (Problem 4 in Section 7). However, this does not appear necessary, because we already know what the answer should be from our experience with the Cartesian coordinate system.

5. Harmonic Oscillators in the Four-Dimensional Euclidean Space

Since Euclidean space is easier for us to visualize, it is not uncommon to use this space when we study physics and mathematics formulated in

Minkowskian space. There are many formalisms and mathematical theorems concerning the relationship between the four-dimensional Minkowskian and Euclidean spaces. However, there are not many concrete examples which illustrate the relation between these two important coordinate systems.

The harmonic oscillator wave functions discussed in Sections 2–4 are normalizable in the four-dimensional Euclidean space of x, y, z and t. The solutions of the Lorentz-invariant oscillator equation can be written as linear combinations of the solutions of the differential equation:

$$\left(\frac{1}{2}\right)\left[-\left(\nabla^2 + \left(\frac{\partial}{\partial t}\right)^2\right) + (\mathbf{x})^2 + t^2\right] u(x) = (\sigma + 2) u(x). \tag{5.1}$$

This differential equation is also separable in many different coordinate systems. It is separable in the Cartesian coordinate system, and the procedure for constructing wave functions is identical to that for the wave functions of Section 4 with $\beta = 0$. Indeed the solution takes the form

$$u(x) = f_a(x) f_b(y) f_n(z) f_k(t), \tag{5.2}$$

where f_a, f_b, ... are normalized one-dimensional oscillator wave functions, and can be written as

$$f_n(z) = [1/(\sqrt{\pi}\, 2^n n!)]^{1/2} H_n(z) \exp(-z^2/2). \tag{5.3}$$

The above solutions form a complete orthonormal set. The eigenvalue σ takes the integer values

$$\sigma = a + b + n + k. \tag{5.4}$$

Thus, for a given value of σ, there are only a finite number of possible combinations of the quantum numbers a, b, n and k.

Let us next consider the effect of Lorentz transformations. The differential equation is not Lorentz-invariant. For this reason, the oscillator wave functions of Sections 2–4 with non-zero β are no longer solutions of the differential equation of Equation (5.1). However, it is still possible to write wave functions as linear combinations of the solutions given in Equation (5.2), because they form a complete orthonormal set. The linear forms in Equation (4.8) and Equation (4.15) are precisely those expansions. Indeed, all of the normalizable solutions discussed in Sections 2–4 are linear combinations of the solutions in the four-dimensional Euclidean space (Kim et al., 1980).

The differential equation of Equation (5.1) is not Lorentz-invariant, but is invariant under rotations in the four-dimensional Euclidean space. In this space, it is convenient to use the coordinate variables ρ, α, θ and ϕ, related to the Cartesian variables by

$$t = \rho \cos \alpha,$$
$$z = \rho \sin \alpha \cos \theta,$$
$$x = \rho \sin \alpha \sin \theta \cos \phi, \quad (5.5)$$
$$y = \rho \sin \alpha \sin \theta \sin \phi.$$

In terms of the new variables, we can write the solution of the differential equation given in Equation (5.1) as (Problem 2 in Section 7)

$$u_{\mu n}^{\ell m}(x) = S_\mu^n(\rho) \, Z_{n+1}^{\ell m}(\alpha, \theta, \phi), \quad (5.6)$$

where

$$Z_{n+1}^{\ell m}(\alpha, \theta, \phi) = (\sin \alpha)^{-1/2} \, P_{n+3/2}^{-n-1/2}(\cos \alpha) \, Y_\ell^m(\theta, \phi).$$

$S_\mu^n(\rho)$ satisifes the radial differential equation:

$$\frac{1}{2}\left[-\frac{d^2}{d\rho^2} - \frac{3}{\rho}\frac{d}{d\rho} + \frac{n(n+2)}{\rho^2} + \rho^2\right] S_\mu^n(\rho) = \sigma \, S_\mu^n(\rho). \quad (5.7)$$

$P_{n+3/2}^{-n-1/2}(\cos \alpha)$ is an associated Legendre function commonly known as the Gegenbauer polynomial (Morse and Feshbach, 1953). The index n in the above expression should not be confused with the one used in Equations (5.2) and (5.3), and is the maximum value the quantum number ℓ can take:

$$\ell = 0, 1, 2, \ldots, n. \quad (5.8)$$

The solution of the radial equation is

$$S_\mu^n(\rho) = (2(\mu!)/[\Gamma(\mu + n + 2)]^3)^{1/2} \, \rho^n \, L_\mu^{n+1}(\rho^2) \, e^{-\rho^2/2}, \quad (5.9)$$

satisfying the orthonormality condition:

$$\int_0^\infty \rho^3 \, S_\mu^n(\rho) \, S_\nu^n(\rho) \, d\rho = \delta_{\mu\nu}. \quad (5.10)$$

$L_\mu^{n+1}(\rho^2)$ is the associated Laguerre polynomial (Arfken, 1985). The eigenvalue σ in this case is

$$\sigma = (2\mu + n). \quad (5.11)$$

While the radial wave functions for the three-dimensional harmonic oscillator are commonly available in various textbooks, it is not easy to find solutions for the four-dimensional problem. Let us write down explicitly the wave functions in the $O(4)$ coordinate system for some low eigenvalues of σ. If $\sigma = 0$, the system is in the ground state, and the wave function is

$$u_{00}^{00}(x) = \left(\frac{1}{\pi}\right) \exp(-\rho^2/2). \tag{5.12}$$

In terms of the Cartesian variables, it is easy to construct the four first excited-state wave functions:

$$(\sqrt{2}/\pi) \, x_i \exp(-\rho^2/2), \tag{5.13}$$

where $x_i = x, y, z$ or t. In terms of the $O(4)$ coordinate variables, when $\sigma = n = 1$, the wave functions are written as

$$u_{01}^{\ell m}(x) = \rho \exp(-\rho^2) \, Z_2^{\ell m}(\alpha, \theta, \phi), \tag{5.14}$$

with

$$Z_2^{00}(\alpha, \theta, \phi) = \frac{\sqrt{2}}{\pi} \cos \alpha,$$

$$Z_2^{1m}(\alpha, \theta, \phi) = \left(\frac{8}{3\pi}\right)^{1/2} \sin \alpha \, Y_1^m(\theta, \phi), \quad m = 1, 0, -1.$$

Here again the wave function is four-fold degenerate. The above wave functions can be written as linear combinations of the Cartesian forms given in Equation (5.3).

For $\sigma = 2$, there are four Cartesian wave functions of the form

$$(1/\pi\sqrt{2}) \, (2 \, x_i^2 - 1) \exp(-\rho^2/2), \tag{5.15}$$

and six wave functions of the form

$$(2/\pi) \, x_i \, x_j \exp(-\rho^2/2), \quad \text{with } i \neq j. \tag{5.16}$$

If we use the $O(4)$ coordinate system, we have to consider two different values of the radial quantum number μ:

(a) $\mu = 1$, with $n = 0$,
(b) $\mu = 0$, with $n = 2$.

If $\mu = 1$ and $n = 0$, the wave function becomes

$$u_{10}^{00}(x) = (1/\sqrt{2}\pi) \, (2 - \rho^2) \exp(-\rho^2/2). \tag{5.17}$$

If, on the other hand, $\mu = 0$ and $n = 2$, the wave function is

$$u_{02}^{\ell m}(x) = (1/\sqrt{3}) \, \rho^2 \exp(-\rho^2/2) \, Z_3^{\ell m}(\alpha, \theta, \phi). \tag{5.18}$$

For $\ell = 0$, there is one wave function with

$$Z_3^{00} = \frac{\sqrt{2}}{\pi}(4\cos^2\alpha - 1). \tag{5.19}$$

For $\ell = 1$, there are three wave functions with

$$Z_3^{\ell m} = (16/\pi)^{1/2} \cos\alpha \sin\alpha \; Y_\ell^m(\theta, \phi). \tag{5.20}$$

For $\ell = 2$, there are five wave functions with

$$Z_3^{2m} = (16/5\pi)^{1/2} \sin^2\alpha \; Y_2^m(\theta, \phi). \tag{5.21}$$

We have given above enough explicit calculations to show how to construct the harmonic oscillator wave functions using the "$O(4)$ harmonics", and how they can be converted to the Cartesian forms. The above calculation also illustrates that the solutions diagonal in the Casimir operators of the $O(4)$ group are not diagonal in those of the Poincaré group discussed in Section 4. Since, however, the $O(4)$ solutions form a complete set, we can construct the wave functions needed for the representations of the Poincaré group by making suitable linear combinations of the $O(4)$ solutions (Problem 5 in Section 7).

6. Moving O(4) Coordinate System

We considered in Section 5 the solutions of the differential equation of Equation (5.1). It was noted that this differential equation is separable in both the Cartesian and $O(4)$ coordinate systems. We observed that the harmonic oscillator formalism can serve as an illustrative example of how rotations in the $O(4)$-coordinate system are different from Lorentz transformations, while the Lorentz transformation of the normalizable oscillator wave functions can be regarded as a unitary transformation in the $O(4)$-space. The difference is of course due to the fact that the differential equation of Equation (5.1) is not Lorentz-invariant. The linear expansion of Equation (4.11), although realizable in the $O(4)$ space, is not a transformation which preserves the eigenvalue σ which remains invariant under the $O(4)$ rotations.

Mathematically, the difference between the above-mentioned two differential equations is very simple. We can obtain one from the other by replacing the variable t by (it). This is in fact one of the most commonly used procedures in physics, especially in connection with the Bethe-Salpeter equation (Wick, 1954; Nakanishi, 1972). However, the physics of this procedure requires further explanation.

In order to complete the discussion of this problem, let us go back to the Lorentz-invariant differential equation of Equation (2.1) written in terms of the x variables for the hadron at rest and that of Equation (2.3) written in the Lorentz-boosted variables x' for the hadron moving along the z direction. Because of the Lorentz invariance, both Equation (2.1) and Equation (2.3) are the same differential equation. In Section 5, we discussed only the $O(4)$ counterpart of Equation (2.1). The $O(4)$ counterpart of Equation (2.3) can be written simply as

$$\frac{1}{2}\left[-\left(\nabla'^2 + \left(\frac{\partial}{\partial t'}\right)^2\right) + (\mathbf{x}')^2 + (t')^2\right] u'(x) = (\sigma' + 2) u'(x). \quad (6.1)$$

It is possible to repeat the calculation of Section 5 and write all the solutions in terms of the primed variables. It is even possible to write the $O(4)$ solution in one Lorentz frame as a linear expansion of those in another Lorentz frame, although the mathematics is rather cumbersome.

However, the important point is that the above differential equation is only a Wick-rotated form of Equation (2.3), and is different from the $O(4)$ counterpart of Equation (2.1) given in Equation (5.1). The only common connection the moving $O(4)$ system has with the rest-frame $O(4)$ system is that the moving system becomes that of the rest frame when the hadronic invariant procedure by carrying out the mathematics (Kim et al., 1980).

In addition to the problem of Lorentz invariance, the procedure of replacing t by (it) has the following complication. Although the differential equation can be obtained by a simple replacement of the variable, the solutions of the differential equation discussed in Section 5 are not an analytic continuation of those given in Section 2. Both solutions are normalizable. However, the analytic continuation from t to (it) or vice versa leads to non-normalizable solutions. This means that the boundary condition on the solutions is an independent condition, and cannot be regarded as a part of the Wick rotation (Problem 10 in Section 7).

In Chapter X, we shall discuss the connection between the solutions in the $O(4)$ coordinate system with those which are diagonal in the Casimir operators of the homogeneous Lorentz group.

7. Exercises and Problems

Exercise 1. It is widely believed that we can obtain momentum wave functions in the harmonic oscillator regime by replacing the coordinate variables by the their conjugate momentum variables. For this reason, there is a tendency to overlook the details. For instance, what happens to phases? Consider the one-dimensional oscillator, and use $f_n(x)$ as the wave function for the n-th excited state. Consider next an arbitrary function $F(x)$ which can be expanded as

$$F(x) = \sum_n A_n f_n(x). \tag{7.1}$$

Then, is it possible to obtain the momentum wave function by simply replacing x by p?

If we write the momentum wave function as a Fourier tranform of the spatial wave function:

$$g_n(p) = (1/2\pi)^{1/2} \int f_n(x) \, e^{-ipx} \, dx. \tag{7.2}$$

Then, for the ground state, we can show

$$g_0(p) = f_0(x). \tag{7.3}$$

However, since the n-th excited-state wave functions are obtained by repeated application of step-up operators on the ground-state wave function, we can show that

$$g_n(p) = (-i)^n f_n(p). \tag{7.4}$$

Thus, the Fourier transformation of $F(x)$ is

$$G(p) = \sum_n (-i)^n A_n f(p). \tag{7.5}$$

For a more detailed treatment of this problem, see K. B. Wolf (1979). This point is important when we analyze the high-energy data using models based on mixed oscillator states. See Le Youanc et al. (1975, 1977).

Exercise 2. Throughout this chapter, we have been using the harmonic oscillator equations and wave functions in configuration space and time. Discuss the oscillator formalism using momentum-energy coordinates.

The mathematics of the harmonic oscillator is the same for both configuration space and momentum space. The Lorentz transformation of the momentum-energy four vector is also identical to that for the space-time four vector. Therefore, we expect that the analysis in momentum-energy space be identical to that for space-time configuration space.

However, in view of the phase factors of the momentum wave function discussed in Exercise 1, the momentum wave functions are either real or imaginary. For this reason, we have to be careful when we take the inner product of two wave functions as in the case of Equation (2.14). In addition, we have to modify the expansion formula given in Equation (4.8). The easiest way to approach this problem would be to note that $(i)(-i) = 1$, and make the momentum wave function absorb an appropriate power of i, while the coefficient $A_k^n(\beta)$ can absorb the same power of $(-i)$.

The momentum-energy wave function can be obtained from the Fourier transformation of the space-time wave function only when the Fourier integral exists.

Exercise 3. While mesons are believed to be two-body bound states of a quark and antiquark, baryons such as protons and neutrons are bound states of three quarks. Assuming that the interaction between each pair of quarks is like that of harmonic oscillator, write down the covariant oscillator equation for a hadron consisting of three quarks, and carry out the separation of variable suitable for the description of massive hadrons.

If we denote the space-time coordinates of three quarks by x_a, x_b, and x_c, the three-particle oscillator equation can be written as

$$\left(-3\left[\left(\frac{\partial}{\partial x_{a\mu}}\right)^2 + \left(\frac{\partial}{\partial x_{b\mu}}\right)^2 + \left(\frac{\partial}{\partial x_{c\mu}}\right)^2\right] + \frac{1}{36}[(x_a - x_b)^2 \right.$$

$$\left. + (x_b - x_c)^2 + (x_c - x_a)^2] - m_0^2 \right) \psi(x_a, x_b, x_c) = 0. \tag{7.6}$$

This equation can be separated if we use the variables X, r and s, defined as

$$x_a = X + r - \sqrt{3}\, s,$$
$$x_b = X + r + \sqrt{3}\, s, \tag{7.7}$$
$$x_c = X - 2r.$$

X is the hadronic coordinate, and r and s are separable normal coordinates.

In terms of these new variables, the wave function in Equation (7.6) can be written as

$$\psi(x_a, x_b, x_c) = f(X)\, R(r)\, S(s), \tag{7.8}$$

where $f(X)$, $R(r)$ and $S(s)$ satisfy the following equations:

$$\left[-\left(\frac{\partial}{\partial X}\right)^2 + m_0^2 + \lambda_r + \lambda_s\right] f(X) = 0,$$

$$\frac{1}{2}\left[-\left(\frac{\partial}{\partial r_\mu}\right)^2 + r_\mu^2\right] R(r) = \lambda_r\, R(r), \tag{7.9}$$

$$\frac{1}{2}\left[-\left(\frac{\partial}{\partial s_\mu}\right)^2 + s_\mu^2\right] S(s) = \lambda_s\, S(s).$$

Here again, $f(X)$ is a solution of the free-particle Klein-Gordon equation with four-momentum P satisfying

$$-P^2 = m_0^2 + \lambda_r + \lambda_s. \tag{7.10}$$

We can eliminate time-like excitations in both the r and s coordinate systems by imposing the conditions:

$$P^\mu \left(r_\mu - \frac{\partial}{\partial x^\mu} \right) R(r) = 0,$$

$$P^\mu \left(s_\mu - \frac{\partial}{\partial s^\mu} \right) S(s) = 0. \tag{7.11}$$

If we use p_1, p_2 and p_3 for the four-momenta for the first, second and third quarks respectively, and P, q and k for the momentum variables conjugate to X, r and s respectively, then

$$p_1 = \tfrac{1}{3}P + \tfrac{1}{6}q - \tfrac{1}{2}(1/\sqrt{3})k,$$
$$p_2 = \tfrac{1}{3}P + \tfrac{1}{6}q + \tfrac{1}{2}(1/\sqrt{3})k, \tag{7.12}$$
$$p_3 = \tfrac{1}{3}P - \tfrac{1}{3}q.$$

The oscillator wave functions in the momentum space are identical to those in the space-time coordinates, except the phase factors discussed in Exercise 2. See Feynman *et al.* (1971) and Kim and Noz (1977b).

Exercise 4. Derive the result of Equation (4.10) by evaluating the integral of Equation (4.9).

Let us replace the Hermite polynomials in the integrand by their respective generating functions:

$$G(r, z) = \exp(-r^2 + 2rz)$$

$$= \sum_{m=0}^{\infty} \frac{r^m}{m!} H_m(z). \tag{7.13}$$

The the calculation is reduced to the evaluation of the integral:

$$I = \int dt\, dz\, G(r, z)\, G(s, t)\, G(r', z')$$

$$\times \exp[-(z^2 + t^2 + z'^2 + t'^2)/2]. \tag{7.14}$$

The integrand in the above expression becomes one exponential function whose argument is quadratic in z and t. This quadratic form can be diagonalized, and the result is

$$I = \pi(1 - \beta^2)^{1/2} \exp(2\beta rs) \exp[2rr'(1 - \beta^2)^{1/2}]. \tag{7.15}$$

By expanding the above exponential factors and choosing the term containing the r^{n+k}, s^k and $r'^{n'}$ for $H_{n+k}(z)$, $H_k(t)$, and $H_{n'}(z')$ respectively, we arrive at the result of Equation (4.10). If we set $k = 0$, and $n = n'$, the result is Equation (4.10). See Ruiz (1974), and Kim *et al.* (1980).

Exercise 5. Calculate the elements of the infinite-by-infinite matrix of Equation (4.15) for Lorentz transformation of the covariant harmonic oscillators.

This exercise is an extension of Exercise 3 to include excitations in the t' variable. Thus we have to consider the transformation of the function

$$\psi(z, t) = f_n(z) f_m(t), \tag{7.16}$$

where

$$f_n(z) = [1/(\sqrt{\pi}\, 2^n\, n!)]^{1/2} H_n(z)\, e^{-z^2/2}.$$

This means that we have to compute the coefficients in the following linear expansion:

$$f_n(z') f_m(t') = \sum_k \sum_j A_{nm}^{kj} f_k(z) f_j(t). \tag{7.17}$$

Thus

$$A_{mn}^{kj} = \int dz\, dt\, f_k(z) f_j(t) f_n(z') f_m(t').$$

Again we can use the generating function of Hermite polynomials to evaluate the above integral. This time, we have to evaluate the integrand involving four generating functions:

$$\int dt\, dz\, G(r, z)\, G(s, t)\, G(r', z')\, G(s', t')$$
$$\times \exp[-(z^2 + t^2 + z'^2 + t'^2)/2]. \tag{7.18}$$

The procedure for evaluation of this integral is similar to that for Exercise 4. However, the calculation is far more complicated. Rotbart (1981) carried out this calculation and obtained explicit expressions for all the elements of the infinite-by-infinite matrix given in Equation (4.15). The result of Rotbart's calculation is

$$A^{kj}_{nm} = \beta^{k-n} [n!m!k!j!]^{1/2} \delta_{j-m,k-n}$$

$$\times \sum_{s=0}^{s_0} \frac{[-\beta^2/(1-\beta^2)]^s}{s!(m-s)!(n-s)!(k-n+1)!}, \qquad (7.19)$$

where s_0 is the smaller of n or m.

Problem 1. Derive the relations given in Equation (4.22) using Equation (4.19), Equation (4.20), and the recurrence relation for the spherical harmonics.

Problem 2. Write down the Casimir operators of the $O(4)$ group and the differential equation given in Equation (5.1) in terms of the ρ, α, θ and ϕ variables. The $O(4)$ group has been extensively discussed in connection with the symmetry of the hydrogen atom. See L. C. Biedenharn (1961) for a complete and thorough discussion of this problem. See G. Domokos and P. Suranyi (1964) for a discussion of the Klein-Gordon equation in the $O(4)$ coordinate system.

Problem 3. The degeneracies of the $O(4)$ harmonics given in Section 5 are like those of the Rydberg energy levels of the hydrogen atom. Starting from the Hamiltonian of the hydrogen atom, show that the energy levels should exhibit the $O(4)$ symmetry. This problem is extensively discussed in the literature. See, for instance, the above-mentioned paper of Biedenharn (1961). See also Gilmore (1974).

Problem 4. From the transformation matrix of Equation (4.15) based on the Cartesian form of normalizable wave functions, we know that the derivatives of the right-hand sides of Equation (4.23) should be replaced in terms of the Laguerre functions in r and the Hermite polynomials in t. In order to solve this problem, we have to know the recurrence relations for the associated Laguerre functions and those for the Hermite polynomials. The recurrence relations for the Hermite polynomials are readily available, while those for the Laguerre functions needed for solving this problem are not yet available in standard mathematical tables. However, it is possible to write the associated Laguerre polynomials in finite power series. From these power series, derive the recurrence relation for the covariant harmonic oscillator wave functions starting from Equation (4.23).

Problem 5. Discuss degeneracies of the harmonic oscillator wave functions in the four-dimensional isotropic space using the Cartesian coordinate system.

Find the unitary matrix which will transform the solutions given in Section 5 into the Cartesian forms.

Problem 6. The nonrelativistic harmonic oscillator serves as an illustrative example for many branches of physics (Moshinsky, 1969). It is therefore quite natural for physicists to have developed a group theory of harmonic oscillators. Let us consider a one-dimensional harmonic oscillator. We are quite familiar with the step-up and step-down operators:

$$a^\dagger = (1/\sqrt{2})\left(x - \frac{\partial}{\partial x}\right),$$

$$a = (1/\sqrt{2})\left(x + \frac{\partial}{\partial x}\right),$$

(7.20)

and the number operator

$$N = a^\dagger a.$$

Write down the commutation relations for $(ia^\dagger)$, (ia), (iN). Show that the (iI) has to be added to form a closed Lie algebra. The group generated by these four operators is called the harmonic oscillator group. Show that the following four-parameter matrix forms one of the representations of the harmonic oscillator group (Miller, 1972).

$$\begin{bmatrix} 1 & ce^d & a & d \\ 0 & e^d & b & 0 \\ 0 & 0 & 1 & 0 \\ 0 & 0 & 0 & 1 \end{bmatrix}.$$

(7.21)

Problem 7. The problem becomes more complicated in the case of nonrelativistic three-dimensional isotropic oscillators. The additional symmetry we have to consider is that of three-dimensional rotation. What is the resulting Lie algebra? See Gilmore (1974).

Problem 8. If we introduce special relativity to the three-dimensional harmonic oscillator, what is the resulting Lie algebra? Translate the treatment of Section 3 into the language of this algebra. See Shapiro (1968).

Problem 9. Discuss the Lie algebra for the four-dimensional isotropic harmonic oscillator using the appropriate step-up and step-down operators. Discuss the connection between this algebra and that of Problem 8. This could be a future research problem.

Problem 10. Compare the Wick rotation applicable to the Bethe-Salpeter equation and wave function to those of the harmonic oscillator equations and wave functions discussed in this chapter.

Problem 11. In this chapter we have been primarily concerned with the use of harmonic oscillators in relativistic physics. Another area of physics in which the harmonic oscillator plays the central role is the coherent state representation. This representation is useful in many branches of modern physics, including optics (Glauber, 1963; Klauder and Sudarshan, 1968; Goldin, 1982), quantum mechanics (Carruthers and Nieto, 1965; Glauber, 1966; Moncrief, 1978; Klauder, 1979), and high-energy scattering theory (Carruthers and Shih, 1983; Carruthers and Zachariasen, 1983). In order to understand the basic mathematics of the coherent state representation, let us consider the following problems starting from the one-dimensional harmonic oscillator.

(a) In Dirac's notation, the n-th excited oscillator is written as $|n\rangle$. The coherent state is defined as

$$|\alpha\rangle = e^{-|\alpha|^2/2} \sum_n \frac{\alpha^n}{\sqrt{n!}} |n\rangle, \qquad (7.22)$$

where α is a complex number. Show that the above form is an eigenstate of the step-down operator a defined in Equation (7.20) with eigenvalue α.

(b) Use the generating function for the Hermite polynomials to show that the above form is a Gaussian function whose origin is $x = \alpha$, and therefore that it satisfies the minimum uncertainty relation.

(c) Show that the inner product of the two coherent states satisfies the relation:

$$|\langle \alpha | \alpha' \rangle|^2 = e^{-|\alpha - \alpha'|^2}. \qquad (7.23)$$

(d) Replace α by $\alpha\, e^{-i\omega t}$. Show then that the probability density of the coherent state describes a simple harmonic oscillation of the Gaussian distribution around $x = 0$. See Goldin (1982).

Chapter VI

Dirac's Form of Relativistic Quantum Mechanics

The purpose of this chapter is to give a physical interpretation to the covariant harmonic oscillator formalism by incorporating Dirac's ideas on relativistic quantum mechanics. Dirac's form of relativistic quantum mechanics serves a very useful purpose in understanding Lorentz deformation properties of relativistic extended hadrons, such as hadrons in the quark model.

As early as in 1927, Dirac observed that the uncertainty relation applicable to the time and energy variables is different from Heisenberg's uncertainty relations applicable to the position and momentum variables, and that this space-time asymmetry is one of the problems we have to face in the process of making quantum mechanics relativistic. In 1945, Dirac considered the possibility of using the four-dimensional harmonic oscillator with normalizable time-like wave functions in connection with relativistic Fock space. The covariant harmonic oscillator wave functions discussed in Chapter V are of Dirac's type. Indeed, the use of the covariant oscillator formalism is perfectly consistent with Dirac's overall plan to construct a relativistic quantum mechanics.

Dirac's plan to construct a "relativistic dynamics of atom" using "Poisson brackets" is contained in his 1949 paper entitled 'Forms of Relativistic Dynamics.' Here Dirac emphasizes that the task of constructing a relativistic dynamics is equivalent to constructing a representation of the inhomogeneous Lorentz group.

In an attempt to find a three-dimensional space in which nonrelativistic quantum mechanics is valid, Dirac in his 1949 paper considered three constraint conditions, each of which reduces the four-dimensional Minkowskian space-time into a three-dimensional Euclidean space. Dirac called the forms of relativistic quantum mechanics with these constraints "instant form", "front form", and "point form." the purpose of this chapter is to show that the covariant harmonic oscillator formalism for massive hadrons based on the $O(3)$-like little group constitutes Dirac's "instant form" quantum mechanics, while displaying the space-time asymmetry in uncertainty relations.

In addition, we shall discuss in detail the kinematics of the front and point forms. In his "front form" quantum mechanics, Dirac introduced the light-cone coordinate system. The beauty of the light-cone coordinate system is that the longitudinal and time-like coordinate variables undergo only scale changes under Lorentz boosts. We shall study in this chapter how Dirac's light-cone coordinate system can serve a useful purpose in describing Lorentz deformation of relativistic extended hadrons. Furthermore, we shall show that the constraint applicable to Dirac's "point form" quantum mechanics is very useful in describing the conservation of probability under Lorentz transformations.

Dirac's approach to relativistic quantum mechanics allows us to construct relativistic bound-state wave functions which can be Lorentz-transformed. These wave functions are needed in understanding relativistic hadrons which are believed to be bound states of quarks and/or antiquarks. We shall study in this chapter Lorentz deformation properties of relativistic hadrons. In Chapter XII, we shall discuss some observable consequences of the Lorentz deformation, particularly the peculiarities observed in Feynman's parton picture (Feynman, 1969).

The ultimate purpose of studying the models of relativistic quantum mechanics is to study how the time-energy uncertainty relation can be combined with the position-momentum uncertainty, in a manner consistent with special relativity and with what we observe in the experimental world. We shall study this question using extensively space-time and momentum-energy diagrams. Indeed, the light-cone coordinate system is an effective language for this pictorial approach.

In Section 1, we study the present form of the uncertainty relation applicable to the time and energy variables. It is noted that there exists an uncertainty relation between the time and energy variables. However, since the time variable is a c-number (Dirac, 1927), there is no Hilbert space associated with this variable (Blanchard, 1982). We shall call this "c-number time-energy uncertainty relation." We then study how this form of the uncertainty relation is applicable to the time separation variable in the quark model, using the harmonic oscillator formalism.

In Section 2, we examine closely how the covariant harmonic oscillator model can serve as a solution of Dirac's "Poisson brackets" in his "instant form" quantum mechanics. It is noted that the use of the c-number time-energy relation and the concept of off-mass-shell particles allows us to avoid the difficulties mentioned in Dirac's 1949 paper. We study in detail the manner in which the time-energy uncertainty relation is combined covariantly with the usual position-momentum uncertainty relation.

Also in his 1949 paper, Dirac introduced the light-cone coordinate system in order to simplify the mathematics of the Lorentz transformation. In Section 3, the geometry of the light-cone coordinate system is discussed in detail. The light-cone coordinate system allows us to regard the Lorentz

Dirac's Form of Relativistic Quantum Mechanics

transformation as a simple scale transformation preserving the area in the two-dimensional coordinate system of longitudinal and time-like variables.

In Section 4, we discuss the covariant harmonic oscillator formalism using the light-cone coordinate system. In Section 5, it is shown that, in spite of the Lorentz deformation of hadronic distribution, it is still possible to define the uncertainty relations in a Lorentz-invariant manner.

Section 6 contains exercises and problems which illustrate some immediate applications of Dirac's form of relativistic quantum mechanics. Chapter XII will deal with observable consequences of the Lorentz-Dirac deformation.

1. C-Number Time-Energy Uncertainty Relation

The time-energy uncertainty relation in the form of $(\Delta t)(\Delta E) \simeq 1$ was known to exist even before the present form of quantum mechanics was formulated. However, the treatment of this subject in the existing quantum mechanics textbooks is not adequate. Students, as well as physicists, are likely to be confused on the following three issues:

(a) Is the time-energy uncertainty relation a consequence of the time-dependent Schrödinger equation? Or, is this relation expected to hold even in systems which cannot be described by the Schrödinger equation?

(b) While there exists the time-energy uncertainty relation in the real world, possibly with the form $[t, H] = -i$ (Heisenberg, 1927), this commutator is zero in the case of Schrödinger quantum mechanics. As was noted by Dirac in 1927, the time variable is a c-number. Then, is the c-number time-energy uncertainty relation universal, or true only in nonrelativistic quantum mechanics?

(c) If the time variable is a c-number and the positive variables are q-numbers, then the coordinate variables in different Lorentz frames are mixtures of c and q numbers. This cannot be consistent with special relativity, as was also pointed out by Dirac (1927).

The reason why we are not able to get satisfactory answers to these questions from the existing literature is very simple. While the uncertainty relation is to be formulated from experimental observations (Bohr, 1958; Heisenberg, 1975), there are not many experimental phenomena which can be regarded as direct manifestations of the time-energy uncertainty relation. In fact, the connection between the life-time and the energy-width of unstable states is the only direct application of this important relation (Wigner, 1972).

It is widely believed that off-mass-shell intermediate particles in quantum field theory are due to the time-energy uncertainty relation. (Heitler, 1954; Schweber, 1961). Since, however, those off-mass-shell particles are not observable, the relation between those two concepts is not directly observable and requires further investigation (Han *et al.*, 1981c).

We study in this section the relativistic quark model as a physical example in which the time-energy uncertainty relation leads to a directly observable effect. In this model, hadrons are bound states of quarks. For a hadron consisting of two quarks whose space-time coordinates are x_a and x_b, we can define new variables:

$$X = (x_a + x_b)/2,$$
$$x = (x_a - x_b)/2\sqrt{2}, \qquad (1.1)$$

as we did in Chapter V. As we noted there, X and x correspond respectively to the overall hadronic coordinate and space-time separation between the quarks. The spatial component of the four-vector X specifies where the hadron is, and its time component tells how old the hadron and the quarks become. The spatial components of the four vector x specify the relative spatial separation between the quarks. Its time component is the time interval or separation between the quarks, as is illustrated in Figure 1.1 (Wick, 1954; Kim and Noz, 1973; Goto, 1977; Dominici and Longhi, 1977; Lukierski and Oziewics, 1977; Jersak and Rein, 1980; Sogami and Yabuki, 1980).

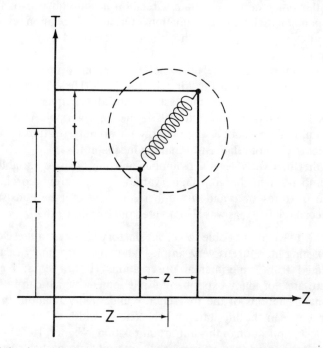

Fig. 1.1. Hadronic and internal coordinate systems. Each of the two quarks in the hadron has its own space-time coordinate. These two coordinates can be translated into the hadronic coordinate (Z, T) and the separation coordinate (z, t).

Because the time-separation variable is not contained in nonrelativistic quantum mechanics, the quark model provides an excellent testing ground to examine whether there exists the time-energy uncertainty relation which does not come from the Schrödinger equation. We are interested in whether there exists an uncertainty relation along the time separation coordinate. We would like then to know whether this time-energy uncertainty relation is the same as the currently accepted form largely based on nonrelativistic quantum mechanics.

According to the currently accepted version (Blanchard, 1982), the time variable is a c-number or

$$[t, H] = 0, \tag{1.2}$$

where H is the Hamiltonian or the energy operator. In nonrelativistic quantum mechanics, this Hamiltonian does not depend on time and commutes with the time variable. Because the commutator vanishes, the Robertson procedure (Robertson, 1929) applicable to Heisenberg's position-momentum uncertainty relation does not work here. Classically, this corresponds to the fact that t and H are not canonically conjugate variables. In quantum mechanics, the above commutator means that there is no Hilbert space in which t and $i\partial/\partial t$ act as operators. However, it is important to note that there still exists a "Fourier" relation between time and energy which limits the precision to $(\Delta t)(\Delta E) \simeq 1$ (Dirac, 1927; Heitler, 1954; Blanchard, 1982).

In order to discuss the uncertainty relations, we need momentum-energy variables in addition to the space-time coordinates. Let us use the four momenta:

$$P = p_a + p_b, \quad q = \sqrt{2}\,(p_a - p_b), \tag{1.3}$$

which were introduced in Chapter V. p_a and p_b are the four-momenta of the first and second quarks resepctively. The spatial and time-like components of the four-vector P is the sum of the momenta and the energy of the two quarks respectively. We assume here that the system is in an eigenstate of this four-momentum, and that every component of P is a sharply defined number. The spatial component of q measures the momentum difference between the quarks. The time-like component of q is the energy difference between the quarks. The concept of this energy separation does not exist in nonrelativistic quantum mechanics. If the wave functions can be Fourier-transformed, P and q are conjugate to X and x respectively. We are particularly interested in the uncertainty relation applicable between the time-like components of four-vectors x and q.

If the hadron has a definite four-momentum and moves along the z direction with velocity parameter β, it is possible to find the Lorentz frame in which the hadron is at rest. We shall use x', y', z', and t' to denote the space-time separations in this frame, and q'_x, q'_y, q'_z, q'_0 for momentum-energy separations. In the Lorentz frame where the hadron is at rest, the uncertainty principle applicable to the space-time separation of quarks is expected to be the same as the presently accepted form largely based on nonrelativistic quantum mechanics. The usual Heisenberg uncertainty relation holds for each of the three spatial coordinates:

$$[x', q'_x] = i,$$
$$[y', q'_y] = i, \qquad (1.4)$$
$$[z', q'_z] = i.$$

On the other hand, the time-separation variable is a c-number and therefore does not cause quantum excitations. The commutator of Equation (1.2) in this case takes the form

$$[t', q'_0] = 0. \qquad (1.5)$$

However, according to the presently accepted form of time-energy uncertainty relation, this commutator still allows the existence of the "Fourier relation" between time and energy variables (Blanchard, 1982) resulting in

$$(\Delta t')(\Delta q'_0) \simeq 1. \qquad (1.6)$$

The crucial question is how these uncertainty relations appear to an observer in the laboratory frame with the space-time separation variables x, y, z and t.

We assume here that the hadron moves along the z axis. Thus the x and y coordinates are not affected by boosts, and the first two commutation relations of Equation (1.4) remain invariant:

$$[x, q_x] = i,$$
$$[y, q_y] = i. \qquad (1.7)$$

If we consider only the ground state, the third or longitudinal commutator of Equation (1.4) can be quantified as

$$\langle \Delta z' \rangle \langle \Delta q'_z \rangle \simeq 1. \qquad (1.8)$$

Then the uncertainty relations associated with both the longitudinal and time-separation variables will lead to a distribution centered around the origin in the $z't'$ coordinate system, as is illustrated in Figure 1.2. It is clear from Figure 1.2 that the circular distribution, with its simpler mathematical form, will share with the arbitrarily-shaped region all qualitative properties associated with the uncertainty relations.

Next, we should examine this localization region in the laboratory frame

Dirac's Form of Relativistic Quantum Mechanics 141

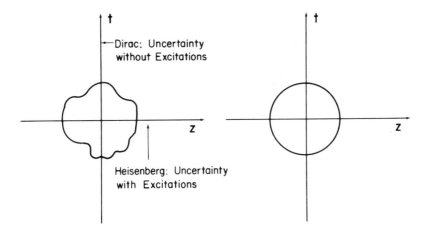

Fig. 1.2. The hadronic distribution dictated by the uncertainty relations. z and t are the longitudinal and time-like separations between two quarks inside a hadron. The distribution is localized around the origin. There can be excited states along the longitudinal axis, while excitations are forbidden along the time-like direction. The circular distribution, with its simple mathematics, is expected to share all qualitative features with the arbitrary distribution.

where the hadron moves along the z direction with the velocity parameter β. The Lorentz transformation from the hadronic rest frame to the lab frame takes the form

$$z = (z' + \beta t')/(1 - \beta^2)^{1/2},$$
$$t = (t' + \beta z')/(1 - \beta^2)^{1/2}, \tag{1.9}$$

when the hadron moves along the z direction with velocity parameter β. The above form is the inverse of the transformation given in Equation (V.2.2).

In studying the connection between the localized region and the commutation relations, the harmonic oscillator is the standard language (Robertson, 1929). In order to study the localization region in the two different Lorentz frames, we use the normalizable covariant harmonic oscillator wave functions which we discussed in Chapter V. While the exact form for the hadronic wave function diagonal in the Casimir operators of the Poincaré group is somewhat complicated, the essential element in the wave function takes the form

$$\psi_\beta^{n,k}(x) = [1/(\pi 2^{n+k} n! k!)]^{1/2} H_n(z') H_k(t') \exp[-(z'^2 + t'^2)/2], \tag{1.10}$$

where we have suppressed all the factors which are not affected by the Lorentz transformation along the z axis. This is possible because the oscillator wave functions are separable in both the Cartesian and spherical coordinate systems. z' and t' are the longitudinal and time-like coordinate variables respectively in the hadronic rest frame.

In terms of the standard step-up and step-down operators:

$$a_\mu = (1/\sqrt{2})(x_\mu + \partial/\partial x^\mu),$$
$$a_\mu^\dagger = (1/\sqrt{2})(x_\mu - \partial/\partial x^\mu), \quad (1.11)$$

the oscillator wave function of Equation (1.10) satisfies the differential equation:

$$a_\mu^\dagger a^\mu \psi(x) = (\lambda + 1)\psi(x), \quad (1.12)$$

where the eigenvalue λ, together with transverse excitations, determines the (mass)^2 of the hadron.

The operators defined in Equation (1.11) satisfy the algebraic relation

$$[a_\mu, a_\nu^\dagger] = g_{\mu\nu}. \quad (1.13)$$

This commutation relation is Lorentz-invariant (Horwitz and Piron, 1973; Sogami and Yabuki, 1980). The time-like component of the above commutator is -1 in every Lorentz frame. This allows time-like excitations. Indeed, Rotbart (1981) discussed the covariant Hilbert space of harmonic oscillator wave functions in which time-like excitations are allowed in all Lorentz frames. The above commutation relation allows also to construct representations diagonal in the Casimir operators of the $O(3,1)$ group. These issues have been discussed in Chapter V.

On the other hand, there is no evidence to indicate the existence of such time-like excitations in the real world. This is perfectly consistent with the fact that the basic space-time symmetry of confined quarks is that of the $O(3)$-like little group of the Poincaré group. We can suppress time-like excitations in the hadronic rest frame by imposing the subsidiary condition:

$$P^\mu a_\mu^\dagger \psi_{nk}(x) = 0, \quad (1.14)$$

where P_μ is the hadronic four-momentum. Then only the solutions with $k = 0$ are allowed, and the commutator given in Equation (1.13) is not consistent with the above subsidiary condition.

How can we then construct a covariant commutator consistent with Equation (1.14)? In order to attack this problem, let us divide the four-dimensional Minkowskian space-time into the one-dimensional time-like space parallel to the hadronic four momentum and the three-dimensional space-like hyperplane perpendicular to the four momentum (Fleming, 1965 and 1966). This hyerplane accommodates the internal space-time symmetry dictated by the $O(3)$-like little group. This leads us to consider the operator

$$b_\mu = a_\mu - (P_\mu P^\nu/M^2)\, a_\nu. \quad (1.15)$$

Then b_μ satisfies the constraint condition:

$$P^\mu b_\mu = P^\mu b_\mu^\dagger = 0, \quad (1.16)$$

and has only three independent components. Thus, for b_μ and $b_\mu^\dagger$, we can write the covariant commutation relation:

$$[b_\mu, b_\nu^\dagger] = -g_{\mu\nu} + P_\mu P_\nu/M^2. \tag{1.17}$$

The right-hand side of the above expression is symmetric in μ and ν, and satisfies the relation:

$$P^\mu(-g_{\mu\nu} + P_\mu P_\nu/M^2) = 0. \tag{1.18}$$

Therefore, the covariant commutation relation given in Equation (1.17) is consistent with the subsidiary condition of Equation (1.14).

The covariant form of Equation (1.17) represents the usual Heisenberg uncertainty relations on the three-dimensional space-like hypersurface perpendicular to the hadronic four momentum. This form enables us to treat separately the uncertainty relation applicable to the time-like direction, without destroying covariance. The existence of the t' distribution due to the ground-state wave function in Equation (1.10) restricted by Equation (1.14) allows us to write the time-energy uncertainty relation in the form

$$(\Delta t')(\Delta E') \simeq 1, \tag{1.19}$$

without postulating the commutation relation. E' in this case is the energy separation between the quarks in the Lorentz frame in which the hadron is at rest. As we shall see in Chapter XII, the parton phenomenon, together with other high-energy features in the quark model, indicates clearly the existence of this uncertainty relation.

As has been expected, the uncertainty relation exists between the time and energy separation variables in the quark model. The remarkable fact is that this uncertainty relation is just like the one expected in all other physical phenomena (Weisskopf and Wigner, 1930; Moshinsky, 1951a, b and 1952a, b; Landau and Lifschitz, 1958; Aharonov and Bohm, 1961 and 1964; Fock, 1962; Eberly and Singh, 1973; Bauer and Mello, 1976 and 1978; Papp, 1977; Recami, 1977; Rayski and Rayski, 1977; Fujiwara et al., 1980; Pauri, 1980; Prugovecki, 1981; Han et al., 1981c; Blanchard, 1982).

2. Dirac's Form of Relativistic Theory of "Atom"

Dirac's atom in modern language is a hadron that is a bound state of quarks and/or antiquarks. We have defined in Section 1 the kinematical variables of the quarks and hadrons. In order to embed quantum mechanics into relativistic space-time, Dirac in his 1949 paper was interested in constructing a three-dimensional subspace of the four-dimensional Minkowskian space in which nonrelativistic quantum mechanics is valid.

Dirac considered three possible constraint conditions which are called "instant form", "front form" and "point form". Each of these forms imposes one constraint condition which reduces the four-dimensional Minkowskian space into a three-dimensional Euclidean space. In his "instant form", Dirac considered the condition

$$x_0 \simeq 0, \tag{2.1}$$

whose covariant form is

$$P_\mu x^\mu \simeq 0, \tag{2.2}$$

where P is the total momentum of the hadron. Equation (2.2) becomes Equation (2.1) when the hadron is at rest. Dirac avoided using the exact numerical equality in writing down the above constraint in order to allow further physical interpretations consistent with quantum mechanics and special relativity. In particular, Dirac had in mind the possibility of the left-hand side becoming an operator acting on state vectors.

After introducing constraints, Dirac emphasizes that the relativistic dynamical equations should consist of transformation operators which generate space-time translations, rotations, and Lorentz boosts. He points out further that those transformation operators should be generators of the Poincaré group. Dirac then writes down the "Poisson bracket" relations these generators should satisfy.

$$\begin{aligned} &[P_\mu, P_\nu] = 0, \\ &[M_{\mu\nu}, M_{\rho\sigma}] = g_{\mu\rho} P_\nu - g_{\nu\rho} P_\mu, \\ &[M_{\mu\nu}, M_{\rho\sigma}] = g_{\mu\rho} M_{\nu\sigma} - g_{\nu\rho} M_{\mu\sigma} + g_{\mu\sigma} M_{\rho\nu} - g_{\nu\sigma} M_{\rho\mu}, \end{aligned} \tag{2.3}$$

where P_μ and $M_{\mu\nu}$ are the generators of space-time translations and Lorentz transformations respectively. Here again, we are using the same notation for the operator and eigenvalue of the hadronic four-momentum. If we translate the above brackets into the language of quantum mechanics, they become the commutators for the generators of the Poincaré group discussed in Chapter III.

We are thus led to Section 3 of Chapter V where we constructed representations of the Poincaré group using the covariant harmonic oscillator. This procedure has been thoroughly discussed in Chapter V.

The only remaining step in constructing Dirac's dynamical system is therefore to make the constraint condition consistent with the generators of the Poincaré group. Dirac noted in particular that the constraint condition of Equation (2.2) can be an operator equation, and that its "Poisson brackets" with other dynamical variables should be zero or become zero in the manner in which the right-hand side of Equation (2.2) vanishes.

Dirac's Form of Relativistic Quantum Mechanics 145

In terms of the step-up and step-down operators introduced in Section 1, we can rewrite the harmonic oscillator equation given in Equation (V.1.3) as

$$\left\{\left(\frac{\partial}{\partial X_\mu}\right)^2 + \frac{1}{2} a_\mu^\dagger a^\mu + m_0^2\right\} \phi(X, x) = 0, \qquad (2.4)$$

with the subsidiary condition given in Equation (1.14). The (mass)^2 of the hadron P^2 is constructed by the eigenvalue of the oscillator differential equation of

$$P^2 = \lambda + m_0^2$$
$$= m_0^2 + \tfrac{1}{2} a_\mu^\dagger a^\mu. \qquad (2.5)$$

As was pointed out before, Dirac was interested in a solution of his "Poisson bracket" equations consistent with the "instant form" constraint of Equation (2.2). The key question is whether the subsidiary condition of Equation (1.14) meets this requirement.

What Dirac wanted from his conditional equality of Equation (2.2) was to freeze the motion along the time separation variable in a manner consistent with quantum mechanics and relativity. This means that we can allow a time-energy uncertainty along this time-like axis without excitations, in accordance with the c-number time-energy uncertainty relation discussed in Section 1. The c-number in matrix language is one-by-one matrix, and is the ground state with no excitations in the harmonic oscillator system.

As we have observed in Section 2 of Chapter V, the subsidiary condition of Equation (1.14) becomes

$$(a_0')^\dagger \psi(x) = 0, \qquad (2.6)$$

which forbids time-like excitations in the hadronic rest frame. We can therefore conclude that the subsidiary condition of Equation (1.14) is a quantum mechanical form of Dirac's "instant form" constraint given in Equation (2.2).

In order that the dynamical system be completely consistent, the subsidiary condition should commute with the generators of the Poincaré group:

$$[P_\alpha, P^\mu a_\mu^\dagger] = 0,$$
$$[M_{\alpha\beta}, P^\mu a_\mu^\dagger] = 0. \qquad (2.7)$$

The above equations follow immediately from the fact that the operator $P^\mu a_\mu^\dagger$ is invariant under translations and Lorentz transformations.

Since the Casimir operators are constructed from the generators of the Poincaré group, we are tempted to say that the constraint operator commutes with the invariant Casimir operators. However, we have to note that the operator P^2 also takes the form of Equation (2.5). Therefore, in order that the dynamical system be completely consistent, the operator $P^\mu a_\mu^\dagger$ should also commute with the form of P^2 given in Equation (2.5). Let us see whether this is the case. If we compute the commutator using the expression of Equation (2.5) for P^2,

$$[P^2, P^\mu a_\mu^\dagger] = P^\mu a_\mu^\dagger. \tag{2.8}$$

The right-hand side of this commutator does not vanish. This however should not alarm us. Because of the subsidiary condition of Equation (1.14), the right-hand side vanishes in the applicable Hilbert space.

The constraint condition of Equation (1.14) and its commutator with other operators produce zero either identically or in the manner in which Equation (1.14) is zero. Therefore, the subsidiary condition of Equation (1.14) satisfies all the requirements of the "instant form" constraint of Equation (2.2). Indeed, the covariant harmonic oscillator formalism, while serving as the basis for the representations of the Poincaré group, is a solution of Dirac's "Poisson bracket" equations consistent with the "instant form" constraint (Han and Kim, 1980 and 1981b).

In his 1949 paper, Dirac notes a "real difficulty" associated with making his dynamical models consistent with the energy-momentum relation for each particle participating in the interacting system. Let us examine why we do not have this difficulty in the harmonic oscillator formalism. The basic difference between Dirac's original paper and the present case is that, for each quark in the dynamics, Dirac uses the free-particle energy-momentum relation:

$$E_a = [(\mathbf{p}_a)^2 + m_a^2]^{1/2}, \tag{2.9}$$

because the concept of off-mass-shell particles was not firmly established in 1949.

In the harmonic oscillator formalism, the Casimir operators of the Poincaré group indicate clearly that the mass of the hadron is a Poincaré-invariant constant, but they do not tell us anything about the masses of the constituent particles. In fact, the (mass)^2 operator for an individual quark $(p_a)^2$ does not commute with P^2 of Equation (2.5) (Problem 7 in Section 6):

$$[p_a^2, P^2] \neq 0,$$
$$[p_b^2, P^2] \neq 0. \tag{2.10}$$

The above non-vanishing commutators would have caused very serious difficulties in 1949 when the concept of off-mass-shell particles did not exist.

The above commutators do not cause any difficulty today, because it is by now a well-accepted view that particles in the interacting field or within a bound state are not necessarily on their mass shells. We know also that those particles can become unphysical or virtual particles due to the time-energy uncertainty relation. It is interesting to note that Dirac's 1949 difficulty can be resolved by the c-number time-energy uncertainty relation which Dirac discussed in 1927 (Kim and Noz, 1979a; Han and Kim, 1981b).

3. Dirac's Light-Cone Coordinate System

The most uncomfortable feature in Lorentz transformation is to deal with a skew symmetric coordinate system. Many attempts have been made in the past to rectify this situation. Among them, the most common practice has been to use a four-dimensional Euclidean geometry with three real spatial coordinates and one imaginary time variable. This practice has not been helpful in understanding the fundamental aspects of the subject.

In an attempt to formulate "a relativistic dynamics of atom", Dirac introduced a coordinate system which allows us to use a rectangular coordinate system commonly known as the light-cone coordinate system. In 1949, Dirac observed that it would be more convenient to use the variables:

$$u = (z + t)/\sqrt{2},$$
$$v = (z - t)/\sqrt{2}, \quad (3.1)$$

for the system being boosted along the z direction. The above combinations of z and t are commonly called the light-cone variables.

In order to see how these light-cone variables are Lorentz transformed, let us write down the light-cone variables in the hadronic rest frame u' and v' defined as

$$u' = (z' + t')/\sqrt{2},$$
$$v = (z' - t')/\sqrt{2}. \quad (3.2)$$

Then the transformation from u' and v' to u and v can be written as

$$u = \left[\frac{1+\beta}{1-\beta}\right]^{1/2} u',$$

$$v = \left[\frac{1-\beta}{1+\beta}\right]^{1/2} v'. \quad (3.3)$$

Under the Lorentz transformation, one axis becomes elongated while the other goes through a contraction so that the product uv will stay constant (Kim and Noz, 1982):

$$uv = u'v'$$
$$= (t^2 - z^2)/2 = [(t')^2 - (z')^2]/2. \quad (3.4)$$

This transformation property is illustrated in Figure 3.1. If the stationary space-time region is a square, then the moving region will appear like a rectangle with the same area.

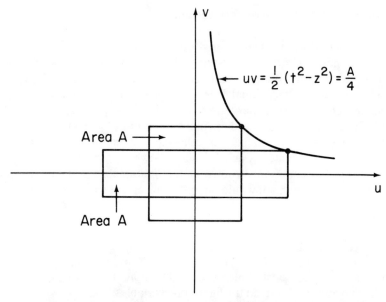

Fig. 3.1. Space-time diagram of a Lorentz boost in the light-cone coordinate system. The coordinate system remains orthogonal. The space-time point (z, t) traces a hyperbola, and the area of the rectangle is $2(t^2 - z^2)$ and remains invariant.

In his 1949 paper, Dirac discussed his "point form" constraint, which imposes the restriction

$$x^\mu x_\mu = \text{const.} \tag{3.5}$$

In terms of the light-cone variables, Equation (3.4) and Figure (3.1) illustrate the content of "point form" constraint.

The Lorentz deformation property in the $q_z q_0$ plane is the same as that for the zt plane, and it is possible to use the light-cone coordinate system for these variables:

$$\begin{aligned} q_+ &= (q_z + q_0)/\sqrt{2}, \\ q_- &= (q_z - q_0)/\sqrt{2}, \end{aligned} \tag{3.6}$$

with the transformation property:

$$\begin{aligned} q_+ &= \left[\frac{1+\beta}{1-\beta}\right]^{1/2} q'_+, \\ q_- &= \left[\frac{1-\beta}{1+\beta}\right]^{1/2} q'_-. \end{aligned} \tag{3.7}$$

Dirac's Form of Relativistic Quantum Mechanics

The Lorentz deformation property in the momentum-energy plane can also be seen in Figure 3.1.

The basic advantage of using the light-cone variables is that the coordinate system remains orthogonal. It is not difficult to visualize the deformation of the regions given in Figure 3.2 due to the boost. The circular region in the hadronic rest frame will appear as an ellipse to the lab-frame observer.

With this understanding, let us use the Gaussian wave function which has the circular uncertainty distribution in the hadronic rest frame:

$$\psi_\beta(z, t) = (1/\sqrt{\pi}) \exp(-(1/2) [(z')^2 + (t')^2]). \tag{3.8}$$

This is the ground-state space-time wave function in the covariant oscillator formalism discussed in Chapter V. Since

$$(z')^2 + (t')^2 = (u')^2 + (v')^2$$

$$= \left(\frac{1-\beta}{1+\beta}\right) u^2 + \left(\frac{1+\beta}{1-\beta}\right) v^2, \tag{3.9}$$

the Gaussian form of Equation (3.8) can be written as

$$\psi_\beta(z, t) = \sqrt{1/\pi} \exp\left(-\frac{1}{4}\left[\frac{1-\beta}{1+\beta}(z+t)^2 + \frac{1+\beta}{1-\beta}(z-t)^2\right]\right). \tag{3.10}$$

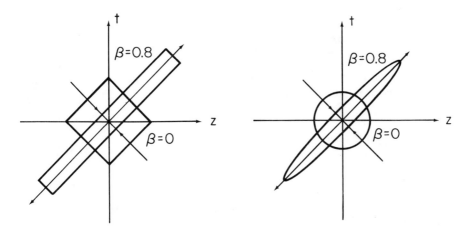

Fig. 3.2. Lorentz deformation in the longitudinal and time-like coordinate system. The left-hand side of this figure is identical to Figure 3.1. It is clear that the circular region in the zt plane will undergo an elliptic deformation.

The momentum-energy wave function becomes

$$\phi_\beta(q_z, q_0) = (1/2\pi) \int \exp[i(q_z z - q_0 t)]\, \psi(z, t)\, dt\, dz$$

$$= (1/\sqrt{\pi}) \exp(-[(q'_z)^2 + (q'_0)^2]/2). \tag{3.11}$$

This wave function can also be written as

$$\phi_\beta(q_z, q_0) = (1/\sqrt{\pi}) \exp\left(-\frac{1}{4}\left[\frac{1-\beta}{1+\beta}(q_z + q_0)^2 + \frac{1+\beta}{1-\beta}(q_z - q_0)^2\right]\right). \tag{3.12}$$

Because $q^\mu x_\mu$ takes the form

$$q^\mu x_\mu = q_x x + q_y y + q_z z - q_0 t$$
$$= q_x x + q_y y + \tfrac{1}{2}[(q_z - q_0)(z + t) + (q_z + q_0)(z - t)]. \tag{3.13}$$

The Fourier relations between the space-time and momentum-energy coordinates are (Kim and Noz, 1979a)

$$q_u = q_- = -i\frac{\partial}{\partial u},$$
$$q_v = q_+ = -i\frac{\partial}{\partial v}. \tag{3.14}$$

This means that the major and minor axes of the momentum-energy coordinates are the "Fourier conjugates" of the minor and major axes of the space-time coordinates respectively. Thus we have the following Lorentz invariant relations.

$$(\Delta u)(\Delta q_-) = (\Delta u')(\Delta q'_-) \simeq 1,$$
$$(\Delta v)(\Delta q_+) = (\Delta v')(\Delta q'_+) \simeq 1. \tag{3.15}$$

This is indeed a Lorentz-invariant statement of the c-number time-energy uncertainty relation combined with Heisenberg's position-momentum uncertainty relations.

The above statement was based on the uncertainty relation along the t' direction and that associated with the ground state along the longitudinal direction. However, we have to take into account the fact that there are excitations along the z' direction. We discuss this problem in Section 4.

4. Harmonic Oscillators in the Light-Cone Coordinate System

We have so far been interested in Lorentz deformation properties of the ground-state harmonic oscillator wave function using the light-cone coordinate system. In this section, we shall study the harmonic oscillator formalism including all excited states.

It was noted in Sections 5 and 6 of Chapter V that the covariant harmonic oscillator wave functions forming the basis for the $O(3)$-like little group can be written as a linear sum of the orthonormal solutions in the $O(4)$ coordinate system. It was noted also that, although not Lorentz-invariant, the $O(4)$-invariant differential equation is covariant in the sense that its $O(3, 1)$ counterpart is invariant under Lorentz transformations.

Because the light-cone coordinate system is only a 45° rotation of the $O(4)$ coordinate system in the subspace of longitudinal and time-like coordinates, we can write the $O(4)$-invariant oscillator equation in terms of the light-cone variables as

$$\frac{1}{2}\left[-\left(\frac{\partial^2}{\partial x^2} + \frac{\partial^2}{\partial y^2} + \frac{\partial^2}{\partial u^2} + \frac{\partial^2}{\partial v^2}\right) + (x^2 + y^2 + u^2 + v^2)\right] g(x) = (\lambda + 2) g(x). \tag{4.1}$$

When the system is boosted along the z direction, u and v are to be replaced by u' and v' respectively. Since we are not interested in the transverse components which remain invariant under Lorentz boosts along the z direction, the above differential equation can now be written as (Kim et al., 1980)

$$\frac{1}{2}\left[-\left(\frac{\partial^2}{\partial u'^2} + \frac{\partial^2}{\partial v'^2}\right) + (u'^2 + v'^2)\right] g'(u, v) = (\lambda + 1) g'(u, v). \tag{4.2}$$

The above differential equation is separable in the u and v variables, and remains separable under the boost because the Lorentz boost does not mix u and v. It is therefore easy to construct the normalizable solutions of the above differential equation. For a given integer value of $\lambda = n$, we can write

$$g_n(x) = \sum_{m=0}^{n} C_n^m f_m(u') f_{n-m}(v'), \tag{4.3}$$

where $f_m(u)$ is the normalized one-dimensional harmonic oscillator wave function defined in Section 2 of Chapter V.

If we insist on the subsidiary condition of Equation (1.14) forbiding time-like oscillations in the hadronic rest frame, the oscillator wave function should have excitations only along the z' direction, or along $(u' + v')/\sqrt{2}$. In terms of the u' and v' variables, the Hermitian polynomial $H_n(z')$ can be written as (Magnus et al., 1949)

$$H_n(z') = H_n([u' + v']/\sqrt{2})$$
$$\times \left(\frac{1}{2}\right)^{n/2} \sum_{m=0}^{n} \binom{n}{m} H_{n-m}(u') H_m(v'). \quad (4.4)$$

Thus the explicit form of the physical wave function becomes (Kim et al., 1980).

$$\psi_\beta^n(z, t) = \left(\frac{1}{2}\right)^n \left(\frac{1}{\pi n!}\right)^{1/2} \left[\sum_{m=0}^{n} \binom{n}{m} H_{n-m}(u') H_m(v')\right]$$
$$\times \exp[-(u'^2 + v'^2)/2]. \quad (4.5)$$

In order to express this in terms of the oscillator functions of the u and v variables, we still have to make an expansion in an infinite series. The power of Hermite polynomials remains the same. However, the Gaussian factor requires an infinite sum of the wave functions in terms of the u and v variables. This calculation remains as a future research problem.

5. Lorentz-Invariant Uncertainty Relations

As was pointed out in Section 1, the uncertainty principle applicable to the space-time separation of quarks in the hadronic rest frame is the same as the currently accepted form based on existing theories and observations. The usual Heisenberg uncertainty relation holds for each of the three spatial coordinates. The time-separation variable is a c-number and therefore does not cause quantum excitations. However, this does not forbid the existence of the "Fourier relation" between time and energy variables resulting in the time-energy uncertainty relation given in Equation (1.19).

Also in Section 1, it was noted that this peculiar time-energy uncertainty can be combined with Heisenberg's position-momentum uncertainty relation to give a covariant form for the internal space-time separation variables in the quark model within the framework of the covariant harmonic oscillator formalism. It was shown then that the uncertainty relations describing the oscillator system can be generalized into a *covariant commutator* form. This means that the uncertainty principle which holds in the oscillator formalism is applicable to all other models designed to explain hadronic structures in a covariant manner.

Furthermore, as is shown in Section 4, the covariant form of the uncertainty relations given in Equation (1.17) and Equation (1.19) takes a *Lorentz-invariant form* in the light-cone coordinate system, as is shown in Equation (3.15). However, the crucial question is how these uncertainty relations appear to an observer in the laboratory frame with the space-time separation variables z and t. Another important question is whether the uncertainty relations in the observer's frame can be stated in a Lorentz-invariant manner.

If we consider only the ground-state wave function, then the localization dictated by the uncertainty relations associated with both space and time separation variables will led to a distribution centered around the origin in the $z't'$ coordinate system, as is illustrated in Figure 1.2. It is clear from Figure 1.2 that the circular distribution, with its simpler mathematical form, will share with the arbitrarily-shaped region all qualitative properties associated with the uncertainty relations.

The question then is how this arbitrarily-shaped region appears to the lab-frame observer, while the coordinates of the two different frames are related by the Lorentz transformation of Equation (1.9). It is very easy to describe the Lorentz deformation of this region in the light-cone coordinate system, in which the Lorentz transformation is simply a scale transformation along the light cones. If we use the Lorentz transformation parameter η, where

$$\sinh \eta = \beta/(1 - \beta^2)^{1/2}. \tag{5.1}$$

then the Lorentz transformation of Equation (3.3) can be written as

$$u = [\exp(\eta)] \, u', \quad v = [\exp(-\eta)] \, v'. \tag{5.2}$$

Lorentz transformation in the light-cone coordinate system is illustrated in Figure 3.1., where the area of the rectangle remains the same as that of the square.

The basic advantage of using the light-cone variables is that the coordinate system remains orthogonal. It is not difficult to visualize the deformation of the regions given in Figure 3.2 due to the boost. The circular region in the hadronic rest frame will appear as an ellipse to the observer in the laboratory frame. It is also possible to determine accurately the Lorentz deformation of the arbitrarily-shaped region. It is therefore sufficient to study the circular region in Figure 1.2 in order to study the transformation properties of more complicated regions. With this understanding, let us go back to the oscillator formalism which has the circular uncertainty distribution in the hadronic rest frame. We can use the space-time wave function of Equations (3.8) and (3.10) and the momentum-energy wave function of Equations (3.11) and (3.12).

Again, in terms of the light-cone variables, we write the transformation of the momentum and energy variables as

$$q_+ = \exp(\eta) \, q'_+, \quad q_- = \exp(-\eta) \, q'_-. \tag{5.3}$$

Within the framework of the harmonic oscillator formalism, the momentum-energy wave function has the same Lorentz-transformation property as space-time wave function.

It is then not difficult to arrive at the Lorentz-invariant uncertainty products given in Equation (3.15). However, we are interested in what implications these invariant relations have on the z and t coordinates. We started with the uncertainty relations:

$$(\Delta z')(\Delta q_z') \simeq 1 \quad \text{and} \quad (\Delta t')(\Delta q_0') \simeq 1. \tag{5.4}$$

in the hadronic rest frame. The question is whether or not we are allowed to write

$$(\Delta z)(\Delta q_z) \simeq 1 \quad \text{and} \quad (\Delta t)(\Delta q_0) \simeq 1, \tag{5.5}$$

in the laboratory frame. In order to answer this question, let us look carefully at the deformation properties described in Figure 3.2.

If we measure the z distribution along the z axis where $t = 0$, the distribution is contracted by

$$\langle \Delta z \rangle_c = \langle \Delta z' \rangle [\cosh(2\eta)]^{-1/2}. \tag{5.6}$$

On the other hand, if we look at the distribution along the light-cone axis and its projection on the z direction, then the distribution is elongated by

$$\langle \Delta z \rangle_e = \langle \Delta z' \rangle [\cosh(2\eta)]^{1/2}. \tag{5.7}$$

The geometry is the same along the t axis, and also for the momentum-energy coordinate system. We can therefore write

$$\begin{aligned} (\Delta z)_c(\Delta q_z)_e \simeq 1 \quad &\text{and} \quad (\Delta z)_e(\Delta q_z)_c \simeq 1, \\ (\Delta t)_c(\Delta q_0)_e \simeq 1 \quad &\text{and} \quad (\Delta t)_e(\Delta q_0)_c \simeq 1. \end{aligned} \tag{5.8}$$

The uncertainty relations of Equation (5.5) are basically correct if they are interpreted in the above manner.

As is illustrated in Figure 5.1., the above interpretation requires the concept that the hadron is contracted if looked at in one way and elongated if looked at in another way along the longitudinal and time-like directions. This should be true also in the momentum-energy coordinate system. The important question is whether this dual nature of Lorentz deformation is consistent with the real world. The concept of "Lorentz contraction" of hadronic matter is not new, and there are several physical models based on the idea that the longitudinal dimension of the hadron becomes shortened (Byers and Yang, 1966; Chou and Yang, 1968; Gursey and Orfanidis, 1972).

The concept of elongation is also familiar to us, because it is built into the parton model. First of all, it has been firmly established that the momentum distribution becomes wide-spread as the hadron moves very fast. This is precisely the elongation in the momentum-energy coordinate system. Since the Lorentz deformation property in the space-time coordinate system is identical to that in the momentum-energy coordinate system, there should also be an elongation in the space-time coordinate system. Then, does this effect manifest itself in high-energy laboratories? We shall discuss this and other related questions in Chapter XII.

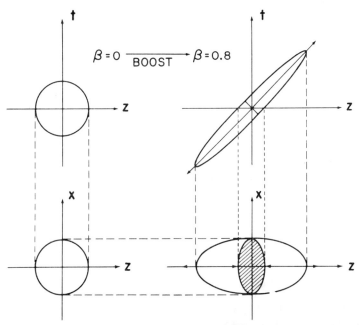

Fig. 5.1. Lorentz deformation of a spherical hadron. The upper and lower parts represent its projection on the zt and zx planes respectively. The zt view is a Lorentz deformation of the circular part of Figure 1.2 due to the Lorentz transformation of Figure 3.2. The lower part shows that the longitudinal component undergoes Lorentz contraction *and* Lorentz elongation, while the transverse components remain unchanged. The Lorentz contraction is the longitudinal measure of the hadron along the z axis with $t = 0$. The Lorentz elongation is the projection of the hadronic size along the major light-cone axis to the z axis.

6. Exercises and Problems

Exercise 1. In Section 3, we introduced the light-cone variables u and v and discussed their Lorentz transformation properties. how are their derivatives transformed? How does the wave equation appear in terms of the light-cone variables?

Because u and v variables do not mix with each other when boosted along the z direction, their derivatives undergo simple scale transformations. From Equation (5.2),

$$\frac{\partial}{\partial u'} = e^{\eta}\frac{\partial}{\partial u}, \quad \text{and} \quad \frac{\partial}{\partial v'} = e^{-\eta}\frac{\partial}{\partial v}. \tag{6.1}$$

The above form is consistent with the definition of the conjugate variables in Equation (3.14) and the transformation property of the q_+ and q_- variables. It is clear from the above expression that the second derivative $\partial^2/\partial u \partial v$ is a Lorentz-invariant operator. In terms of u and v, the differential operators in the conventional z and t variables can be written as

$$\frac{\partial}{\partial z} = (1/\sqrt{2})\left(\frac{\partial}{\partial u} + \frac{\partial}{\partial v}\right), \quad \text{and} \quad \frac{\partial}{\partial t} = (1/\sqrt{2})\left(\frac{\partial}{\partial u} - \frac{\partial}{\partial v}\right). \tag{6.2}$$

Then, after a simple calculation,

$$\frac{\partial^2}{\partial u \partial v} = \left(\frac{\partial}{\partial z}\right)^2 - \left(\frac{\partial}{\partial t}\right)^2. \tag{6.3}$$

Both sides of the equation are Lorentz-invariant. The immediate application of the above relation is the wave equation, which can now be written as

$$\frac{\partial^2}{\partial u \partial v} \psi(u, v) = 0. \tag{6.4}$$

This equation tells us that solutions of the wave equation depend either on u or v, not on both. Thus the most general form of the solution takes the form

$$\psi(u, v) = f(u) + g(v). \tag{6.5}$$

$f(u)$ and $g(v)$ depend only on $(z + t)$ and $(z - t)$ respectively. See Bacry and Cadilhac (1981), and Kim and Noz (1981).

Exercise 2. In the Lorentz frame where the hadron is at rest, the uncertainty principle applicable to the space-time separation of quarks is expected to be the same as the presently accepted form largely based on nonrelativistic quantum mechanics. The usual Heisenberg uncertainty relation holds for each of the three spatial coordinates:

$$[x', q'_x] = i, \quad [y', q'_y] = i, \quad [z', q'_z] = i. \tag{6.6}$$

On the other hand, the time-separation variable is a *c*-number and therefore does not cause quantum excitations. The commutator in this case takes the form

$$[t', q'_0] = 0, \tag{6.7}$$

accompanied by the "Fourier relation":

$$(\Delta t')(\Delta q'_0) \simeq 1. \tag{6.8}$$

How would these uncertainty relations appear to an observer in the laboratory frame?

Since the hadron moves along the z axis, the x and y coordinates are not affected by boosts, and the first two commutation relations of Equation (6.6) remain invariant:

$$[x, q_x] = i, \quad [y, q_y] = i. \tag{6.9}$$

As for the third commutation relation of Equation (6.6) for the longitudinal direction, we have to consider the Lorentz transformation of the coordinate system:

$$z = (z' + \beta t')/(1 - \beta^2)^{1/2},$$
$$t = (t' + \beta z')/(1 - \beta^2)^{1/2}. \tag{6.10}$$

Likewise, the transformation equations for the momentum-energy variables can be written as

$$q_z = (q'_z + \beta q'_0)/(1 - \beta^2)^{1/2},$$
$$q_0 = (q'_0 + \beta q'_z)/(1 - \beta^2)^{1/2}. \tag{6.11}$$

In terms of variables in the laboratory frame, the uncertainty relation $[z', q'_z] = i$ of Equation (6.6) and the time-energy relation of Equation (6.7) can be written as

$$[z, q_z] = i/(1 - \beta^2),$$
$$[t, q_0] = i\beta^2/(1 - \beta^2). \tag{6.12}$$

In addition, because the Lorentz transformation is not an orthogonal transformation, the commutation relations between z and q_0 and between t and q_z do not vanish:

$$[z, q_0] = i\beta/(1 - \beta^2),$$
$$[t, q_z] = i\beta/(1 - \beta^2). \tag{6.13}$$

The commutation relations of Equations (6.9), (6.12) and (6.13) can now be combined into a covariant form:

$$[x_\mu, q_\nu] = -g_{\mu\nu} + P_\mu P_\nu/M^2, \tag{6.14}$$

with the covariant form of Equation (6.8):

$$(\Delta(P \cdot x/M))(\Delta(P \cdot q/M)) \simeq 1, \tag{6.15}$$

where M is the hadronic mass. The above form of the uncertainty relations have two troublesome features. According to Equation (6.13), Planck's constant appears frame-dependent. According to Equation (6.13), there is a lack of the orthogonality relation among independent variables. If we use the light-cone coordinate system, we can see how these two problems cancel each other out, and how the concept of the uncertainty relations becomes Lorentz-invariant (Hussar et al., 1985).

Problem 1. Discuss time dilation, space contraction and clock synchronization using the light-cone coordinate system (Parker and Schmieg, 1970).

Problem 2. Write down Maxwell's equations and derive the wave equations using the light-cone coordinate system (Parker and Schmieg, 1970).

Problem 3. Discuss the Lorentz transformation properties of electromagnetic potentials and fields in terms of the light-cone variables. See Chapter X.

Problem 4. Show that the time-energy uncertainty relation in quantum field theory is the same as the c-number time-energy uncertainty relation we used in this chapter. See Han *et al.* (1981).

Problem 5. It was observed in Chapter V that the (mass)^2 spectrum has no lower bound if excitations along the time-like axis are allowed. Show that there is no lower energy bound in nonrelativistic quantum mechanics if the time variable becomes a q number. See Prugovečki (1981).

Problem 6. Since it is believed that the time-energy uncertainty relation is independent of the Schrödinger equation, there has been a tendency to formulate the problem in terms of other quantities. Show that the time-energy uncertainty can be stated in terms of the density operators (Eberly and Singh, 1973; Blanchard, 1982).

Problem 7. Show that the individual quark mass cannot be simultaneously diagonalized with the hadronic mass. Prove Equation (2.10). See Han and Kim (1980).

Problem 8. The solution of the oscillator equation given in Equation (4.3) is essentially a harmonic oscillator wave function in a two-dimensional space. One-dimensional and three-dimensional harmonic oscillators are discussed extensively in the literature. However, two-dimensional oscillators are seldom discussed. Show that the Lorentz deformation is essentially a problem of two-dimensional harmonic oscillators.

Problem 9. Describe the localization of light waves using the light-cone coordinate system.

Chapter VII

Massless Particles

We shall study in this chapter the internal space-time symmetries of massless particles. As was noted in Chapter III, the little group for massless particles is locally isomorphic to the group of Euclidean transformations on a two-dimensional plane which is often called $E(2)$. Therefore, the study of massless particles requires a careful investigation of the $E(2)$ group or its unimodular subgroup $SE(2)$, and its relation to the little group. We already know that the generators of these two groups satisfy the same set of commutations relations.

Among many massless particles, photons occupy the most prominent position. Indeed, the development of modern physics has been the process of understanding properties of photons. The role of photons in the development of quantum theory is well known. The role of electromagnetism in the development of relativity is also well known. Quantum electrodynamics which serves as the basis for all field theoretic models is a theory of photons interacting with charged point particles.

In spite of this overwhelming role in physics, we have been and are still behind in understanding the internal space-time symmetry of photons. One of the difficulties associated with studying photons has been that the four-vector form for the photon wave function is not unique and is dependent on gauge degrees of freedom which are not measurable. It is of course possible to bypass four-potentials and use electric and magnetic fields which are independent of gauge parameters. However, this does not simplify our problem.

Unlike $SO(3)$, $SE(2)$ is a non-compact group, and its unitary representations are infinite-dimensional. Certainly, this does not make the photon problem easier for us. In fact, the study of finite subsets of unitary representations of the $SE(2)$ group in the past led to the restriction to the $O(2)$ subgroup of $SE(2)$ (Wigner, 1939). This restriction of course freezes the translational degrees of freedom in the $SE(2)$ group (Weinberg, 1964a).

A closer look at the four-by-four matrices of the $E(2)$-like little group applicable to the photon four-vector indicates that they form a finite-dimensional non-unitary representation. Therefore, instead of restricting ourselves to unitary representations of the $SE(2)$ group, we should study also its finite-dimensional non-unitary representations, and examine how these non-unitary repreesntations will help us in understanding unitary representations of physical transformations. As is indicated in Table 1, it was not until 1982 that the desired non-unitary representations were discussed in the physics literature.

TABLE 1

Representations of the $SE(2)$ group. In order that representations be useful in describing massless particle with definite helicity, they should be diagonal in the generators of rotations.

Diagonal in	Unitary infinite dimensional	Non-unitary finite dimensional
P_1 and P_2	Wigner (1939)	Trivial
L_3	Inonu and Wigner (1953)	Han et al. (1982)

Many of the properties of the $SE(2)$ group have already been discussed in Chapters I and II. The present chapter is devoted to the study of finite-dimensional non-unitary representations of the $E(2)$-like little group for massless particles and their relationship with the correspondig representations of the $SE(2)$ group. We shall study first photons in detail. We shall then extend the formalism to include neutrinos. Gravitons are discussed in Chapter VIII.

In spite of what we said above, we are eventually interested in constructing unitary representations of physical transformations. Then what do we do with the non-unitary representations of the $E(2)$-like little group which we study in this chapter? It is not difficult to associate rotations of the $SE(2)$ group with the helicity. We shall see in this chapter that the translational degrees of freedom correspond to gauge transformations. Since the gauge transformation is not an observable transformation, its representation does not have to be unitary.

On the other hand, Lorentz transformations which change the four-momentum are physical transformations. Among those Lorentz transformations, there are rotations around and boosts along the three orthogonal directions. Rotations can be represented by unitary matrices. Boost matrices applicable to four-vectors are non-unitary. Therefore, when we boost four-potentials, we should consider the possibility of constructing a unitary representation by combining the boost and gauge transformations.

In Section 1, we study the $SE(2)$ group in terms of transformations on the two-dimensional Euclidean plane by summarizing what we have already discussed in Chapters I and II. Section 2 is devoted to a detailed study of the little group for photons. It is shown that the four-by-four transformation matrices of the little group can be reduced to the three-by-three transformation matrices of the $SE(2)$ group. It is shown also that the translation-like degrees of freedom in the $E(2)$-like little group are gauge degrees of freedom.

Section 3 deals with Lorentz transformation properties of the four-potential. Because the choice of gauge is arbitrary, we usually choose a specific gauge for convenience. However, there is a unique gauge which makes the four-potential diagonal in the helicity operator. This gauge condition in one Lorentz frame is not always respected in other Lorentz frames. For this reason, we often say that the electromagnetic four-vector is not manifestly covariant. We study this problem in detail and show that it is possible to perform Lorentz transformations on the four-potential in a manner which preserves this gauge condition. It is shown in Section 4 that this gauge-preserving transformation is a unitary transformation with one-dimensional trivial representation of the little group.

In Section 5, we study the little group for neutrinos. Since the discovery of parity violation in weak interaction (Lee and Yang, 1957; Wu et al., 1957) and the discovery of neutrino polarization (Goldhaber et al., 1958), one of the most outstanding questions has been what the physical or kinematical origin of neutrino polarization is. We shall see in this section that the polarization of neutrinos is due to the requirement of the invariance under the $E(2)$-like little group for massless particles, which in the case of photons corresponds to gauge invariance. We shall see also that the gauge dependence of the photon four-potential can be traced to the spinor variance under the little group transformation.

For both photons and neutrinos, the internal space is purely a spin space, and is not derivable from the coordinate variables. Since the internal variable in the covariant harmonic oscillator discussed in Chapter V is directly derivable from the coordinate variables, we are led to consider whether such a study is possible. With this point in mind, we discuss in Section 6 the covariant harmonic oscillator wave functions and their $E(2)$-like symmetry for massless hadrons. While Section 7 lists exercises and problems, the primary purpose of this section is to discuss the items which are not contained in the main sections.

1. What is the E(2) Group?

As we noted in Chapters I and II, the two-dimensional Euclidean group, often called $SE(2)$, consists of rotations and translations on a two-dimensional Euclidean plane. The coordinate transformation takes the form

$$\begin{aligned} x' &= x\cos\alpha - y\sin\alpha + u, \\ y' &= x\sin\alpha + y\cos\alpha + v. \end{aligned} \quad (1.1)$$

This transformation can be written in matrix form as

$$\begin{bmatrix} x' \\ y' \\ 1 \end{bmatrix} = \begin{bmatrix} \cos\alpha & -\sin\alpha & u \\ \sin\alpha & \cos\alpha & v \\ 0 & 0 & 1 \end{bmatrix} \begin{bmatrix} x \\ y \\ 1 \end{bmatrix}. \qquad (1.2)$$

the algebraic properties of the above transformation matrix has been discussed in Chapters I and II.

The three-by-three matrix in Equation (1.2) can be exponentiated as

$$E(u, v, \alpha) = \exp[-i(uN_1 + vN_2)] \exp(-i\,\alpha L_3), \qquad (1.3)$$

where L_3 is the generator of rotations, and N_1 and N_2 are the generators of translations. These generators in the case of the above coordinate transformation take the form

$$L_3 = \begin{bmatrix} 0 & -i & 0 \\ i & 0 & 0 \\ 0 & 0 & 0 \end{bmatrix},$$

$$N_1 = \begin{bmatrix} 0 & 0 & i \\ 0 & 0 & 0 \\ 0 & 0 & 0 \end{bmatrix}, \quad N_2 = \begin{bmatrix} 0 & 0 & 0 \\ 0 & 0 & i \\ 0 & 0 & 0 \end{bmatrix}. \qquad (1.4)$$

These generators satisfy the following commutation relations:

$$[N_1, N_2] = 0,$$
$$[L_3, N_1] = iN_2, \qquad (1.5)$$
$$[L_3, N_2] = -iN_1.$$

The transformation described in Equations (1.1) and (1.2) generated by the matrices of Equation (1.4) is "active" in the sense that it transforms the object.

Let us next consider transformations of functions of x and y, and continue to use N_1 and N_2 as the generators of translations and L_3 as the generator of rotations. These generators take the form

$$N_1 = -i\frac{\partial}{\partial x},$$
$$N_2 = -i\frac{\partial}{\partial y}, \qquad (1.6)$$
$$L_3 = -i\left[x\frac{\partial}{\partial y} - y\frac{\partial}{\partial x}\right].$$

The above operators satisfy the commutation relations of Equation (1.5). The transformation through these differential operators is "passive" in the sense that it is achieved through a coordinate transformation which is inverse to that for the active transformation given in Equation (1.1).

As in the case of the rotation group, the standard method of studying this group is to find an operator which commutes with all three of the above generators. It is easy to check that N^2, defined as

$$N^2 = N_1^2 + N_2^2, \tag{1.7}$$

commutes with all three generators. Thus one way to construct representations of the $SE(2)$ group is to solve the equation

$$[N_1^2 + N_2^2]\, \psi(x, y) = k^2\, \psi(x, y), \tag{1.8}$$

using the differential forms of N_1 and N_2 given in Equation (1.6). This partial differential equation can be separated in the polar, Cartesian, parabolic or elliptic coordinate system (Winternitz and Fris, 1965).

In his original paper (Wigner, 1939), Wigner used the Cartesian coordinate system diagonal in N_1 and N_2. Inonu and Wigner (1953) later studied the differential equation using the polar coordinate system. If we solve the above partial differential equation in the Cartesian coordinate system, the solution is

$$\psi(x, y) = \exp[-i(k_1 x + k_2 y)], \tag{1.9}$$

where

$$k^2 = k_1^2 + k_2^2.$$

However, in the polar coordinate system, the solution becomes

$$\psi(r, \phi) = J_m(kr)\, e^{\pm im\phi}, \tag{1.10}$$

where

$$r = [x^2 + y^2]^{1/2},$$

$$\phi = \tan^{-1}(y/x),$$

and $J_m(kr)$ is the Bessel function of order m. These solutions are diagonal in N^2. The Cartesian solution of Equation (1.9) is diagonal in N_1 and N_2, while the polar solution of Equation (1.10) is diagonal in L_3. These results are summarized in Table 1.1.

It is easy to see that the Cartesian solution given in Equation (1.9) forms the representation basis for the infinite-dimensional unitary representation of the $SE(2)$ group. The application of $\exp(-iuN_1)$ or $\exp(-ivN_2)$ induces a solution different from the original one, and this process can be repeated to produce infinitely many different wave functions. The dimensionality problem for the polar solution is discussed in Exercise 1 in Section 7.

TABLE 1.1
Representation spaces for the $SE(2)$ group.

Diagonal in	Unitary infinite dimensional	Non-unitary fintie dimensional
N_1 and N_2	$\exp[-i(k_1 x + k_2 y)]$	1 (trivial)
L_3	$J_m(kr)\, e^{\pm im\phi}$	$r^m\, e^{\pm im\phi}$

In addition to unitary representations, there are also finite-dimensional non-unitary representations. The coordinate transformation matrix of Equation (1.2) is a three-by-three non-unitary matrix. If we compute N^2 using the generators given in Equation (1.4), it vanishes. We are thus led to consider the fundamental differential equation of Equation (1.8) with vanishing k^2 (Han et al., 1982a):

$$\left\{\left(\frac{\partial}{\partial x}\right)^2 + \left(\frac{\partial}{\partial y}\right)^2\right\} \psi(x, y) = 0. \tag{1.11}$$

This is a two-dimensional Laplace equation, and its solutions are quite familiar to us. The analytic solution of this equation takes the form

$$\psi = r^m \exp[\pm im\phi]$$
$$= (x \pm iy)^m. \tag{1.12}$$

This is an eigenstate of L_3 or a rotation around the origin.
  The effect of the rotation operator

$$R(\alpha) = \exp[-i\alpha L_3] \tag{1.13}$$

on ψ of Equation (1.12) is that

$$R(\alpha)\, \psi = r^m\, e^{\pm im(\phi - \alpha)}.$$

If we translate the expression of Equation (1.12) by applying the operator

$$T(u, v) = \exp[-i(uN_1 + vN_2)], \tag{1.14}$$

then the translated form becomes

$$T(u, v)\, \psi(x, y) = [(x - u) \pm i(y - v)]^m. \tag{1.15}$$

This is no longer an eigenstate of rotation around the origin but around the point at $x = u$ and $y = v$. The rotation operator in this case is

$$R_{uv}(\alpha) = T(u, v)\, (e^{-i\alpha L_3})\, T^{-1}(u, v). \tag{1.16}$$

In Section 3 of Chapter I, we discussed the explicit matrix representation for the coordinate transformations in the two-dimensional plane. All rotations by

Massless Particles

the same angle but not necessarily around the same point belong to the same eigenvalence class.

If $m = 1$, we are dealing with a linear form in x and y. Thus the transformation described above is simply a coordinate transformation. In order to establish the connection between the above analysis and the matrix transformation given in Equation (1.2), let us consider the form:

$$V_1 = \begin{bmatrix} x + iy \\ x - iy \\ 1 \end{bmatrix} = \begin{bmatrix} r e^{i\phi} \\ r e^{-i\phi} \\ 1 \end{bmatrix}. \tag{1.17}$$

Then the passive transformation (Exercise 2 in Section 6 of Chapter II) applicable to this vector is

$$E(u, v, \alpha) = \exp(-i\alpha L_3) \exp(-iuN_1 - ivN_2), \tag{1.18}$$

with

$$L_3 = \begin{bmatrix} 1 & 0 & 0 \\ 0 & -1 & 0 \\ 0 & 0 & 0 \end{bmatrix},$$

$$N_1 = \begin{bmatrix} 0 & 0 & -i \\ 0 & 0 & -i \\ 0 & 0 & 0 \end{bmatrix}, \quad N_2 = \begin{bmatrix} 0 & 0 & 1 \\ 0 & 0 & -1 \\ 0 & 0 & 0 \end{bmatrix}. \tag{1.19}$$

The N matrices of Equation (1.4) or (1.19) satisfy the commutation relations for the $SE(2)$ group given in Equation (1.5). In addition, these three-by-three matrices satisfy

$$N_1^2 + N_2^2 = 0, \tag{1.20}$$

which is a reflection of the differential equation with $k^2 = 0$ in Equation (1.11). In addition, N_1 and N_2 satisfy

$$N_1^2 = N_2^2 = N_1 N_2 = 0, \tag{1.21}$$

reflecting the fact that

$$\left(\frac{\partial}{\partial x}\right)^2 (x \pm iy) = \left(\frac{\partial}{\partial y}\right)^2 (x \pm iy) = \frac{\partial^2}{\partial x \partial y} (x \pm iy) = 0. \tag{1.22}$$

Because of Equations (1.20) and (1.21), the power series expansion for $E(u, v, 0)$ terminates:

$$E(u, v, 0) = \exp(-iuN_1 - ivN_2)$$
$$= I - (iuN_1 + ivN_2). \tag{1.23}$$

The series expansion truncates because of Equation (1.21).

It is of interest to see how this procedure can be extended for higher values of m. This problem will is discussed in Chapter VIII.

2. E(2)-like Little Group for Photons

Because the four-potential for the electromagnetic field is non-unique and dependent on gauge degrees of freedom, many attempts have been made in the past to construct an alternative form for the photon wave function. These attempts have been aimed at constructing a simpler form than the four-vector form. The most obvious alternative is the use of electric and magnetic fields which are directly measurable. However, we then have to deal with a quantity which behaves like a tensor under Lorentz transformations. Furthermore, we do not yet know any form other than the four-potential which is more convenient in describing the interaction of photons with charged particles.

It is therefore more productive to face the problem of gauge degrees of freedom than to avoid it. In approaching this problem, we note that the translational-like degrees of freedom were left unexplained in Wigner's original paper (1939). We note also that Wigner was only interested in constructing unitary representations even in his later papers (Inonu and Wigner, 1953; Wigner, 1962b). In this section, we shall study the photon problem by filling in these gaps.

Let us consider a single free photon moving along the z direction. Then we can write the photon wave function as

$$A^\mu(x) = A^\mu \, e^{i\omega(z-t)}, \tag{2.1}$$

where

$$A^\mu = (A_1, A_2, A_3, A_0).$$

The momentum four-vector is clearly

$$p^\mu = (0, 0, \omega, \omega). \tag{2.2}$$

Then, as we discussed in Chapter III, the little group applicable to the photon four-potential is generated by

$$L_3 = \begin{bmatrix} 0 & -i & 0 & 0 \\ i & 0 & 0 & 0 \\ 0 & 0 & 0 & 0 \\ 0 & 0 & 0 & 0 \end{bmatrix},$$

$$N_1 = \begin{bmatrix} 0 & 0 & -i & i \\ 0 & 0 & 0 & 0 \\ i & 0 & 0 & 0 \\ i & 0 & 0 & 0 \end{bmatrix}, \quad N_2 = \begin{bmatrix} 0 & 0 & 0 & 0 \\ 0 & 0 & -i & i \\ 0 & i & 0 & 0 \\ 0 & i & 0 & 0 \end{bmatrix}. \tag{2.3}$$

These matrices satisfy the commutation relations:

$$[L_3, N_1] = iN_2,$$
$$[L_3, N_2] = -iN_1, \tag{2.4}$$
$$[N_1, N_2] = 0.$$

From the generators, we can construct the transformation matrices

$$D(u, v, \alpha) = D(u, 0, 0) D(0, v, 0) D(0, 0, \alpha) \tag{2.5}$$

where

$$D(u, 0, 0) = D_1(u) = \exp[-iuN_1],$$
$$D(0, v, 0) = D_2(v) = \exp[-ivN_2],$$
$$D(0, 0, \alpha) = R(\alpha) = \exp[-i\alpha L_3].$$

We can expand the above formulas in power series. The expansion for $D(0, 0, \alpha)$ is a very familiar procedure for us, and the result is

$$R(\alpha) = \begin{bmatrix} \cos \alpha & -\sin \alpha & 0 & 0 \\ \sin \alpha & \cos \alpha & 0 & 0 \\ 0 & 0 & 1 & 0 \\ 0 & 0 & 0 & 1 \end{bmatrix}, \tag{2.6}$$

If we expand the $D(0, u, 0)$ and $D(0, 0, v)$

$$D_1(u) = 1 - iuN_1 - (uN_1)^2/2,$$
$$D_2(v) = 1 - ivN_2 - (vN_2)^2/2, \tag{2.7}$$

with

$$(N_1)^3 = (N_2)^3 = N_1(N_2)^2 = N_2(N_1)^2 = 0.$$

Therefore, the four-by-four matrices for $D_1(u)$ and $D_2(v)$ are quadratic in u and v respectively, and

$$D_1(u) = \begin{bmatrix} 1 & 0 & -u & u \\ 0 & 1 & 0 & 0 \\ u & 0 & 1 - u^2/2 & u^2/2 \\ u & 0 & -u^2/2 & 1 + u^2/2 \end{bmatrix}, \tag{2.8}$$

$$D_2(v) = \begin{bmatrix} 1 & 0 & 0 & 0 \\ 0 & 1 & -v & v \\ 0 & v & 1 - v^2/2 & v^2/2 \\ 0 & v & -v^2/2 & 1 + v^2/2 \end{bmatrix}. \tag{2.9}$$

The above D matrices commute with each other, and the product $D(u, v, 0) = D_1(u)\,D_2(v)$ becomes the $T(u, v)$ matrix in Equation (III.2.6).

The algebraic properties of the above D matrices have already been discussed in Section 2 of Chapter III. It was noted there that these matrices leave the four-momentum given in Equation (2.2) invariant. It was shown in Exercise 1 in Section 7 of Chapter III that each of the above matrices can be written as a product of one boost matrix and one rotation followed by another boost matrix. Therefore it is a transformation matrix belonging to the group of Lorentz transformations.

The above D matrices have the same algebraic property as that for the E matrices discussed in Section 1. Then, why do they look so different? In the case of the $O(3)$-like little group, the four-by-four matrices of the little group can be reduced to a block diagonal form consisting of the three-by-three rotation marix and one-by-one unit matrix. Is it then possible to reduce the D matrices to the extent that they can be directly compared with the three-by-three E matrices?

One major problem in bringing the D matrix to the form of the E matrix is that the D matrix is quadratic in the u and v variables. We thus have to get rid of these quadratic terms. In order to attack this problem, let us note that we are allowed to impose the Lorentz condition

$$\frac{\partial}{\partial x^\mu}(A^\mu(x)) = p^\mu A_\mu(x) = 0. \tag{2.10}$$

resulting in

$$A_3 = A_0.$$

Since the third and fourth components are identical, the N_1 and N_2 matrices of Equation (2.3) can be replaced respectively by

$$N_1 = \begin{bmatrix} 0 & 0 & 0 & 0 \\ 0 & 0 & 0 & 0 \\ i & 0 & 0 & 0 \\ i & 0 & 0 & 0 \end{bmatrix}, \quad N_2 = \begin{bmatrix} 0 & 0 & 0 & 0 \\ 0 & 0 & 0 & 0 \\ 0 & i & 0 & 0 \\ 0 & i & 0 & 0 \end{bmatrix}. \tag{2.11}$$

At the same time the D matrices become (Han et al., 1982c)

$$D(u, v, 0) = \begin{bmatrix} 1 & 0 & 0 & 0 \\ 0 & 1 & 0 & 0 \\ u & v & 1 & 0 \\ u & v & 0 & 1 \end{bmatrix}, \tag{2.12}$$

It is now clear from the expressions for L_3 of Equation (2.3) and for N_1 and N_2 of Equation (2.11), and from the above form of D that the fourth row and column are redundant. The matrices of Equations (2.3) and (2.11) can therefore be reduced further to

$$\bar{L}_3 = \begin{bmatrix} 0 & -i & 0 \\ i & 0 & 0 \\ 0 & 0 & 0 \end{bmatrix},$$

$$\bar{N}_1 = \begin{bmatrix} 0 & 0 & 0 \\ 0 & 0 & 0 \\ i & 0 & 0 \end{bmatrix}, \quad \bar{N}_2 = \begin{bmatrix} 0 & 0 & 0 \\ 0 & 0 & 0 \\ 0 & i & 0 \end{bmatrix}. \tag{2.13}$$

Thus

$$\bar{D}(u, v, 0) = \begin{bmatrix} 1 & 0 & 0 \\ 0 & 1 & 0 \\ u & v & 1 \end{bmatrix}, \tag{2.14}$$

The above L_3 is identiccal to that for the $SE(2)$ group given in Equation(1.4). However, $\bar{N}_1$ and $\bar{N}_2$ are the negative of the Hermitian conjugates of the corresponding three-by-three matrices in Equation (1.4). The commutation relations for the $SE(2)$ group remain unchanged under Hermitian conjugation or the sign change in the N_1 and N_2 operators. Consequently, the rotation matrix is the same as that for the $SE(2)$ group. However, the $\bar{D}(u, v, 0)$ matrix is related to $E(u, v, 0)$ by

$$\bar{D}(u, v, 0) = [E^{-1}(-u, -v, 0)]^\dagger. \tag{2.15}$$

We can also write the reduced matrix for the rotation. Then

$$\bar{D}(0, 0, \alpha) = E(0, 0, \alpha). \tag{2.16}$$

Since the rotation matrix is orthogonal,

$$\bar{D}(0, 0, \alpha) = [E^{-1}(0, 0, \alpha)]^\dagger. \tag{2.17}$$

The inverse matrices are multipled in the inverse order. So are the Hermitian conjugations. Thus

$$\bar{D}(0, 0, \alpha) = [E^{-1}(-u, -v, \alpha)]^\dagger, \tag{2.18}$$

and the algebraic property of the D matrix is identical to that of the E matrices for the $SE(2)$ group.

Consequently, all algebraic properties of the $SE(2)$ group are directly applicable to the $E(2)$-like little group for photons. For instance, the concept of equivalence class in $SE(2)$ is directly applicable to the u and v transformations in the little group (see Wightman, 1960). This transformation has

3. Transformation Properties of Photon Polarization Vectors

The use of scalar and vector potentials for electromagnetic fields plays an important role in all branches of physics. There are two important properties associated with these physical quantities. First, the electromagnetic fields derivable from these potentials are invariant under gauge transformations. Second, the scalar and vector potentials form a four-vector, and the electromagnetic field derivable from a Lorentz-transformed four vector potential is identical to that derivable from the Lorentz-transformed electromagnetic field tensor.

This means that we are free to choose a particular gauge condition to serve our convenience. The usual gauge condition for free electromagnetic waves is that the longitudinal and time-like components of the four-potential vanish in the observer's Lorentz frame. Do we use this condition only for convenience? The answer to this question is "No." In order that the photon states be diagonal in the helicity operator, it is essential that the non-transverse components vanish. This condition remains invariant in Lorentz frames which move along the direction of the photon momentum. However, the time-like component does not remain zero when we perform a Lorentz transformation along the direction perpendicular to the photon momentum.

With this point in mind, let us write down the photon wave functions diagonal in the helicity operator L_3. The four-vector A^μ given in Equation (2.1) should take the form

$$\varepsilon_\pm^\mu = (1, \pm i, 0, 0), \tag{3.1}$$

where ε_+^μ and ε_-^μ represent the photon states with positive and negative helicities respectively. In order that the system be diagonal in L_3, it is essential that the third and fourth components vanish. We shall call these four-component vectors the photon *polarization vectors* (Wightman, 1960). We can normalize the polarization vector of Equation (3.1) by dividing every element by $\sqrt{2}$. However, this procedure is trivial and appears unnecessary.

We obtain the above polarization vector by imposing the transversality condition

$$\nabla \cdot \mathbf{A}(x) = 0, \quad \mathbf{p} \cdot \mathbf{A} = 0, \quad \text{or} \quad A_3 = 0, \tag{3.2}$$

in addition to the Lorentz condition of Equation (2.10). These two conditions reduce the number of independent components in Equation (2.1) from four

Massless Particles

to two. Indeed, if we combine the above condition with the Lorentz condition of Equation (2.10),

$$A_3 = A_0 = 0. \tag{3.3}$$

Since the above codnition is so essential for the photon four-vector to be an eigenstate of the helicity operator, we shall call the combined effect of the Lorentz and transversality conditions the *helicity gauge*. The Lorentz condition is Lorentz-invariant. However, the transversality condition of Equation (3.2) is not Lorentz-invariant. Therefore we are faced with the question of whether helicity is a Lorentz-invariant concept in the case of photons.

With this point in mind, let us first examine the effect of $D(u, v, 0)$ transformations on the above polarization vectors. If we use the expression of Equation (2.12) and perform a straightforward calculation:

$$\begin{bmatrix} \zeta_{\pm 1} \\ \zeta_{\pm 2} \\ \zeta_{\pm 3} \\ \zeta_{\pm 0} \end{bmatrix} = \begin{bmatrix} 1 & 0 & 0 & 0 \\ 0 & 1 & 0 & 0 \\ u & v & 1 & 0 \\ u & v & 0 & 1 \end{bmatrix} \begin{bmatrix} 1 \\ \pm i \\ 0 \\ 0 \end{bmatrix}, \tag{3.4}$$

the result is (Weinberg, 1964b; Han and Kim, 1981a)

$$\zeta_{\pm}^{\mu} = D(u, v, 0)\, \varepsilon_{\pm}^{\mu} = \varepsilon_{\pm}^{\mu} + \left(\frac{u \pm iv}{\omega}\right) p^{\mu}$$

$$= (1, \pm i, u \pm iv, u \pm iv), \tag{3.5}$$

where p^{μ} is the energy-momentum four-vector given in Equation (2.2). Thus $D(u, v, 0)$ applied to the polarization vector results in the addition of a term which is proportional to the four-momentum. $D(u, v, 0)$ therefore performs a gauge transformation on ε^{μ}.

Because $D(u, v, 0)$ is a Lorentz transformation,

$$\zeta^{\mu}\zeta_{\mu} = \varepsilon^{\mu}\varepsilon_{\mu}, \tag{3.6}$$

and because $D(u, v, 0)$ leaves the four-momentum invariant,

$$p^{\mu}\zeta_{\mu} = p^{\mu}\varepsilon_{\mu} = 0. \tag{3.7}$$

The third component of ζ^{μ} is equal to its fourth component, but they do not vanish. If we calculate the electric and magnetic fields from this four-potential [Problem 2 in Section 7], then they are identical to those obtained from ε^{μ}. ζ^{μ} is indeed a gauge transformation of ε^{μ}.

The four-component vector ζ^{μ} is no longer an eigenstate of L_3, but is diagonal in

$$L_3(u, v) = D(u, v, 0)\, L_3\, D^{-1}(u, v, 0). \tag{3.8}$$

The rotations by the same angle generated by L_3 and $L_3(u, v)$ belong to the same eigenvalue class. the gauge transformation is therefore a transformation within an equivalence class (Wightman, 1960).

Let us next examine Lorentz transformations of the polarization vector ε^μ given in Equation (3.1). If we boost this vector along the z direction, it remains invariant. However, if we boost it along the x direction,

$$\begin{bmatrix} A'_{\pm 1} \\ A'_{\pm 2} \\ A'_{\pm 3} \\ A'_{\pm 0} \end{bmatrix} = \begin{bmatrix} \cosh \eta & 0 & 0 & \sinh \eta \\ 0 & 1 & 0 & 0 \\ 0 & 0 & 1 & 0 \\ \sinh \eta & 0 & 0 & \cosh \eta \end{bmatrix} \begin{bmatrix} 1 \\ \pm i \\ 0 \\ 0 \end{bmatrix}, \tag{3.9}$$

or

$$A'^\mu_\pm = (\cosh \eta, \pm i, 0, \sinh \eta). \tag{3.10}$$

Under this Lorentz boost, the momentum four-vector becomes

$$p'^\mu = (\omega \sinh \eta, 0, \omega, \omega \cosh \eta), \tag{3.11}$$

The four-vectors p'^μ and $A'^\mu_\pm$ satisfy the Lorentz condition:

$$p'^\mu A'_{\pm\mu} = 0. \tag{3.12}$$

However, the transversality condition is no longer satisfied:

$$\mathbf{p}' \cdot \mathbf{A}'_\pm = \omega \cosh \eta \sinh \eta, \tag{3.13}$$

which does not vanish. $\mathbf{A}'_\pm$ is not a helicity state.

On the other hand, if we boost the four vector $\zeta^\mu_\pm$ given in Equation (3.5) along the x direction:

$$\begin{bmatrix} \zeta'_{\pm 1} \\ \zeta'_{\pm 2} \\ \zeta'_{\pm 3} \\ \zeta'_{\pm 0} \end{bmatrix} = \begin{bmatrix} \cosh \eta & 0 & 0 & \sinh \eta \\ 0 & 1 & 0 & 0 \\ 0 & 0 & 1 & 0 \\ \sinh \eta & 0 & 0 & \cosh \eta \end{bmatrix} \begin{bmatrix} \zeta_{\pm 1} \\ \zeta_{\pm 2} \\ \zeta_{\pm 3} \\ \zeta_{\pm 0} \end{bmatrix}, \tag{3.14}$$

the result is (see Han and Kim, 1981a)

$$\zeta'_{\pm 1} = \cosh \eta + (u \pm iv) \sinh \eta,$$
$$\zeta'_{\pm 2} = \pm 1, \quad \zeta'_{\pm 3} = (u \pm iv), \tag{3.15}$$
$$\zeta'_{\pm 0} = \sinh \eta + (u \pm iv) \cosh \eta.$$

The Lorentz condition $p^\mu \zeta_{\pm\mu} = 0$ is maintained. However,

$$\mathbf{p}' \cdot \boldsymbol{\zeta}'_\pm = (\omega \cosh \eta) [\sinh \eta + (u \pm iv) \cosh \eta]. \tag{3.16}$$

Since u and v are gauge parameters which do not affect the observable quantities, we are led to examine whether they can be chosen in such a way that $\zeta'_{\pm 0}$ of Equation (3.15) and $\mathbf{p}' \cdot \boldsymbol{\zeta}'_{\pm}$ of Equation (3.16) vanish simultaneously. The answer to this question is definitely "Yes". Indeed, both of them vanish if

$$u \pm iv = -\tanh \eta. \tag{3.17}$$

Thus

$$u = -\tanh \eta, \quad \text{and} \quad v = 0. \tag{3.18}$$

The Lorentz boost of the polarization vector $\varepsilon^\mu_\pm$ along the x direction does not lead to another polarization vector. However, it is still possible to construct the new polarization vector $\varepsilon'^\mu_\pm$ by boosting $\varepsilon^\mu_\pm$ after the $D(u, v, 0)$ transformation whose parameters are determined by the transversality condition of Equation (3.17) or Equation (3.18). Again if we use the form of Equation (3.1), the multiplication of the gauge transformation matrix of Equation (3.4) and the boost matrix of Equation (3.14) leads to

$$\begin{bmatrix} \varepsilon'_{\pm 1} \\ \varepsilon'_{\pm 2} \\ \varepsilon'_{\pm 3} \\ 0 \end{bmatrix} = \begin{bmatrix} 1/\cosh \eta & 0 & 0 & \sinh \eta \\ 0 & 1 & 0 & 0 \\ -\tanh \eta & 0 & 1 & 0 \\ 0 & 0 & 0 & \cosh \eta \end{bmatrix} \begin{bmatrix} 1 \\ \pm i \\ 0 \\ 0 \end{bmatrix}, \tag{3.19}$$

Thus

$$\varepsilon'^\mu_\pm = (1/\cosh \eta, \pm i, -\tanh \eta, 0). \tag{3.20}$$

This polarization vector satisfies the transversality condition

$$\mathbf{p}' \cdot \boldsymbol{\varepsilon}'_\pm = 0. \tag{3.21}$$

In addition, the transformation of Equation (3.19) is a norm-preserving transformation:

$$|\varepsilon'_{\pm 1}|^2 + |\varepsilon'_{\pm 2}|^2 + |\varepsilon'_{\pm 3}|^2 = |\varepsilon_{\pm 1}|^2 + |\varepsilon_{\pm 2}|^2 + |\varepsilon_{\pm 3}|^2. \tag{3.22}$$

We can carry out a similar calculation when the system is boosted along the y direction. The gauge parameters in this case become

$$u = 0, \quad \text{and} \quad v = -\tanh \eta. \tag{3.23}$$

and the matrix which transforms the polarization vector becomes

$$\begin{bmatrix} 1 & 0 & 0 & 0 \\ 0 & 1/\cosh \eta & 0 & \sinh \eta \\ 0 & -\tanh \eta & 1 & 0 \\ 0 & 0 & 0 & \cosh \eta \end{bmatrix}. \tag{3.24}$$

We can also obtain this matrix by rotating the matrix of Equation (3.19) by 90°.

4. Unitary Transformation of Photon Polarization Vectors

We have seen in Section 3 that Lorentz boosts on photon polarization vectors are not in general helicity-preserving transformations. However, it is still possible to preserve the helicity gauge by performing a transformation of the $E(2)$-like little group before (or after) the boost. The net effect was a product of two non-unitary matrices. The resulting transformation matrix such as the one given in Equation (3.19) is not unitary.

However, the result given in Equation (3.22) is clearly that of a unitary transformation. Then there must be a way to transform the matrix of Equation (3.19) to a unitary matrix. Can this matrix be decomposed further into a product of a unitary matrix and a non-unitary matrix which does not change the photon polarization vector. For example, the boost matrix along the z direction is not unitary, but does not change the photon polarization vector of Equation (3.1).

In order to tackle this problem, let us use the notation $B_x(\eta)$ for the four-by-four matrix we used in Equations (3.9) and (3.14). Under this boost operation, the four momentum p became p'. However, in order to achieve the same purpose, we can consider a boost along the z direction followed by a rotation around the y axis as is shown in Figure 4.1. In this case, the boost and rotation matrices are (See Exercise 2 in Section 7 of Chapter III)

$$B_z(\eta) = \begin{bmatrix} 1 & 0 & 0 & 0 \\ 0 & 1 & 0 & 0 \\ 0 & 0 & \frac{1}{2}(1+\cosh^2 \eta)/\cosh \eta & \frac{\sinh \eta}{2} \tanh \eta \\ 0 & 0 & \frac{\sinh \eta}{2} \tanh \eta & \frac{1}{2}(1+\cosh^2 \eta)/\cosh \eta \end{bmatrix}, (4.1)$$

$$R_y(\eta) = \begin{bmatrix} 1/\cosh \eta & 0 & \tanh \eta & 0 \\ 0 & 1 & 0 & 0 \\ -\tanh \eta & 0 & 1/\cosh \eta & 0 \\ 0 & 0 & 0 & 1 \end{bmatrix}, \quad (4.2)$$

with

$$(1/\cosh \eta)^2 + (\tanh \eta)^2 = 1.$$

The application of the transformation $[R_y(\eta)B_z(\eta)]$ on the four-momentum gives the same effect as that of the application of $B_x(\eta)$. Indeed, the matrix

$$D_x(\eta) = [B_x(\eta)]^{-1} R_y(\eta) B_z(\eta) \qquad (4.3)$$

Massless Particles

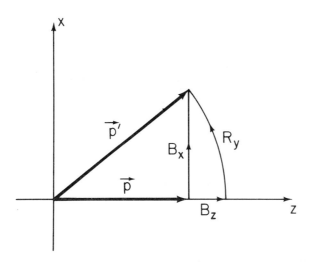

Fig. 4.1. Lorentz boost along the x direction. The four-momentum can be boosted either directly by B_x or through the rotation R_y preceded by B_z along the z direction. These operators produce two different four-vectors when appled to the polarization vector. However, they are connected by a gauge transformation.

leaves the four-momentum invariant, and is therefore an element of the $E(2)$-like little group for photons. The four-by-four matrix form for $D_x(\eta)$ is complicated, but it becomes very simple once the Lorentz condition of Equation (2.10) is taken into account. Since $B_z(\eta)$ leaves $\varepsilon_\pm^\mu$ invariant, we arrive at the conclusion that

$$\varepsilon_\pm^{\prime\mu} = R_y(\eta)\, \varepsilon_\pm^\mu. \tag{4.4}$$

Indeed, the Lorentz boost $B_x(\eta)$ on $\varepsilon_\pm^\mu$ preceded by the gauge transformation $D_x(\eta)$ leads to the pure rotation $R_y(\eta)$. This rotation is a finite-dimensional unitary transformation.

The above result raises the following delicate mathematical question. The non-compact group of Lorentz transformations does not have Abelian invariant subgroups, and therefore there cannot be finite-dimensional unitary representations of this group. However, this theorem does not apply to trivial one-dimensional representations. The point is that, if we impose the helicity gauge, the little group has only one degree of freedom which in this case is the rotation around the momentum. The transformation with one-dimensional internal space can be described by a trivial representation (Wigner, 1939 and 1962; Mackey, 1968; Lipsman, 1974).

It is possible to approach the above problem through the construction of infinite-dimensional unitary representations. We note that the $E(2)$-like little group for photons is non-compact, and its unitary representation is necessarily infinite-dimensional (Wigner, 1962). However, its translation-like subgroup governing gauge degrees of freedom is an Abelian invariant subgroup. For this reason, we can write

$$D(u, v, \alpha) = D(0, 0, \alpha) D(u', v', 0), \qquad (4.5)$$

where the gauge parameters u' and v' are determined from u, v, and α (Problem 4 in Section 7).

If the above operator is applied to the state which does not depend on the gauge parameters:

$$N_1|\text{state}\rangle = N_2|\text{state}\rangle = 0, \qquad (4.6)$$

then $D(u', v', 0)$ can be replaced by 1. The representation $D(u, v, \alpha)$ is simply that of the rotation around the momentum. This rotation is generated by the helicity operator. Its representation is that of $O(2)$ and is one-dimensional. Thus, again, we end up with the trivial representation.

Let us summarize what we have done so far in this chapter. The electromagnetic four-potential is commonly used in quantum electrodynamics (QED) as well as in classical physics. The usual gauge condition is either the Lorentz gauge of Equation (2.9) or the Coulomb gauge of Equation (3.2), but not both. In either case, photons are not in eigenstate of the helicity operator.

If the photon is in a helicity eigenstate, it can be represented by a *trivial representation* of a unitary representation, or equivalently by the four-potential satisfying both the Lorentz and Coulomb gauge conditions. This situation can be summarized in the Table 4.1.

TABLE 4.1
Representations of the $E(2)$-like little group for photons. If we freeze gauge degrees of freedom, both unitary infinite-dimensional and non-unitary finite-dimensional representations become the one dimensional trivial representation.

Reps. of the little group	Free photons with definite helicity	Coupled photons in QED
Unitary infinite-dimensional	One-dimensional trivial representation	
Non-unitary finite-dimensional	$\frac{\partial}{\partial x_\mu} A_\mu = \nabla \cdot \mathbf{A} = 0$	Lorentz or Coulomb gauge

5. Massless Particles with Spin 1/2

As we noted before, for a massless particle moving along the z direction, the little group is generated by

$$N_1 = K_1 - J_2, \quad N_2 = K_2 + J_1, \quad J_3, \qquad (5.1)$$

where the above two-by-two matrices were introduced in Section 2 of Chapter IV. The N operators generate gauge transformations in the case of photons. What is then the physical significance of these operators for massless particles with spin-1/2?

The commutation relations for the generators of the Lorentz group remain invariant under the sign change in K_i. We can determine the sign of K_i unambiguously when they are applicable to the space-time coordinate variables and photon four-vectors. In the $SL(2, c)$ regime, we can choose $S_i = i/2\,\sigma_i$ as the generators of rotations. However, the boost generators can take two different signs: $K_i = \pm i\tfrac{1}{2}\sigma_i$. We thus have to consider both $N_i^{(+)}$ and $N_i^{(-)}$:

$$N_1^{(+)} = \begin{bmatrix} 0 & i \\ 0 & 0 \end{bmatrix}, \quad N_2^{(+)} = \begin{bmatrix} 0 & 1 \\ 0 & 0 \end{bmatrix}. \tag{5.2}$$

We can obtain $N_1^{(-)}$ and $N_2^{(-)}$ by taking the Hermitian conjugation of the above expressions respectively. Then the transformation matrices become

$$D^{(+)}(u, v) = \exp(-i[uN_1^{(+)} + vN_2^{(+)}]) = \begin{bmatrix} 1 & u - iv \\ 0 & 1 \end{bmatrix},$$

$$D^{(-)}(u, v) = \exp(-i[uN_1^{(-)} + vN_2^{(-)}]) = \begin{bmatrix} 1 & 0 \\ -u - iv & 1 \end{bmatrix}. \tag{5.3}$$

Since these matrices correspond to gauge transformation matrices for photons, we shall hereafter call them the gauge transformation matrices in the $SL(2, c)$ regime (Wigner, 1939; Han et al., 1982b).

There are two sets of spinors in $SL(2, c)$. For spinors whose boosts are generated by $K_i = i/2\,\sigma_i$, we shall use the usual Pauli notation α and β for positive and negative helicity states respectively. For those whose boosts are generated by $K_i = -i/2\,\sigma_i$, we shall use $\dot\alpha$ and $\dot\beta$. Then, the spinors are gauge-invariant in the sense that

$$D^{(+)}(u, v)\,\alpha = \alpha, \quad D^{(-)}(u, v)\,\dot\beta = \dot\beta. \tag{5.4}$$

On the other hand, the SL(2, c) spinors are gauge-dependent in the sense that

$$D^{(+)}(u, v)\,\beta = \beta + (u - iv)\,\alpha, \quad D^{(-)}(u, v)\,\dot\alpha = \dot\alpha - (u + iv)\,\dot\beta. \tag{5.5}$$

The gauge-invariant spinors of Equation (5.4) appear as polarized neutrinos in the real world. However, where do the above gauge-dependent spinors stand in the physics of spin-1/2 particles? Are they really responsible for the gauge dependence of electromagnetic four-potentials when we construct a four-vector by taking a bilinear combination of spinors?

As we did in Equations (4.1) and (4.2) of Chapter IV, we can construct the following unit vectors in the Minkowskian space by taking the direct products of two $SL(2, c)$ spinors:

$$-\alpha\dot{\alpha} = (1, i, 0, 0), \quad \beta\dot{\beta} = (1, -i, 0, 0),$$
$$\alpha\beta = (0, 0, 1, 1), \quad \beta\dot{\alpha} = (0, 0, 1, -1). \tag{5.6}$$

For $D(u, v)$ to be consistent with Equation (5.6), we should choose

$$D(u, v) = D^{(+)}(u, v) \, D^{(-)}(u, v), \tag{5.7}$$

where $D^{(+)}$ and $D^{(-)}$ are applicable to the first and second spinors of Equation (5.6) respectively. Then

$$D(u, v)(-\alpha\dot{\alpha}) = -\alpha\dot{\alpha} + (u + iv) \, \alpha\beta,$$
$$D(u, v) \, \beta\dot{\beta} = \beta\dot{\beta} + (u - iv) \, \alpha\beta, \tag{5.8}$$
$$D(u, v) \, \alpha\beta = \alpha\beta.$$

The plane-wave photon four-potential does not depend on $\beta\dot{\alpha}$. The first two equations of the above expression correspond to the gauge transformations on the photon polarization vectors. The third equation corresponds to the effect of the D transformation on the four-momentum, confirming the fact that $D(u, v)$ is an element of the little group.

The remaining question is how the above analysis can be translated into the language of the Dirac equation. We shall study this problem in Chapter VIII.

6. Harmonic Oscillator Wave Functions for Massless Composite Particles

Indeed, the study of the internal space-time symmetry of the composite particles in the harmonic oscillator regime is relegated to finding suitable separable coordinate systems. In Chapter V, we were concerned with the representations of the $O(3)$-like little group for massive particles. The little group for massless particles is locally isomorphic to $SE(2)$, and its generators are different from those for the $O(3)$-like little group for massive particles. It is therefore of interest to see whether there exists a separable coordinate system for studying internal space-time symmetries of massless composite particles.

In one of the thirty four coordinate systems discussed by Kalnins and Miller (1977), three of the space-time coordinate variables are conjugate to the generators of the $SE(2)$-like little group for the massless composite particle. We can therefore construct representations of the $SE(2)$-like little group by solving the differential equation using this Kalnins-Miller coordinate system.

Massless Particles

Let us go back to the oscillator differential equation discussed in Chapter V. If the composite particle is massless, this means that m_0 and λ given in Section 1 of Chapter V are such that p^2 is zero:

$$p^2 = m_0^2 + \lambda = 0. \tag{6.1}$$

We assume without loss of generality that the momentum of the massless composite particle is in the z direction. Then the little group is generated by L_3, N_1 and N_2, where

$$N_1 = K_1 + L_2$$
$$= -i\left(x\frac{\partial}{\partial t} + t\frac{\partial}{\partial x}\right) - i\left(z\frac{\partial}{\partial x} - x\frac{\partial}{\partial z}\right), \tag{6.2}$$

$$N_2 = K_2 - L_1$$
$$= -i\left(y\frac{\partial}{\partial t} + t\frac{\partial}{\partial y}\right) + i\left(z\frac{\partial}{\partial y} - y\frac{\partial}{\partial z}\right).$$

These generators satisfy the commutation relations for the generators of the $SE(2)$ group given in Equation (2.4); L_3 in this case is the helicity operator. In constructing representations, we should note that there are two maximal commuting sets of operators in the Lie algebra. They are (Han et al., 1982a and 1983a)

(a) $N^2, N_1, N_2,$
(b) $N^2, L_3,$ \hfill (6.3)

where $N^2 = N_1^2 + N_2^2$. Because we are interested in states with definite helicities, we have to construct wave functions which are diagonal in L_3 and N^2.

In order to construct solutions of the oscillator wave equation, we have to make a judicious choice of the coordinate system in which the differential equation is separable. We shall call this the Kalnins-Miller coordinate system.

If x is time-like with positive t, the Kalnins-Miller coordinate variables ρ, ξ, α, and ϕ are related to x, y, z, and t by

$$x = \rho\, e^{-\alpha}\, \xi \cos\phi,$$
$$y = \rho\, e^{-\alpha}\, \xi \sin\phi,$$
$$z = (\rho/2)[e^\alpha + (\xi^2 - 1)\, e^{-\alpha}], \tag{6.4}$$
$$t = (\rho/2)[e^\alpha + (\xi^2 + 1)\, e^{-\alpha}].$$

These equations can also be written as

$$\rho = (t^2 - z^2 - r^2)^{1/2},$$
$$\xi = r/(t - z),$$
$$\alpha = -\ln\{(t - z)/(t^2 - z^2 - r^2)^{1/2}\}, \quad (6.5)$$
$$\phi = \tan^{-1}(y/x).$$

where

$$r = (x^2 + y^2)^{1/2}.$$

In terms of the Kalnins-Miller variables, L_3, N_1, N_2 and N^2 take the form

$$L_3 = -i\partial/\partial\phi,$$
$$N_1 = -i\partial/\partial\xi_1,$$
$$N_2 = -i\partial/\partial\xi_2, \quad (6.6)$$
$$N^2 = -(\partial/\partial\xi)^2 - (1/\xi)(\partial/\partial\xi) - (1/\xi^2)(\partial/\partial\phi)^2,$$
$$= -\{(\partial/\partial\xi_1)^2 + (\partial/\partial\xi_2)^2\},$$

where

$$\xi_1 = \xi \cos\phi, \quad \xi_2 = \xi \sin\phi.$$

The oscillator differential equation can then be written as

$$\left(\frac{1}{\rho}\right)^3 \frac{\partial}{\partial\rho}\left[\rho^3 \frac{\partial}{\partial\rho} \psi(x)\right] + \left(\frac{1}{\rho}\right)^2 \left[\left(\frac{\partial}{\partial\alpha}\right)^2 - 2\left(\frac{\partial}{\partial\alpha}\right)\right]\psi(x)$$

$$- \left[\left(\frac{1}{\rho}\right)^2 e^{-2\alpha} N^2 + \rho^2\right]\psi(x) = 2\lambda\,\psi(x). \quad (6.7)$$

In order to separate this differential equation, let us write $\psi(x)$ in the form

$$\psi(x) = G(\rho, \alpha) F(\xi, \phi). \quad (6.8)$$

If $F(\xi, \phi)$ satisfies the eigenvalue equation

$$N^2 F(\xi, \phi) = b^2 F(\xi, \phi), \quad (6.9)$$

then, $G(\rho, \alpha)$ should satisfy the differential equation

$$\left(\frac{1}{\rho}\right)^3 \frac{\partial}{\partial\rho}\left(\rho^3 \frac{\partial}{\partial\rho} G\right) + \left(\frac{1}{\rho}\right)^2 \left\{\left(\frac{\partial}{\partial\alpha}\right)^2 - 2\left(\frac{\partial}{\partial\alpha}\right)\right.$$

$$\left. - e^{-2\alpha} b^2\right\} G - \rho^2 G = 2\lambda G. \quad (6.10)$$

The eigenvalue equation of Equation (6.9) is a two-dimensional Helmholtz equation if b^2 does not vanish. It is a Laplace equation if $b^2 = 0$. As is discussed fully in Section 1, b^2 has to vanish in order that the system be physically interesting. The solution then becomes

$$F(\xi, \phi) = \xi^m \exp(\pm im\phi), \qquad (6.11)$$

where m is an integer and is the magnitude of the angular momentum. Since $b^2 = 0$, the differential equation of Equation (6.10) becomes

$$\left(\frac{1}{\rho}\right)^3 \frac{\partial}{\partial \rho}\left(\rho^3 \frac{\partial}{\partial \rho} G\right) + \left(\frac{1}{\rho}\right)^2 \left[\left(\frac{\partial}{\partial \alpha}\right)^2 - 2\left(\frac{\partial}{\partial \alpha}\right)\right] G - \rho^2 G = 2\lambda G. \qquad (6.12)$$

The solution of the above differential equation will then take the form

$$G_{\mu n}(\rho, \alpha) = [\rho^n \exp(-\rho^2/2)] L_\mu^{(n+1)}(\rho^2) A_n^{(\pm)}(\alpha), \qquad (6.13)$$

where

$$A_n^{(+)}(\alpha) = \exp[(n+2)\alpha],$$
$$A_n^{(-)}(\alpha) = \exp(-n\alpha).$$

$L_\mu^{(n+1)}(\rho^2)$ is the associated Laguerre function. The eigenvalue in Equation (6.12) is

$$\lambda = -(n + 2\mu + 1), \qquad (6.14)$$

where n and μ take integer values. In order that the composite particle be massless, the above eigenvalue and m_0^2 should satisfy the condition for massless particles:

$$p^2 = m_0^2 + \lambda = 0. \qquad (6.15)$$

We have considered so far only the case where x is time-like with positive values of t. If t is negative, we can reverse the sign of the Cartesian coordinate variables given in Equations (6.4) and (6.5). If x is a space-like vector, z and t of Equation (6.4) have to be modified to

$$z = (\rho/2)[e^\alpha - (\xi^2 - 1)e^{-\alpha}],$$
$$t = (\rho/2)[e^\alpha - (\xi^2 + 1)e^{-\alpha}]. \qquad (6.16)$$

Consequently, three of the equations in Equation (6.5) are modified to

$$\rho = (r^2 + z^2 - t^2)^{1/2},$$
$$\xi = r/(z-t), \qquad (6.17)$$
$$\alpha = -\ln\{\rho/(z-t)\}.$$

The process of separating and solving the differential equation is the same as in the case of time-like region. We can use the form of Equation (6.11) for $F(\xi, \phi)$, and Equation (6.13) for $G(\rho, \alpha)$. However, the eigenvalue λ in this case takes the values:

$$\lambda = n + 2\mu + 1. \tag{6.18}$$

It is important that the masslessness condition of Equation (6.15) be satisfied for the above values of λ. Since λ's for the time-like and space-like regions have opposite signs, m_0^2 will also have different signs. This does not cause any conceptual difficulty, because the time-like region never mixes with the space-like region under Poincaré transformations.

In studying space-time symmetries of the solution of the wave equation obtained in Section 1, we note that the internal wave function $\psi(x)$ is a product of $G(\rho, \alpha)$ and $F(\xi, \phi)$, as is given in Equation (6.8). This allows us to deal with F and G separately.

Let us first discuss the F function. It is not difficult to see that ϕ in Equation (6.11) is the angle variable specifying the rotation around the z axis with angular momentum $\pm m$. This is known as the helicity for the massless particle. Since physically observable states are expected to be helicity eigenstates, they are invariant under the rotation around the z axis.

In terms of the ξ_1 and ξ_2 variables defined in Equation (6.6), F can be written as

$$F(\xi, \phi) = F(\xi_1, \xi_2) = (\xi_1 \pm i\xi_2)^m. \tag{6.19}$$

with

$$\xi_1 = x/(t-z), \quad \xi_2 = y/(t-z),$$

for the time-like region. The operators N_1 and N_2 given in Equation (6.2) now generate translations in the $\xi_1 \xi_2$ plane. Since both of these "translation" operators commute with N^2, the differential equation of Equation (6.7) is invariant under this transformation. We can replace ξ_1 and ξ_2 in $F(\xi_1, \xi_2)$ of Equation (6.19) by ξ_1' and ξ_2' respectively, where

$$\xi_1' = \xi_1 - \nu_1,$$

$$\xi_2' = \xi_2 - \nu_2. \tag{6.20}$$

without changing the differential equation.

As was shown in Sections 2 and 3, the above-mentioned N_1 and N_2 transformations are equivalent to gauge transformations. Then what is this gauge transformation in terms of the conventional space-time variables? The "translation" of Equation (6.20) causes the following changes in the ξ and ϕ variables.

$$\xi \to \xi' = [(\xi_1 - \nu_1)^2 + (\xi_2 - \nu_2)^2]^{1/2},$$

$$\phi \to \phi' = \tan^{-1}[(\xi_2 - \nu_2)/(\xi_1 - \nu_1)]. \tag{6.21}$$

Another way to interpret the above transformation is to regard Equation (6.19) as a rotation around the origin in the $\xi_1 \xi_2$ plane. Then the transformation of Equation (6.21) shifts the center of rotation from the origin to the coordinate point (v_1, v_2).

In order to see the effect of the N_1 and N_2 transformations in terms of the Cartesian space-time variables, let us write Equation (6.4) for the time-like region as

$$r = \rho \xi \, e^{-\alpha},$$
$$y/x = \tan \phi,$$
$$t + z = \rho [e^\alpha + \xi^2 \, e^{-\alpha}], \quad (6.22)$$
$$z - t = -\rho \, e^{-\alpha}.$$

It is apparent that the $(t - z)$ variable remains invariant under the N transformation. The effects of Equation (6.10) or (6.11) on other variables are

$$y'/x' = \tan \phi',$$
$$r'/r = \xi'/\xi, \quad (6.23)$$
$$(z' + t')/(z + t) = (e^{2\alpha} + \xi'^2)/(e^{2\alpha} + \xi^2).$$

The variable $\rho = (t^2 - z^2 - r^2)^{1/2}$ is a N-invariant quantity.

Since $(t - z)$ is invariant under the N_1 and N_2 transformations, it is clear from Equation (6.19) that the x and y coordinate variables are directly proportional to ξ_1 and ξ_2 respectively. Indeed, for $b^2 = 0$, the differential equation given in Equation (6.9) can be written as

$$[(\partial/\partial x)^2 + (\partial/\partial y)^2] \, F(x, y) = 0, \quad (6.24)$$

with the solution

$$F(x, y) = (x \pm iy)^m.$$

The mathematics of this form is quite familiar to us, and does not require any further explanation. The point is that the N transformation parameters are now directly related to the x and y coordinate variables, and the spin of the massless composite particle is indeed due to the above orbit-like form. The N transformation in this case is a translation of the rotation axis from the origin to another point in the xy plane.

As for the normalization of $F(\xi, \phi)$, the ϕ dependence is just like the case of hydrogen atom. The Hilbert space and the normalization of the wave function associated with this variable are well known. The ξ dependence is not normalizable, and there is no Hilbert space associated with this variable. As was noted before, this is due to the fact that the N transformation is not measurable.

Let us next disscuss properties of the $G(\rho, \alpha)$ function given in Equation

(6.13). This function is a product of two separate functions. The ρ dependence is normalizable, and the wave function is concentrated within a hyperbolic region near the light cones. On the other hand, the α-dependence, which measures the $(t - z)$ variable for fixed ρ, is not normalizable. However, this does not introduce any additional difficulty to the overall wave function which is not normalizable due to the non-observability of N transformations.

The above discussion has so far been restricted to x in the forward light cone. By changing the sign of the Cartesian variables given in Equation (6.4), we can give the same reasoning for the backward light cone. By replacing the z and t variables by those given in Equation (6.16), we can give a similar treatment for the space-like region.

In this section, we have studied in detail the part of the solutions of the oscillator equation for massless particles containing the space-time symmetries of the $E(2)$-like little group. This part is independent of the form of potential.

Unlike the case of massive composite particles, wave functions are not normalizable. However, this should not alarm us. The transverse coordinates in this case are proportional to the parameters of the N transformation. According to the discussions given in Sections 2, 3, and 5, the N transformation can be identified as a gauge transformation which is not observable. In the case of massless composite particles, the fact that the N transformation is not observable is translated into the lack of a Hilbert space associated with the transverse coordinate variables.

We do not have answers to the fundamental question of whether massless composite particles exist in nature, or whether the existing massless particles such as photons and gravitons are ultimately composite. Yet, it is of interest to note that there are solutions of the harmonic oscillator differential equation which allow us to study the $E(2)$-like symmetry for massless particles in terms of the conventional space-time variables.

7. Exercises and Problems

Exercise 1. In Section 1, we noted that the two-dimensional Helmholtz equation has solutions with vanishing k^2 as well as with non-vanishing k^2. If k^2 does not vanish, the representation is unitary and infinite-dimensional. On the other hand, if $k^2 = 0$, the representation is finite-dimensional and non-unitary. Explain these.

The fact that the representation is finite-dimensional and non-unitary for $k^2 = 0$ has been seen in Section 1. If k^2 does not vanish, the solution of the Helmholtz equation is

$$\psi(x, y) = e^{im\phi} J_m(\mathrm{k}r). \tag{7.1}$$

Unlike the solutions of Laplace's equation which describe the finite-dimensional representations, the above solutions form a complete orthonormal basis for the two-dimensional space.

The rotation operation on this function is trivial. However, if we translate the origin of the coordinate system, so that (See Inonu and Wigner, 1953)

$$x' = x - u, \quad \text{and} \quad y' = y - v.$$

Then

$$e^{im\phi} J(kr) = \sum_{m'=-\infty}^{\infty} A_{m'}^{m} e^{im'\phi'} J_{m'}(kr'). \tag{7.2}$$

The summation contains an infinite number of terms. Thus the form given in Equation (7.1) is the basis for the infinite-dimensional unitary representation. We can use the Bessel function identity (Miller, 1968):

$$e^{im\phi} J_m(kr) = \sum_{m'=-\infty}^{\infty} (e^{-i(m-m')\phi''} J_{m-m'}(kr'')) e^{im\phi'} J_{m'}(kr'), \tag{7.3}$$

with

$$u + iv = r'' e^{i\phi''},$$

to determine the coefficient $A_{m'}^{m}$ in Equation (7.2).

Exercise 2. Show that the D matrices for massless spin-1/2 particles given in Equations (5.3) can also be obtained from the combined effect of boost and rotation described for photons in Section 4.

Let us start with the boost along the x axis:

$$B_x^{(\pm)}(\eta) = \begin{bmatrix} \cosh(\eta/2) & \pm\sinh(\eta/2) \\ \pm\sinh(\eta/2) & \cosh(\eta/2) \end{bmatrix}. \tag{7.4}$$

The rotation around the y axis takes the form

$$R_y(\eta) = [1/\cosh(\eta)]^{1/2} \begin{bmatrix} \cosh(\eta/2) & -\sinh(\eta/2) \\ \sinh(\eta/2) & \cosh(\eta/2) \end{bmatrix}. \tag{7.5}$$

Finally, the boost along the z axis is

$$B_z^{(\pm)}(\eta) = \begin{bmatrix} [\cosh \eta]^{\pm 1/2} & 0 \\ 0 & [\cosh \eta]^{\mp 1/2} \end{bmatrix}. \tag{7.6}$$

We can now calculate

$$D_x^{(\pm)}(\eta) = [B_x^{(\pm)}(\eta)]^{-1} R_y(\eta) B_z^{(\pm)}(\eta). \tag{7.7}$$

After matrix multiplications, we arrive at

$$D_x^{(+)}(\eta) = \begin{bmatrix} 1 & -\tanh \eta \\ 0 & 1 \end{bmatrix},$$

$$D_x^{(-)}(\eta) = \begin{bmatrix} 1 & 0 \\ \tanh \eta & 1 \end{bmatrix},$$
(7.8)

which are gauge transformation matrices given in Equation (5.3).

Exercise 3. Show that a Lorentz boost on the D-invariant spinor is a helicity preserving transformation, while the helicity is not preserved for a Lorentz boost of a spinor which is not D-invariant.

In order to see the effect of these transformations, let us apply the boost operator $B_x^{(+)}(\eta)$ of Equation (7.4) on a right-handed massless particle moving along the z direction. The resulting spinor is

$$B_x^{(+)}(\eta) \, \alpha = \begin{bmatrix} \cosh(\eta/2) \\ \sinh(\eta/2) \end{bmatrix}$$

$$= [\cosh(\eta/2)]^{1/2} \, \alpha',$$
(7.9)

where α' is the *normalized* positive-helicity spinor along the momentum **p**'. Indeed, when applied to α, $B_x^{(+)}(\eta)$ is a helicity-preserving transformation. However, this is not a unitary transformation.

On the other hand, if we apply the same boost operator on β, the result is

$$B_x^{(+)}(\eta) \, \beta = \begin{bmatrix} \sinh(\eta/2) \\ \cosh(\eta/2) \end{bmatrix}.$$
(7.10)

This spinor is not orthogonal to that of Equation (7.9), and therefore does not represent the negative helicity state. When applied to β, $B_x^{(+)}(\eta)$ is not a helicity-preserving transformation.

The question then is whether there is a transformation which will preserve the negative helicity state. The answer to this question is "yes." One way is to use $B_x^{(-)}(\eta)$, so that

$$B_x^{(-)}(\eta) \, \beta = \begin{bmatrix} -\sinh(\eta/2) \\ \cosh(\eta/2) \end{bmatrix},$$

$$= [\cosh(\eta/2)]^{1/2} \, \beta'.$$
(7.11)

Another approach to this problem is to perform the D transformation before applying $B_x^{(+)}(\eta)$, in analogy to what we did in Section 5. Indeed,

$$B_x^{(+)}(\eta) \, D_x^{(+)}(\eta) \, \beta = [1/\cosh \eta]^{1/2} \, \beta'.$$
(7.12)

Since $D_x^{(+)}(\eta)$ leaves α unchanged, the effect of $[B_x^{(+)}(\eta)\, D_x^{(+)}(\eta)]$ is the helicity preserving transformation applicable to both α and β.

This is indeed the place where the D matrices play the decisive role. As in the case of spin 1, the boost preceded by the D transformation produces the desired spinor:

$$B_x^{(+)}(\eta)\, D_x^{(+)}(\eta)\, \alpha = [\cosh \eta]^{1/2}\, \alpha'. \tag{7.13}$$

The transformation $B_x^{(+)}(\eta)\, D_x^{(+)}(\eta)$ is therefore a helicity preserving transformation for both helicity states. It is not a unitary transformation because of the factors $[\cosh \eta]^{\pm 1/2}$ in Equations (7.12) and (7.13).

This lack of unitarity should not alarm us. Unlike the case of massless particles with spin 1, the helicity-preserving Lorentz boost along the direction of the momentum is not a unitary transformation in the $SL(2, c)$ regime. We are in fact quite familiar with this in the Dirac equation, and we know how to take care of the problem.

Problem 1. Show that Equation (7.1) becomes a transformation of a finite-dimensional non-unitary representation in the $k \to 0$ limit. Calculate the coefficient $A_{mm'}^m$ in this limit using the form

$$J_m(kr) \to \frac{1}{m!}\left(\frac{kr}{2}\right)^m. \tag{7.14}$$

Problem 2. Calculate the electric and magnetic field from the four-vectors of Equation (3.1) and Equation (3.16), and show that these two different four-vectors give the same electric and magnetic fields. See Han and Kim (1981a).

Problem 3. Find solutions of the differential equation given in Equation (6.10) when b^2 does not vanish. Do they form a basis for an infinite-dimensional unitary representation of the $E(2)$-like little group? See Kalnins and Miller (1977).

Problem 4. Because the translation-like subgroup of the $E(2)$-like little group is an invariant subgroup, Equation (4.5) is possible. Calculate the parameters u' and v' in terms of u, v and α.

Problem 5. Consider the solutions of the differential equation of Equation (1.8) which are diagonal in the translation operators. They are not diagonal in the rotation operator, but can be expanded in terms of eigenfunctions of the rotation operator. Explain Equation (4.5) in terms of this expansion.

Problem 6. Throughout this chapter, we concentrated our effort on the electromagnetic four-potentials, and did not discuss the electromagnetic fields. However, since they are derivable from the potentials by differentiation, and since they are independent of gauge parameters, their transformation property is expected to be simpler than that of the four-potential. Is this true? Work out the Lorentz boost along the z and x directions, assuming that the photon propagates along the z direction. See Weinberg (1964a). See also discussions in Chapter X.

Problem 7. It was noted in Chapter IV that the $E(2)$-invariance condition alone leads to the solutions of the Dirac equation for massless particle. Show that the same invariance requirement will lead to gauge-invariant solutions of Maxwell's equations (Weinberg, 1964a).

Problem 8. What is the one-dimensional analog of the transition from Equation (1.8) to Equation (1.19) in the $k^2 \to 0$ limit? See Bowen and Coster (1981).

Problem 9. The three-dimensional Euclidean space is the space in which we conduct our daily life. From the group theoretical point of view, $E(3)$ is a semi-direct product of the three-dimensional rotation and translation groups. Work out the irreducible representations of $E(3)$. See Rno (1985).

Chapter VIII

Group Contractions

Let us summarize what we have learned so far on massive and massless particles.

For a massive relativistic particle, there exists a Lorentz frame in which the particle is at rest. In this Lorentz frame, we can define its intrinsic angular momentum. In covariant language, it is said that the internal space-time symmetry group for massive relativistic particles is locally isomorphic to the three-dimensional rotation group. This $O(3)$-like internal space-time symmetry group or "little group" is generated by the three angular momentum operators in the Lorentz frame in which the particle is at rest.

On the other hand, there are no Lorentz frames in which massless particles are at rest. The internal space-time symmetry group for massless particles is locally isomorphic to the two-dimensional Euclidean group. The content of this isomorphism has been discussed in detail in Chapter VII. The $E(2)$-like little group for massless particles is generated by the helicity operator and the two generators of gauge transformations.

We are quite familiar with the doctrine that a massive particle moving with velocity very close to that of light should appear like a massless particle, especially in view of Einstein's energy-momentum relation:

$$E = [P^2 + M^2]^{1/2}, \tag{1}$$

where E, P and M are the energy, momentum, and mass of the free particles respectively. This doctrine is incomplete without the proof that the internal space-time symmetry of massless particles is a limiting case of that for massive particles.

Since we know how the $O(3)$-like little group for massive particles and the $E(2)$-like little group for massless particles correspond to $O(3)$ and $E(2)$ respectively, we can study the above-mentioned problem by looking closely at how the $E(2)$ group can be regarded as a limiting case of the $O(3)$ group. This particular limiting procedure is one of the examples of the mathematical procedure known as *group contraction* (Segal, 1951; Inonu and Wigner, 1953; Talman, 1968).

Although the mathematics of group contraction is not widely discussed in the established physics curriculum, we very often think of $E(2)$ as a limiting case of $O(3)$. For instance, when we commute from home to school, we are making transformations within a two-dimensional Euclidean space. However, when we travel on the surface of the earth, we know that we are performing rotations. We then wonder how the motions on the $E(2)$ plane can be reconciled with those on the spherical surface.

Another example of group contraction is our belief that a Lorentz transformation becomes a Galilei transformation in the limit of small velocity compared with that of light. Yet, another example is our experience in high-energy scattering processes that, while it is convenient to use the spherical partial waves for analyzing the data in the low-energy region, it is more convenient to use the Bessel-function expansion in the high-energy limit. We shall discuss also these topics in this chapter.

However, our main interest is in the internal space-time symmetries of massive and massless particles. Because a massless particle has one rotational degree of freedom and two gauge degrees of freedom instead of three rotational degrees of freedom for massive particles, and because we firmly believe that the massless particle is an infinite-momentum/zero-mass limit of the massive particles, we have to prove that the gauge transformations are Lorentz-boosted rotations.

In Section 1, we study the geometry of the contraction of $SO(3)$ to $SE(2)$ using a spherical surface. It is shown that this contraction procedure is a flat surface approximation on a sphere with a large radius. Section 2 contains the proof that gauge transformations are Lorentz-boosted rotations in the infinite-momentum/zero-mass limit. In Section 3, the same limiting procedure is applied to the Dirac equation.

In spite of its importance in understanding physics, finite-dimensional representations of the $E(2)$ group have never been systematically discussed in the literature. Thus, in Section 4, we use the group contraction procedure to convert the spherical harmonics into the $E(2)$ harmonics which are essentially solutions of Laplace's equation. Massless particles with spin 2 are discussed in Section 5 as an application of one of the finite-dimensional representations.

In Section 6, the Galilei group is discussed as a contraction of the Lorentz group. As a physical application of the infinite-dimensional unitary representation, we discuss in Section 7 a description of high-energy scattering processes. In Section 8, we discuss other related examples in the form of exercises and problems.

1. SE(2) Group as a Contraction of SO(3)

We studied in Chapter VII the $SE(2)$ group and its connection with the $E(2)$-like little group for massless particles. We can still achieve a better understanding of these groups by studying their connection with the $SO(3)$ group and the $O(3)$-like little group for massless particles. We are particularly interested in studying the $SE(2)$ group as a limiting case of $O(3)$. This procedure is known to physicists as the contraction of $SO(3)$ to $SE(2)$ (Inonu and Wigner, 1953).

This approach starts from the coordinate transformation. One way to deal with this problem is to consider a sphere with large radius, and a small area around the north pole (Talman, 1968; Gilmore, 1974). This area would appear like a flat surface. We can then make Euclidean transformations on this surface, consisting of translations along the x and y directions and rotations around any point within this area. Strictly speaking, however, these Euclidean transformations are $SO(3)$ rotations around the x axis, y axis, and around the axis which makes a very small angle with the z axis.

Let us start with the familiar $SO(3)$ rotation operators which can be written in the form

$$\exp[-i(\xi_1 L_1 + \xi_2 L_2)] \exp(-i\phi L_3). \tag{1.1}$$

L_3 generates rotations around the north pole. L_2 takes the form

$$L_2 = -i\left(z\frac{\partial}{\partial x} - x\frac{\partial}{\partial z}\right). \tag{1.2}$$

Therefore, for large values of the radius R, $z = R$, and

$$N_1 = -i\frac{\partial}{\partial x} = \left(\frac{1}{R}\right) L_2. \tag{1.3}$$

If we rotate the system by angle ξ_2 around the y axis, the resulting translation on the $SE(2)$ plane is

$$u = \xi_2 R \quad \text{or} \quad \xi_2 = \frac{u}{R}. \tag{1.4}$$

Likewise

$$N_2 = -\left(\frac{1}{R}\right) L_1, \tag{1.5}$$

with

$$\xi_1 = \frac{v}{R}.$$

If we write the commutation relations for the $SO(3)$ group as

$$\left[L_3, \left(\frac{1}{R}\right)L_1\right] = i\left(\frac{1}{R}\right)L_2,$$

$$\left[L_3, \left(\frac{1}{R}\right)L_2\right] = -i\left(\frac{1}{R}\right)L_1, \qquad (1.6)$$

$$\left[\left(\frac{1}{R}\right)L_1, \left(\frac{1}{R}\right)L_2\right] = i\left(\frac{1}{R}\right)^2 L_3,$$

they become in the large-R limit the commutation relations for the $SE(2)$ group discussed repeatedly in Chapters I, II, and VII.

Let us translate the above limiting procedure into the language of coordinate transformations. $O(3)$ transformations on a spherical surface are generated by

$$L_1 = \begin{bmatrix} 0 & 0 & 0 \\ 0 & 0 & -i \\ 0 & i & 0 \end{bmatrix}, \quad L_2 = \begin{bmatrix} 0 & 0 & i \\ 0 & 0 & 0 \\ -i & 0 & 0 \end{bmatrix},$$

$$L_3 = \begin{bmatrix} 0 & -i & 0 \\ i & 0 & 0 \\ 0 & 0 & 0 \end{bmatrix}. \qquad (1.7)$$

For the present purpose, we can restrict ourselves to a small region near the north pole, where z is large and equal to the radius of the sphere, and x and y are much smaller than the radius. We can then write

$$\begin{bmatrix} x \\ y \\ z \end{bmatrix} = \begin{bmatrix} x \\ y \\ R \end{bmatrix} = \begin{bmatrix} 1 & 0 & 0 \\ 0 & 1 & 0 \\ 0 & 0 & R \end{bmatrix} \begin{bmatrix} x \\ y \\ 1 \end{bmatrix}. \qquad (1.8)$$

The column vectors on the left- and right-hand sides are respectively the coordinate vectors on which the $SO(3)$ and $SE(2)$ transformations are applicable. We shall use the notation A for the three-by-three matrix on the right-hand side. In the limit of large R,

$$L_3 = A^{-1} L_3 A,$$
$$N_1 = \left(\frac{1}{R}\right) A^{-1} L_2 A, \qquad (1.9)$$
$$N_2 = -\left(\frac{1}{R}\right) A^{-1} L_1 A.$$

This procedure leaves L_3 invariant. However, L_1 and L_2 of the $SO(3)$ group become the N_1 and N_2 matrices whose three-by-three matrix forms are discussed extensively in Chapter VII. We can obtain the N matrices by replacing all the elements in the lower left part of the L matrices by zero.

As for the coordinate vectors, the contraction transforms (x, y, z) to $(x, y, 1)$ as is done in Equation (1.9). If we use the spherical base vectors for $SO(3)$

$$r \sin\theta \, e^{\pm i\phi} \to (x \pm iy),$$
$$r \cos\theta \to 1. \qquad (1.10)$$

with large r and small θ. This limiting process leads to the base vectors of the $SE(2)$ transformation given in Equation (1.6) (Han et al., 1983c).

The form of L_3 is not affected by the similarity transformation of Equation (1.9). For this reason, $\text{Tr}[\exp(-i\alpha L_3)]$ remains invariant throughout the contraction procedure. Thus the operational definition of the equivalence class in $SO(3)$ remains valid in $SE(2)$. Instead of rotations by the same amount around different axes in the $SO(3)$ case, the equivalence class in $SE(2)$ is defined in terms of rotations by a fixed angle around different points in the xy plane.

2. E(2)-like Little Group as an Infinite-momentum/zero-mass Limit of the O(3)-like Little Group for Massive Particles

We have studied in Sections 1–4 of Chapter VII the $SE(2)$ group and determined its role in explaining the internal space-time symmetry of massless particles. In Chapters V and VI, we studied in detail how the $SO(3)$ group can explain the internal space-time symmetry of massive particles. In Section 1 of this chapter, we learned that the $SE(2)$ group can be regarded as a limiting case of the $SO(3)$ group.

Einstein's formula, $E = [(\mathbf{P})^2 + M^2]^{1/2}$, states that the energy-momentum relation for massive particles becomes that for massless particles in the limit of large momentum and/or zero-mass. Therefore, we should expect that the internal symmetry of massive particles becomes that of massless particles in the same limit. We should expect further that this limiting procedure will be very similar to the group contraction process discussed in Section 1.

For simplicity, we shall use the four-by-four coordinate transformation matrices to illustrate this limiting procedure. If a massive particle is at rest, the symmetry group is generated by the angular momentum operators L_1, L_2 and L_3 given in Section 1 of Chapter III. These operators do not change the four-momentum of the particle at rest. If this particle moves along the z direction, L_3 remains invariant, and its eigenvalue is the helicity. However, what happens to L_1 and L_2, particularly in the infinite-momentum limit?

There are no Lorentz frames in which massless particles are at rest. The little group for a massless particle moving along the z direction is generated by L_3, N_1 and N_2 given in Chapter VII. As we noted there, these generators are related to the boost and rotation operators by

$$N_1 = K_1 - L_2, \quad N_2 = K_2 + L_1. \tag{2.1}$$

The explicit four-by-four matrix forms for N_1 and N_2 are also given in Section 2 of Chapter VII. The four-momentum of the massless particle remains invariant under transformations generated by these operators. These generators of course satisfy the commutation relations for the generators of the $SE(2)$. L_3 is like the generator of rotation while N_1 and N_2 are like the generators of translations in the two-dimensional plane. As we discussed in Chapter VII, these translation-like operators generate gauge transformations.

Let us carry out an explicit calculation to justify the above expectation starting with a massive particle at rest whose $O(3)$-like little group is generated by L_1, L_2 and L_3. If we boost this massive particle along the z direction, its momentum and energy will become P and $E = [P^2 + M^2]^{1/2}$ respectively. The boost matrix is

$$B(P) = \begin{bmatrix} 1 & 0 & 0 & 0 \\ 0 & 1 & 0 & 0 \\ 0 & 0 & E/M & P/M \\ 0 & 0 & P/M & E/M \end{bmatrix}. \tag{2.2}$$

P is the momentum along the z direction. Under this boost operation, L_3 will remain invariant:

$$L_3' = BL_3B^{-1} = L_3. \tag{2.3}$$

However, the boosted L_2 and L_1 become

$$L'_2 = (E/M)L_2 - (P/M)K_1,$$
$$L'_1 = (E/M)L_1 + (P/M)K_2. \tag{2.4}$$

Because the Lorentz boosts in Equations (2.3) and (2.4) are similarity transformations, the L' operators still satisfy the $SO(3)$ commutation relations:

$$[L'_i, L'_j] = i\varepsilon_{ijk} L'_k. \tag{2.5}$$

Since the quantities in Equation (2.4) become very large as the momentum increases, we introduce new operators

$$G_1 = -(M/E) L'_2,$$
$$G_2 = (M/E) L'_1. \tag{2.6}$$

In terms of these new operators, we can write the $SO(3)$ commutation relations of Equation (2.5) as

$$[L_3, G_1] = -iG_2,$$
$$[L_3, G_2] = iG_1, \tag{2.7}$$
$$[G_1, G_2] = -(M/E)^2 L_3.$$

The quantity $(M/E)^2$ becomes vanishingly small if the mass becomes small or the momentum becomes very large. In this limit, G_1 and G_2 become N_1 and N_2 respectively, and the above commutation relations become those for the $E(2)$-like little group in the same manner as the $SO(3)$ commutation relations become those for the $SE(2)$ group through the group contraction process in Equation (1.6). The quantity (M/E) acts like the radius of the sphere we used for the model of the contraction of $SO(3)$ to $SE(2)$. For completeness, let us write the commutation relations shared by the $SE(2)$ group and the $E(2)$-like little group:

$$[L_3, N_1] = -iN_2, \quad [L_3, N_2] = iN_1, \quad [N_1, N_2] = 0. \tag{2.8}$$

In Chapter VII, we learned that N_1 and N_2 are the generators of gauge transformations for the photon case. Indeed, rotations around the axes perpendicular to the momentum become gauge transformations in the infinite-momentum/zero-mass limit (Han *et al.*, 1983c and 1984).

We have so far worked out the large-momentum/zero-mass limit of the rotation using the four-by-four matrix applicable to four-vectors. Section 3 will deal with this limiting process for spinors.

3. Large-momentum/zero-mass Limit of the Dirac Equation

Let us start with the three generators of the $SU(2)$ subgroup:

$$S_i = \tfrac{1}{2}\, \sigma_i. \tag{3.1}$$

The Lorentz boost along the z direction is accomplished through the similarity transformation:

$$S'_i = B(P)\, S_i\, B^{-1}(P), \tag{3.2}$$

where the applicable boost matrix is

$$B(P) = \begin{bmatrix} e^{\xi/2} & 0 \\ 0 & e^{-\xi/2} \end{bmatrix}, \tag{3.3}$$

with

$$e^{\xi/2} = \left(\frac{E+P}{E-P} \right)^{1/4}.$$

In the large-(momentum/mass) limit,

$$e^{\xi} \to \frac{2E}{M}. \tag{3.4}$$

Under the similarity transformation of Equation (3.2), S_3 remains invariant. However, S_1 and S_2 become

$$S'_1 = \begin{bmatrix} 0 & \tfrac{1}{2} e^{\xi} \\ \tfrac{1}{2} e^{-\xi} & 0 \end{bmatrix}, \quad S'_2 = \begin{bmatrix} 0 & -\tfrac{i}{2} e^{\xi} \\ \tfrac{i}{2} e^{-\xi} & 0 \end{bmatrix}. \tag{3.5}$$

We can go through a limiting procedure similar to that given in Equation (2.6) and (2.7) to obtain

$$N_1 = -\frac{M}{E} S'_2, \quad N_2 = \frac{E}{M} S'_1, \tag{3.6}$$

in the large-(momentum/mass) limit. The two-by-two matrix forms for N_1 and N_2 have been discussed in Chapters IV and VII. They are upper triangular matrices with zero diagonal elements. The multiplication of the factor $e^{-\xi}$ is exactly like the procedure of getting G_1 and G_2 in Equation (2.6) from L_1' and L_2' respectively.

Unlike the case of coordinate transformations, we have to consider here both signs of the boost generators. The effect of changing the sign is the same as taking the Hermitian conjugate of the above N matrices. These Hermitian-conjugated matrices are lower triangular with vanishing diagonal elements.

Let us now discuss the large-momentum/zero-mass limit of the Dirac equation. In order to accommodate both signs of the boost operators, we can write down the generators of $SL(2, c)$ in the form:

$$J_i = \begin{bmatrix} (1/2)\sigma_i & 0 \\ 0 & (1/2)\sigma_i \end{bmatrix}, \quad K_i = \begin{bmatrix} (i/2)\sigma_i & 0 \\ 0 & (-i/2)\sigma_i \end{bmatrix}. \tag{3.7}$$

These are the generators applicable to the Dirac wave functions in the Weyl representation. Indeed, choosing the sign of the boost operator is equivalent to choosing the sign of γ_5. With this point in mind, we can construct the gauge transformation matrix:

$$D(u, v) = \begin{bmatrix} D^{(+)}(u, v) & 0 \\ 0 & D^{(-)}(u, v) \end{bmatrix}, \tag{3.8}$$

applicable to the Dirac spinors. The two-by-two matrices $D^{(\pm)}(u, v)$ and their effect on the $SL(2, c)$ spinors have been discussed in detail in Section 5 of Chapter IV and Section 5 of Chapter VII.

In order to understand the effect of the above D matrix, let us start with the eigenspinors S_3 of Equation (3.1) for a massive Dirac particle at rest.

$$U(0) = \begin{bmatrix} \alpha \\ \pm \dot{\alpha} \end{bmatrix}, \quad V(0) = \begin{bmatrix} \pm \beta \\ \dot{\beta} \end{bmatrix}. \tag{3.9}$$

The $+$ and $-$ signs in the above expression specify positive and negative energy states respectively. If we boost these spinors along the z axis by applying the boost operator generated by K_3, then

$$U(\mathbf{p}) = \begin{bmatrix} [\exp(+\xi/2)]\,\alpha \\ \pm[\exp(-\xi/2)]\,\dot{\alpha} \end{bmatrix}, \quad V(\mathbf{p}) = \begin{bmatrix} \pm[\exp(-\xi/2)]\,\beta \\ [\exp(+\xi/2)]\,\dot{\beta} \end{bmatrix}. \tag{3.10}$$

In the above expression, $\exp(\xi/2)$ becomes large and $\exp(-\xi/2)$ becomes small. From Equations (5.6) and (5.7) of Chapter VII, we can see that the large components are gauge-invariant while the small components are gauge-dependent. Therefore, in general, spin-1/2 particles with non-zero mass are not invariant under gauge transformations.

In the large-momentum/zero-mass limit, we can renormalize the above spinors, and write them as

$$U(\mathbf{p}) = \begin{bmatrix} \alpha \\ 0 \end{bmatrix}, \quad V(\mathbf{p}) = \begin{bmatrix} 0 \\ \beta \end{bmatrix}. \tag{3.11}$$

These spinors are invariant under the D transformation. As we stated in Chapters IV and VII, this invariance is responsible for the polarization of neutrinos. Indeed, the gauge-dependent spinors disappear in the large-momentum/zero-mass limit. This is precisely why we do not talk about gauge transformations on neutrinos represented by the Dirac equation. As we shall see in Chapter X, this situation is similar to the case in which the electric and magnetic fields (not potentials), which are solutions of Maxwell's equations, are invariant under gauge transformations.

Let us summarize what we did above. We had to go through two different steps in taking the large-momentum/zero-mass limit of the Dirac equation and its solutions. The first step was to work out the limiting procedure for obtaining the two-by-two matrices of $SW(2)$ or the $E(2)$-like subgroup of $SL(2, c)$ from $SU(2)$. The second step was to eliminate the small component in the Weyl representation of the Dirac equation.

4. Finite-dimensional Non-unitary Representations of the SE(2) Group

We have so far discussed the $SE(2)$ group and the $E(2)$-like little group for massless particles with spin 1 and spin 1/2, and how their little groups can be obtained from those of their massive counterparts through group contraction. Thus the methods of group contraction may be useful in constructing the representations of the $E(2)$-like little groups for higher-spin particles. Before doing this, we have to study how the representations of the corresponding $SE(2)$ group can be obtained from those of $SO(3)$ or $SU(2)$.

In carrying out this procedure, we have to be careful about the sign of the boost operators. Representations of $SO(3)$ and $SU(2)$ do not depend on this sign. However, as was amply demonstrated in Sections 2 and 3, the representations of $SE(2)$ and $SW(2)$ depend on the sign of the boost operators. Because of this, our present understanding of the relation between $SE(2)$ and $SW(2)$ for higher spins is not as clear as in the case of $SO(3)$ and $SU(2)$. For this reason, we shall confine ourselves to the discussion of $SE(2)$ as a contraction of $SO(3)$ where this sign problem does not arise.

It was noted in Chapter VII that, among the representations of the $SE(2)$ group, finite-dimensional non-unitary representations are likely to be those relevant to physics. We shall therefore restrict ourselves to finite-dimensional representations. In studying these representations, the first question is what the size of the matrix should be. It would be easy to find the form of the L_3 matrix once its size is known. However, in constructing the N_1 and N_2

matrices, we should note the following similarities and differences between the photon and higher-spin cases. For both cases, the N_1 and N_2 matrices should satisfy the condition:

$$N_1^2 + N_2^2 = 0, \tag{4.1}$$

which is equivalent to Laplace's equation given in Section 1 of Chapter VII for finite-dimensional representations. For higher spin cases, we expect that the size of the matrices be larger than three-by-three.

In addition, as was noted in Chapter VII, the N matrices satisfy $N_1^2 = N_2^2 = N_1 N_2 = 0$ for $m = 1$, which is a reflection of the fact that the second derivatives of the linear function in x and y vanish. For $m = 2$, the second derivatives of $(x \pm iy)^2$ do not vanish, but the third derivatives are zero. We should therefore expect that the N_1 and N_2 matrices will satisfy

$$N_1^3 = N_2^3 = N_1^2 N_2 = N_1 N_2^2 = 0, \tag{4.2}$$

in addition to the constraint of Equation (4.1). For massless particles with spin m, the solution of the Laplace equation given in Section 1 of Chapter VII is a homogeneous polynomial of degree m. Thus, the $(m + 1)$-th derivatives of $(x \pm iy)^m$ should vanish. The matrices should therefore satisfy

$$N_1^{m+1} = N_1^m N_2 = \ldots = N_1 N_2^m = N_2^{m+1} = 0. \tag{4.3}$$

We learned in Section 1 that representations of the $SE(2)$ group can be obtained from those of the rotation group, and noted that L_3 remains unchanged throughout the contraction procedure. Therefore the size of the L_3 matrix for $m = 1$ is three-by-three, and that for $m = 2$ has to be five-by-five. For an arbitrary integer value, the size is $(2m + 1)$-by-$(2m + 1)$. The size of the N matrices should be the same as that of L_3.

Let us carry out a detailed calculation for the $m = 2$ case, starting with the solution of the Laplace equation for the ψ function of Section 1 of Chapter VII:

$$\psi(x) = (x \pm iy)^2. \tag{4.4}$$

This function is an eigenstate of L_3. Under the translation operation,

$$T(u, v) \psi(x, y) = [(x - u) \pm i(y - v)]^2$$
$$= (x \pm iy)^2 - 2(u \pm iv)(x \pm iy) + (u + iv)^2. \tag{4.5}$$

From our experience in Section 1, we can regard $\psi(x, y)$ as a contracted form of $Y_2^{\pm 2}(\theta, \phi)$, through the substitution:

$$\cos\theta \to 1,$$
$$\sin\theta \exp(\pm i\theta) \to (x \pm iy). \tag{4.6}$$

Rotation of $r^2 Y_2^{\pm 2}(\theta, \phi)$ around the z axis is trivial. However, rotation of this function around the x or y axis requires a summation of all possible $Y_2^m(\theta, \phi)$

states. This means that the operation of $T(u, v)$ on $\psi(x, y)$ will result in a linear combination of lower powers of $(x \pm iy)$ as is shown in Equation (4.5). The contracted form of the $Y_2^m(\theta, \phi)$ set can then be written as

$$V_2 = \begin{bmatrix} (x+iy)^2 \\ (x-iy)^2 \\ (x+iy) \\ (x-iy) \\ 1 \end{bmatrix}, \qquad (4.7)$$

with the following five-by-five matrices applicable to this column vector:

$$L_3 = \begin{bmatrix} 2 & 0 & 0 & 0 & 0 \\ 0 & -2 & 0 & 0 & 0 \\ 0 & 0 & 1 & 0 & 0 \\ 0 & 0 & 0 & -1 & 0 \\ 0 & 0 & 0 & 0 & 0 \end{bmatrix},$$

$$N_1 = \begin{bmatrix} 0 & 0 & -2i & 0 & 0 \\ 0 & 0 & 0 & -2i & 0 \\ 0 & 0 & 0 & 0 & -i \\ 0 & 0 & 0 & 0 & -i \\ 0 & 0 & 0 & 0 & 0 \end{bmatrix}, \qquad (4.8)$$

$$N_2 = \begin{bmatrix} 0 & 0 & 2 & 0 & 0 \\ 0 & 0 & 0 & -2 & 0 \\ 0 & 0 & 0 & 0 & 1 \\ 0 & 0 & 0 & 0 & -1 \\ 0 & 0 & 0 & 0 & 0 \end{bmatrix}.$$

The form of L_3 is identical to that of the $SO(3)$ case. N_1 and N_2 are the contractions of the five-by-five L_2 and $-L_1$ matrices respectively. They generate translations when they are applied to V_2 of Equation (4.7). These matrices satisfy the commutation relations given in Equation (2.8) for the generators of the $SE(2)$ group. In addition, they satisfy the dimensionality conditions Equation (4.1) and Equation (4.2). Here again it is quite clear that $\text{Tr}[\exp(\pm i\phi L_3)]$ remains invariant under translations. Therefore the translations are transformations within an equivalence class.

We should note in the expressions given in Equation (4.8) that the square matrices consisting of the last three rows and columns constitute the photon case with $m = \pm 1$. The forms given in Equation (4.7) and Equation (4.8) indicate clearly how we can enlarge the three-by-three matrices for $m = 1$ to the five-by-five matrices for $m = 2$. This tells us also how to construct the

Group Contractions

$(2m + 1)$-by-$(2m + 1)$ matrices for higher values of m, in analogy to the case of a simpler group consisting of one-dimensional multiplication and addition of a real number discussed in Gilmore's book (Gilmore, 1974).

Indeed, for an arbitrary integer m, we can write the basis vector as

$$V_m = \begin{bmatrix} (x+iy)^m \\ (x-iy)^m \\ * * \\ * * \\ (x+iy) \\ (x-iy) \\ 1 \end{bmatrix}. \tag{4.9}$$

The matrix for L_3 should be the same as that for the $SO(3)$ group. The N_1 and N_2 matrices applicable to the above column vector should satisfy the constraints of Equation (4.1) and Equation (4.3) which specify that the representation is finite-dimensional and is $(2m + 1)$-by-$(2m + 1)$. These matrices can now be written as

$$L_3 = \begin{bmatrix} m & 0 & 0 & 0 & 0 & 0 & 0 \\ 0 & -m & 0 & 0 & 0 & 0 & 0 \\ 0 & 0 & * & 0 & 0 & 0 & 0 \\ 0 & 0 & 0 & -* & 0 & 0 & 0 \\ 0 & 0 & 0 & 0 & 1 & 0 & 0 \\ 0 & 0 & 0 & 0 & 0 & -1 & 0 \\ 0 & 0 & 0 & 0 & 0 & 0 & 0 \end{bmatrix},$$

$$N_1 = \begin{bmatrix} 0 & 0 & -im & 0 & 0 & 0 & 0 \\ 0 & 0 & 0 & -im & 0 & 0 & 0 \\ 0 & 0 & 0 & 0 & * & 0 & 0 \\ 0 & 0 & 0 & 0 & 0 & * & 0 \\ 0 & 0 & 0 & 0 & 0 & 0 & -i \\ 0 & 0 & 0 & 0 & 0 & 0 & -i \\ 0 & 0 & 0 & 0 & 0 & 0 & 0 \end{bmatrix}, \tag{4.10}$$

$$N_2 = \begin{bmatrix} 0 & 0 & m & 0 & 0 & 0 & 0 \\ 0 & 0 & 0 & -m & 0 & 0 & 0 \\ 0 & 0 & 0 & 0 & * & 0 & 0 \\ 0 & 0 & 0 & 0 & 0 & * & 0 \\ 0 & 0 & 0 & 0 & 0 & 0 & 1 \\ 0 & 0 & 0 & 0 & 0 & 0 & -1 \\ 0 & 0 & 0 & 0 & 0 & 0 & 0 \end{bmatrix}.$$

Like their smaller counterparts, these matrices satisfy the commutation relations for the $SE(2)$ group given in Equation (2.8) and the dimensionality conditions of Equation (4.1) and Equation (4.3).

5. Polarization Vectors for Massless Particles with Integer Spin

We studied in Section 4 how to construct finite-dimensional representations of the $SE(2)$ group for integer values of m. However, it is important to realize that the column vectors given in Equations (4.7) and (4.9) are not physical wave functions. They only form vector spaces for their respective representations of the $SE(2)$ group. The representations of the $E(2)$-like little group are basically Lorentz-transformation matrices with the same algebraic property as that of the two-dimensional Euclidean group.

If $m = 0$, the representation is a unit matrix for both the $SE(2)$ group and the $E(2)$-like little group. If $m = 1$, the representation of the $SE(2)$ group consists of three-by-three matrices discussed in Section 1 of Chapter VII. However, the representation of the little group in this case consists of four-by-four matrices, which are basically Lorentz-transformation matrices. The precise connection between the representations of these two different groups has also been studied in detail in Section 2 of Chapter VII, and this forms the basis for studying higher-spin cases which are constructed from the symmetric direct Cartesian product of the $m = 1$ four-vector.

It was noted in Sections 2 and 3 of Chapter VII that, if we start from the four-potential:

$$A^\mu = (A_1, A_2, A_3, A_0), \tag{5.1}$$

for the photon whose momentum is along the z direction, then the Lorentz condition reduces this form to

$$\zeta_\pm = (1, \pm i, u \pm iv, u \pm iv). \tag{5.2}$$

The third and fourth components in the above expression can be made to vanish by a gauge transformation. These components should vanish in order that the above four-vectors be eigenstates of the helicity operator. We call the four-vectors which are eigenstates of the helicity operator the polarization vectors. As is shown in Section 4 of Chapter VII, the concept of polarization vector is Lorentz-invariant.

The purpose of the present section is to study how one can construct higher-spin states by taking direct products of the four-vector of the form given in Equation (5.2). In order to tackle this problem, we go back to the $SE(2)$ case. The connection between the four-vector of Equation (5.2) and the corresponding representation of the $SE(2)$ group has been discussed in

Section 2 of Chapter VII. With these points in mind, let us discuss in detail the direct product of two $m = \pm 1$ wave functions.

$$\psi_1 = [(x \pm iy) - (u \pm iv)],$$
$$\psi_2 = [(x \pm iy) - (u' \pm iv')]. \quad (5.3)$$

One way to take a direct product of these two functions is

$$\psi = [(x \pm iy) - (u \pm iv)] [(x \pm iy) - (u' \pm iv')]. \quad (5.4)$$

This form can be brought to a form diagonal in L_3 only if $u' = u$ and $v' = v$. Thus ψ_1 and ψ_2 should take the same form, and therefore

$$\psi = [(x \pm iy) - (u \pm iv)]^2$$
$$= (x \pm iy)^2 - 2(u \pm iv)(x \pm iy) + (u \pm iv)^2. \quad (5.5)$$

The above expression satisfies the two-dimensional Laplace equation, and is suitable for describing a massless particle with helicity ± 2. The addition of two helicities in the same direction is therefore possible. We can consider also a direct product of ψ_1 and ψ_2 with opposite signs of m.

$$\psi = [(x \pm iy) - (u \pm iv)] [(x \mp iy) - (u' \mp iv')]. \quad (5.6)$$

This could correspond to a product of two wave fucntions with opposite helicities. However, the above form does not satisfy the two-dimensional Laplace equation. For this reason, the addition of two helicities in opposite directions is not possible.

For $m = 2$ gravitons, we have to consider the symmetric traceless tensor constructed from a direct product of two four-vectors $\zeta_\pm$ and $\xi_\pm$:

$$h_\pm^{\mu\nu} = (\zeta_\pm^\mu \xi_\pm^\nu + \zeta_\pm^\nu \xi_\pm^\mu)/2. \quad (5.7)$$

The form of $\zeta_\pm$ is given in Equation (5.2). Because of the condition on Equation (5.5), the second four-vector $\xi_\pm$ should be the same as $\zeta_\pm$. This tensor is traceless in the sense that

$$g_{\mu\nu} h^{\mu\nu} = 0, \quad (5.8)$$

and its four-by-four matrix form is (Weinberg, 1972)

$$h_\pm^{\mu\nu} = \begin{bmatrix} 1 & \pm i & u \pm iv & u \pm iv \\ \pm i & -1 & \pm i(u \pm iv) & \pm i(u \pm iv) \\ u \pm iv & \pm i(u \pm iv) & (u \pm iv)^2 & (u \pm iv)^2 \\ u \pm iv & \pm i(u \pm iv) & (u \pm iv)^2 & (u \pm iv)^2 \end{bmatrix}. \quad (5.9)$$

The four-by-four little group matrices applicable to the indices μ and ν are identical. This means that there is only one set of gauge parameters u and v, as is indicated in the above matrix.

Let us now compare the above form with the $SE(2)$ function of Equation

(5.5). The expression $(x \pm iy)^2$ on the right-hand side specifies the polarization and is represented by the two-by-two matrix consisting of the first two rows and columns of the four-by-four matrix of Equation (5.9). The second term on the right-hand side of Equation (5.5) corresponds to the two-by-two matrices consisting of the first two rows and the last two columns and of the first two columns and the last two rows. The last term in Equation (5.5) is represented by the two-by-two matrix consisting of the last two rows and columns.

From our experience with the photon case, we know how to eliminate the dependence on the u and v parameters in order to make the form of Equation (5.9) a helicity eigenstate.

6. Lorentz and Galilei Transformations

The purpose of this section is to indicate that there are many other interesting applications of group contractions. In Section 1, we studied the contraction of $O(3)$ to $E(2)$ using the notion of a plane tangent to a spherical surface. The concept of this tangent plane plays a very important role in many branches of science dealing with curved surfaces. As is illustrated in Figure 6.1, all the curved surfaces pictured contract to the tangent plane.

For example, let us consider the surface of the hyperbola in a three-dimensional space spanned by x, y and t:

$$(ct)^2 - x^2 - y^2 = \text{Const.}, \tag{6.1}$$

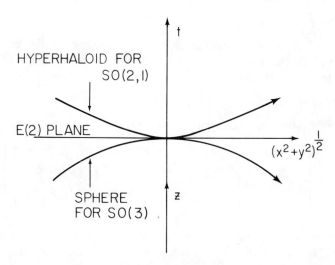

Fig. 6.1. Group contraction and tangent plane. The contraction of $SO(3)$ or $SO(2, 1)$ to $SE(2)$ is essentially a tangent-plane approximation.

where c is a constant and may become very large. This is a description of the $O(2, 1)$ group consisting of Lorentz boosts along the x and y directions and rotations on the xy plane.

We can now consider a plane tangent to the surface at $x = y = 0$. As the constant c becomes very large, the portion of surface in which $(x^2 + y^2)$ is finite becomes flat and coincides with the tangent plane. It is then not difficult to imagine that transformations on this tangent plane are Galilei transformations.

In order to see this point, let us start with coordinate transformations of the group $O(2, 1)$ generated by L_3, K_1, and K_2, where

$$L_3 = \begin{bmatrix} 0 & -i & 0 \\ i & 0 & 0 \\ 0 & 0 & 0 \end{bmatrix},$$

$$K_1 = \begin{bmatrix} 0 & 0 & 0 \\ 0 & 0 & -i \\ 0 & -i & 0 \end{bmatrix}, \quad K_2 = \begin{bmatrix} 0 & 0 & i \\ 0 & 0 & 0 \\ i & 0 & 0 \end{bmatrix}, \qquad (6.2)$$

applicable to the column vector (x, y, ct). K_1 and K_2 are the generators of Lorentz boosts along the x and y directions respectively.

The column vector (x, y, ct) can be written as

$$\begin{bmatrix} x \\ y \\ ct \end{bmatrix} = \begin{bmatrix} 1 & 0 & 0 \\ 0 & 1 & 0 \\ 0 & 0 & c \end{bmatrix} \begin{bmatrix} x \\ y \\ t \end{bmatrix}. \qquad (6.3)$$

Then, as c becomes very large, the circumstance is identical to the case of Equations (1.8) and (1.9). The resulting transformation matrix becomes

$$\begin{bmatrix} x' \\ y' \\ t' \end{bmatrix} = \begin{bmatrix} \cos \alpha & -\sin \alpha & u \\ \sin \alpha & \cos \alpha & v \\ 0 & 0 & 1 \end{bmatrix} \begin{bmatrix} x \\ y \\ t \end{bmatrix}. \qquad (6.4)$$

This form is a rotation on the xy plane followed by Galilei boosts along the x and y directions:

$$x' = x \cos \alpha - y \sin \alpha + ut,$$
$$y' = x \sin \alpha + y \sin \alpha + vt, \qquad (6.5)$$
$$t' = t.$$

The generalization of the above procedure for three-dimensional space is straightforward (Problem 3 in Section 8). Indeed, special relativity becomes Galilei relativity in the limit of large c (Inonu and Wigner, 1953).

Let us use $F(u, v, \alpha)$ for the above three-by-three matrix. This matrix can be decomposed into

$$F(u, v, \alpha) = B(u, v) R(\alpha). \tag{6.6}$$

The mathematical expressions for these matrices have already been given many times in connection with the $SE(2)$ group. $B(u, v)$ is clearly an Abelian invariant subgroup of the homogeneous Galilei group represented by $F(u, v, \alpha)$.

We can write the inhomogeneous Galilei transformation as

$$\begin{bmatrix} x' \\ y' \\ t' \\ 1 \end{bmatrix} = \begin{bmatrix} \cos\alpha & -\sin\alpha & u & x_0 \\ \sin\alpha & \cos\alpha & v & y_0 \\ 0 & 0 & 1 & t_0 \\ 0 & 0 & 0 & 1 \end{bmatrix} \begin{bmatrix} x \\ y \\ t \\ 1 \end{bmatrix}, \tag{6.7}$$

which produces a translation on each of the coordinate variables in addition to the homogeneous Galilei transformation given in Equation (6.5). Let us call the above four-by-four matrix G. Then this matrix can be decomposed to

$$G = T(x_0, y_0, t_0) B(u, v) R(\alpha). \tag{6.8}$$

where $T(x_0, y_0, t_0)$ is the translation matrix, and its form is obvious from the four-by-four matrix in Equation (6.7) and from our experience with the $E(2)$ group.

The group of translation matrices is clearly an Abelian invariant subgroup of G. The boost matrix B is an Abelian invariant subgroup of the homogeneous linear transformation of Equation (6.5). However, B is not an invariant subgroup of G. The group G has the five parameter subgroup TB whose form is

$$TB = \begin{bmatrix} 1 & 0 & u & x_0 \\ 0 & 1 & v & y_0 \\ 0 & 0 & 1 & t_0 \\ 0 & 0 & 0 & 1 \end{bmatrix}. \tag{6.9}$$

The above TB matrix is a semidirect product of T and B, and T is an Abelian invariant subgroup. The inhomogeneous Galilei group is a semidirect product of R and TB. Group TB is not Abelian, but still is an invariant subgroup of G. Therefore, TB is a non-Abelian invariant subgroup of G (see Problem 7 in Section 8).

7. Group Contractions and Unitary Representations of SE(2)

It was noted in Section 6 of Chapter VII that unitary representations of the $SE(2)$ group are infinite dimensional, and their representation space consists of solutions of the Bessel's differential equation. If we look at finite-dimensional representations which are contractions of the spherical harmonics, the dimension increases as the value ℓ increases. For this reason, we expect that the unitary representations of the $SE(2)$ group can be achieved through the contraction of the spherical harmonics with an infinite value of ℓ (Inonu and Wigner, 1953).

Let us start with the spherical harmonics of the form

$$Y_\ell^m(\theta, \phi) = e^{im\phi} P_\ell^m(\cos \theta), \tag{7.1}$$

where we do not worry about the normalization constant. The associated Legendre function $P_\ell^m(\cos \theta)$ satisfies the differential equation:

$$\frac{d}{dz}\left((1-x^2)\frac{d}{dz}P_\ell^m(z)\right) + (\ell(\ell+1) - m^2/(1-z^2)) P_\ell^m(z) = 0, \tag{7.2}$$

where

$$z = \cos \theta.$$

In the limit of small θ, the above differential equation becomes

$$\theta\left(\frac{d}{d\theta}\theta\frac{d}{d\theta}P_\ell^m\right) + (\ell(\ell+1)\theta^2 - m^2) P_\ell^m = 0. \tag{7.3}$$

This is Bessel's differential equation, and its solutions are of course Bessel functions. In the small-θ limit, the spherical harmonics of Equation (7.1) becomes

$$Y_\ell^m(\theta, \phi) \to e^{-im\phi} J_m(k\theta), \tag{7.4}$$

where

$$k^2 = \ell(\ell+1).$$

The above form is just the representation space discussed in Section 6 of Chapter VII. The parameter k or ℓ has to be sufficiently large to make the argument of the Bessel function non-zero in the small-θ limit. When we make the approximation of keeping only the lowest-order term in $k\theta$, the representation becomes finite-dimensional and non-unitary.

The basic difference between the infinite-dimensional unitary representations and finite-dimensional non-unitary representations is that the representation space consisting of Bessel functions is square integrable and all the generators of the group are Hermitian. On the other hand, if the representation is finite-dimensional, the functions are not normalizable. They are only tensor representations of the coordinate variables.

Let us discuss one physical application of this limiting process. It is also a high-momentum limit. In scattering processes in which two incoming particles collide with each other resulting in two particles moving in different directions, we commonly use the Legendre polynomials $P_\ell(\cos\theta)$ to describe the dependence on the scattering angle. The quantum number ℓ is the angular momentum around the scattering center, and can be regarded as a measure of the incoming momentum multiplied by the impact parameter.

When particles move slowly, it is sufficient to consider only two or three lowest values ℓ. On the other hand, when the particles move with speed very close to that of light, the scattering becomes predominantly forward and becomes like the Fraunhoffer diffraction. In this case, we have to deal with large values of ℓ. One way to approach this problem is to start from the operator

$$(\mathbf{L})^2 = L_x^2 + L_y^2 + L_z^2, \tag{7.5}$$

with the eigenvalue $\ell(\ell+1)$. For large values of ℓ, we can ignore L_3 whose eigenvalue is usually not larger than 1, and let

$$\ell(\ell+1) \simeq (\ell+\tfrac{1}{2})^2. \tag{7.6}$$

It is interesting to note (Misra and Maharana, 1976) that, when the scattering angle is very small, we can replace $(L_x^2 + L_y^2)$ by

$$R^2(P_x^2 + P_y^2) = (\ell+\tfrac{1}{2})^2, \tag{7.7}$$

in the spirit of Equations (1.8) and (1.9) with suitable redefinitions for R and P_i. In view of the discussion given in Section 2, we can readily let R be (P_0/M). For a given value of P_0, the eigenvalue of

$$(P_1^2 + P_2^2) \tag{7.8}$$

will give a measure of ℓ. In view of the discussion given in Section 1, this new parameter will be that of the Bessel function. Thus, for large values of ℓ, the Legendre polynomial becomes the Bessel function:

$$P_\ell(\cos\theta) \to J_0(aq\theta), \tag{7.9}$$

where

$$a = P_0/M.$$

The parameter a now measures ℓ, and becomes continuous for large values of P_0 (Blankenbecler and Goldberger, 1962; S. J. Wallace, 1973 and 1974).

The above Bessel-function form is commonly used for studying high-energy data. It is interesting to note that the transition from the use of the Legendre polynomials for low-energy processes to that of the Bessel functions in high-energy scattering is a group contraction of $SO(3)$ to $SE(2)$.

8. Exercises and Problems

Exercise 1. We discussed in Section 3 how $SU(2)$ becomes $SW(2)$ during the contraction process. According to Cartan's criterion, the Cartan determinant given in Section 5 of Chapter II vanishes for $SW(2)$ while it is non-zero for $SU(2)$. Show how the Cartan determinant vanishes during this contraction procedure.

Let us start from the Lie algebra of $SU(2)$:

$$[S_i, S_j] = iC_{ij}^k S_k, \tag{8.1}$$

where

$$C_{ij}^k = \varepsilon_{ijk}.$$

Cartan's determinant is the determinant of the three-by-three matrix:

$$g_{ij} = \varepsilon_{ik}^\ell \varepsilon_{j\ell}^k.$$

The explicit form of this matrix is

$$[g_{ij}] = \begin{bmatrix} -2 & 0 & 0 \\ 0 & -2 & 0 \\ 0 & 0 & -2 \end{bmatrix}. \tag{8.2}$$

The determinant of this matrix does not vanish. According to the group contraction procedure discussed in Sections 2 and 3, we are led to consider the algebra of S_3, N_1 and N_2 given in Equation (3.6). Then

$$C_{12}^3 = -\left(\frac{M}{E}\right)^2, \quad C_{23}^1 = C_{31}^2 = 1, \tag{8.3}$$

and the Cartan matrix becomes

$$\begin{bmatrix} 2(M/E)^2 & 0 & 0 \\ 0 & 2(M/E)^2 & 0 \\ 0 & 0 & -2 \end{bmatrix}, \tag{8.4}$$

where M and E are, as before, the mass and energy of the particle respectively. The determinant of this matrix becomes zero in the high-energy limit.

Exercise 2. The group contraction we studied in this chapter is essentially a flat-surface approximation of curved surfaces. This procedure can prove to be very useful in many branches of physics and engineering. Let us consider here an application in geometric optics. The surface of a lens is approximately spherical. However, it would be more convenient if the light rays can be expressed as linear functions. Find matrix representations of light rays going through a thin lens.

Let us consider a straight line in the xy coordinate system. We can write the equation for this line as

$$y = ax + b. \tag{8.5}$$

The first derivative of this function or the slope of the line is independent of the x variable. Thus for a given x variable, the y variable and the slope completely determine the line. Therefore, we can use the column vector.

$$\begin{bmatrix} y \\ a \end{bmatrix}_x, \tag{8.6}$$

to specify the line. If we wish to move from the coordinate point (x_0, y_0) to (x_1, y_1), we can represent this transformation as

$$\begin{bmatrix} y_1 \\ a \end{bmatrix}_{x_1} = \begin{bmatrix} 1 & (x_1 - x_0) \\ 0 & 1 \end{bmatrix} \begin{bmatrix} y_0 \\ a \end{bmatrix}_{x_0}. \tag{8.7}$$

The matrix on the right-hand side of the above expression is called the translation matrix, and its algebraic property has been studied in Chapter VII in connection with the $E(2)$-like little group for massless particles.

Let us next consider a light ray going through a convex spherical surface of a transparent material whose index of refraction is n, as is specified in Figure 8.1. We assume that the angle between the ray and the axis of the

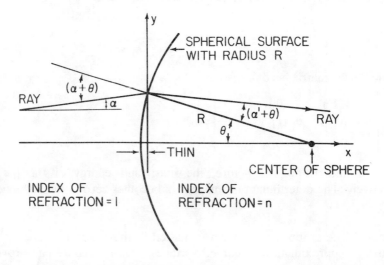

Fig. 8.1. Lens surface. Matrix optics results from the linear approximation of the relation between the slope of the ray and its distance from the axis. The lens surface is spherical or composed of any other curved surface which can be approximated as a spherical surface near the lens axis.

surface is very small and that the distance between the ray and the axis is much smaller than radius of the spherical surface. Then from Snell's law,

$$\sin(\alpha + \theta) = n\sin(\alpha' + \theta), \tag{8.8}$$

where the index of refraction of air is assumed to be 1. Since all the angles in the above expression are small,

$$a = \alpha \quad \text{and} \quad a' = \alpha', \quad \theta = y/R, \tag{8.9}$$

and

$$a + \frac{y}{R} = na' + n\frac{y}{R}. \tag{8.10}$$

This means that the slope undergoes the change through

$$a' = \frac{1}{n}a + \left(\frac{1-n}{nR}\right)y, \tag{8.11}$$

while there is no change in y, a and a' are the slopes of rays in the air and the refractive material respectively. We can ignore the variation in the x coordinate, which is of order $(y/R)^2$, in the spirit of the group contraction discussed in this chapter. Thus the relation between a and a' can be represented by the matrix relation

$$\begin{bmatrix} y \\ a' \end{bmatrix}_x = \begin{bmatrix} 1 & 0 \\ (1-n)/nR & 1/n \end{bmatrix} \begin{bmatrix} y \\ a \end{bmatrix}_x. \tag{8.12}$$

If the ray comes out from the refractive material through another convex surface with radius of curvature R' facing in the opposite direction, then the applicable matrix is

$$\begin{bmatrix} 1 & 0 \\ (1-n)/R' & n \end{bmatrix}. \tag{8.13}$$

If the distance between the two convex surface is small, they constitute a *thin lens*. The overall effect is the multiplication of the matrix of Equation (8.12) from left by the matrix of Equation (8.13). The net effect is the matrix

$$\begin{bmatrix} 1 & 0 \\ -1/f & 1 \end{bmatrix}, \tag{8.14}$$

where

$$\frac{1}{f} = (n-1)\left(\frac{1}{R} + \frac{1}{R'}\right). \tag{8.15}$$

f is called the focal length of the thin lens. The above expression is called the lens makers formula and was discovered first experimentally. The matrix of Equation (8.13) is called the lens matrix. This matrix, together with the translation matrix of Equation (8.7), forms the basic building blocks for a group theoretical approach to modern geometrical optics (Bacry and Cadilhac, 1981). See Blaker (1971) for a more detailed treatment of the matrix approach to geometrical optics.

Problem 1. Throughout this chapter, we used the small area on the north pole of the spherical Earth as a model for group contraction. It is also possible to make approximations on other areas. For example, it is possible to choose a narrow belt around the equator as a basis for approximation. Show in this case that the Legendre function becomes a harmonic oscillator wave function. See Talman (1968) and Gilmore (1974).

Problem 2. Construct the Lie algebra for the three-dimensional (homogeneous) Galilei group and that for the inhomogeneous Galilei group.

Problem 3. It is clear from Equation (6.4) that the two-dimensional Galilei group is isomorphic to $E(2)$. Is the three-dimensional Galilei group isomorphic to $E(3)$?

Problem 4. Show that the Poincaré group can be obtained from the deSitter group $O(4, 1)$ or $O(3, 2)$ through group contraction.

Problem 5. Show that the three-dimensional inhomogeneous Galilei group can also be obtained as a contraction of $O(4, 1)$ or $O(3, 2)$. See Brooke (1978 and 1980).

Problem 6. We discussed in Section 6 how the Galilei transformation in two-dimensional space can be obtained from $O(2, 1)$. Show that the Galilei transformation in three-dimensional space can be obtained from $O(3, 1)$ through group contraction. Show also that we can get the same result by contracting $O(4)$. See Holman (1969).

Problem 7. Show that the subgroup consisting of Galilei boosts and translations in three-dimensional space is a non-Abelian invariant subgroup of the inhomogeneous Galilei group.

Problem 8. Work out the irreducible unitary representations of the inhomogeneous Galilei group applicable to the time-independent Schrödinger wave functions. See Inonu and Wigner (1952) and Bargmann (1954).

Problem 9. We discussed in Chapter V the covariant harmonic oscillator wave functions for massive hadrons. In Section 6 of Chapter VII, we studied the harmonic oscillator wave functions for massless hadrons. Does the Lorentz-boosted oscillator wave function for a massive hadron behave like that of a massless hadron in the high-energy limit?

Problem 10. Discuss the contraction of $SU(2)$ to $SW(2)$ using the conformal mapping for $SL(2, c)$:

$$z' = \frac{az + b}{cz + d}. \tag{8.16}$$

Problem 11. In Section 4 of this chapter, we constructed finite-dimensional representations of the $E(2)$ group for integer values of m, by contracting the spherical harmonics. However, when we start from spinors as in the case of the rotation group, there is a complication due to the fact that the boost operators take two different signs. Is there a systematic approach to this problem? This is a possible future research problem.

Problem 12. In Equation (5.7), we took the symmetric tensor product of two four-vectors. What happens to the antisymmetric product?

Problem 13. Show that the harmonic oscillator group mentioned in Section 7 (Problem 6) of Chapter V is a contraction of $SO(3, 2)$. See Roman and Haavisto (1981).

Problem 14. We learned that $D_1(u)$ and $D_2(v)$ of Equations (2.8) and (2.9) of Chapter VII perform gauge transformations on the photon four-potential. What happens to these matrices when the Lorentz group is contracted to the Galilei group?

Problem 15. The role of the Pauli spin matrices in nonrelativistic quantum mechanics is well known, and they are identical to those used for the generators of rotations in $SL(2, c)$. What happens to the boost operators of $SL(2, c)$ when we contract $SO(3, 1)$ to the three-dimensional Galilei group?

Problem 16. A spherical surface can always be approximated as a parabolic surface. Is it possible to modify the group contraction procedure discussed in this chapter to obtain a parabolic approximation? This procedure is the calculation of aberration.

Chapter IX

SO(2, 1) and SU(1, 1)

After studying the little groups which preserve time-like and light-like four-momenta, we are led to consider the little group which leaves a space-like four-momentum invariant. Unlike the two previous cases, there is no clearly defined physical motivation to study this group as a little group, because we have to deal here with free particles travelling faster than light which are often called tachyons (Feinberg, 1967). Although tachyons are intrinsically interesting from a group theoretical point of view (Schwartz, 1982), there are many other applications of $SO(2, 1)$ of immediate physical and mathematical interest.

First, $SU(1, 1)$ is isomorphic to the simplest symplectic group $Sp(2)$. The word "symplectic group" was introduced by Weyl in 1938 (Weyl, 1946) and this group is relatively new to the physics world. However, this group plays a central role in the current development of physics, including classical mechanics (Arnold, 1978; Abraham and Marsden, 1978), plasma physics (Littlejohn, 1981; Morrison and Greene, 1981; Grebogi and Kaufman, 1983), nuclear physics (Moshinsky and Quesne, 1971; Quesne and Moshinsky, 1971; Moshinsky and Winternitz, 1980), statistical mechanics (Dyson, 1962), general relativity (Faddeev, 1970; Ashteker and Strewbel, 1981), optics (Bacry and Cadilhac, 1981), and high-energy physics (Kim and Noz, 1981, 1982). The reason is very simple. Symplectic transformations preserve skew symmetric products, and there are many skew symmetric products in physics. We are somewhat late in recognizing this important point.

It is not difficult to find discussions of the symplectic group in journals and textbooks. However, in most of the group theory textbooks, the symplectic group is discussed in the context of abstract group theory. In most of the research articles, this group is introduced as a "new device" with which one can generate sophisticated research problems. While the symplectic group plays an increasingly important role in physics research, the existing literature gives the impression that one has to learn completely "new" material in order to understand this group. In this environment, it is gratifying that we can study $Sp(2)$ in conjunction with our effort to understand $SO(2, 1)$ and $SU(1, 1)$.

Second, when we study Lorentz transformations, especially in physical applications, we often use transformations in the xy coordinate system either along the x or along the y axis. We seldom discuss Lorentz transformations in the three-dimensional xyz coordinate system. In spite of this simplification, the conventional method of computing velocity additions and successive Lorentz boosts is still complicated. Therefore, it is of practical interest to see whether there can be a further simplification. For instance, by restricting the problem from $O(3, 1)$ to $O(2, 1)$, we reduce the size of the matrix from four-by-four to three-by-three. We can reduce further the size of matrix to two-by-two by using $Sp(2)$.

Third, the study of $SO(2, 1)$ is essentially that of the Legendre functions with parameters not necessarily confined to those of the rotation group. The Legendre function with complex angular momentum has an important physical application in scattering theory, commonly called the Regge-pole model (Regge, 1959; Chew, 1962; Collins and Squires, 1968).

Fourth, $SU(1, 1)$ and $SO(2, 1)$ constitute a starting point for studying non-compact groups. These groups, like the rotation group, have three-parameters but are general enough to illustrate many of the properties of non-compact groups. Therefore, since the publication of Bargmann's original work (1947), this group has been thoroughly and exhaustively discussed in the literature (Barut and Fronsdal, 1965; Pukanszky, 1961 and 1964; Gel'fand *et al.* 1966; Sally, 1966; Mukunda, 1967, 1968, and 1973; Holman and Biedenharn, 1966 and 1968; Vilenkin, 1968; Kuriyan *et al.*, 1968; Talman, 1968; Miller, 1968; Wolf, 1974; Kalnins and Miller, 1974; Lang, 1975; Basu and Wolf, 1982). As far as $SU(1, 1)$ and $SO(2, 1)$ are concerned, the mathematics is far ahead of physics. We are at present in need of a physical motivation.

In Section 1, we take advantage of the fact that $SU(1, 1)$ is unitarily equivalent to $SL(2, r)$ or $Sp(2)$ consisting of real two-by-two unimodular matrices. In this case, we can study transformation properties of the group using a two-dimensional graph. The two-to-one correspondence with $SO(2, 1)$ is also studied. In Section 2, finite-dimensional non-unitary representations of $SO(2, 1)$ are studied as a preliminary step for later sections. It is illustrated in Section 3 that the study of $SO(2,1)$ is essentially that of the Legendre functions for the magnitude of its argument greater than one. In Section 4, we outline the first step which might be needed in studying unitary irreducible representations of $SU(1, 1)$.

Section 5 consists of exercises and problems. In view of the extensiveness of the coverage of this subject in the mathematical literature, and of the briefness of the discussion given in this section, the list of mathematical exercises can become endless. We therefore choose only those problems which may have physical applications or may serve an illustrative purpose for the items discussed in earlier chapters.

1. Geometry of SL(2, r) and Sp(2)

Matrices representing $SU(1, 1)$ take the form

$$W = \begin{bmatrix} a & b \\ b^* & a^* \end{bmatrix}, \quad \text{with} \quad |a|^2 - |b|^2 = 1, \tag{1.1}$$

This group is generated by

$$S_3 = \frac{1}{2}\sigma_3, \quad K_1 = \frac{i}{2}\sigma_1, \quad K_2 = \frac{i}{2}\sigma_2. \tag{1.2}$$

If we perform a 90° rotation around the y axis followed by another 90° rotation around the x axis, then

$$x \rightarrow z, \quad z \rightarrow y, \quad y \rightarrow x. \tag{1.3}$$

The generators of Equation (1.3) become pure imaginary and take the form

$$S_3 = \begin{bmatrix} 0 & -i/2 \\ i/2 & 0 \end{bmatrix}, \quad K_1 = \begin{bmatrix} i/2 & 0 \\ 0 & -i/2 \end{bmatrix}, \quad K_2 = \begin{bmatrix} 0 & i/2 \\ i/2 & 0 \end{bmatrix}, \tag{1.4}$$

The above generators satisfy the commutation relations:

$$[K_1, K_2] = -iS_3, \quad [K_1, S_3] = iK_3, \quad [K_2, S_3] = iK_1, \tag{1.5}$$

and the matrix W becomes real.

The group of real W matrices generated by the above three imaginary matrices is called $SL(2, r)$ or the two-dimensional symplectic group which is commonly called $Sp(2)$. The W matrix satisfies the condition:

$$\tilde{M}JM = J, \tag{1.6}$$

where

$$J = \begin{bmatrix} 0 & 1 \\ -1 & 0 \end{bmatrix},$$

Because this group consists only of real two-by-two matrices, it is possible to study this group using two-dimensional geometry, which can be sketched easily on a piece of paper. In fact, K_1, K_2 and S_3 generate the following transformation matrices in the two-dimensional xy plane:

$$B(0, \xi) = \exp(-i\xi K_1) = \begin{bmatrix} e^{\xi/2} & 0 \\ 0 & e^{-\xi/2} \end{bmatrix},$$

$$B\left(\frac{\pi}{2}, \eta\right) = \exp(-i\eta K_2) = \begin{bmatrix} \cosh\frac{\eta}{2} & \sinh\frac{\eta}{2} \\ \sinh\frac{\eta}{2} & \cosh\frac{\eta}{2} \end{bmatrix}, \quad (1.7)$$

$$R(\theta) = \exp(-i\theta S_3) = \begin{bmatrix} \cos\frac{\theta}{2} & -\sin\frac{\theta}{2} \\ \sin\frac{\theta}{2} & \cos\frac{\theta}{2} \end{bmatrix}.$$

The R matrix in the above expression represents a rotation by angle θ. $B(0, \xi)$ is the elongation/contraction along the x/y axes respectively. $B(\pi/2, \eta)$ is the elongation/contraction along the 45° direction. The elongation matrix along the $\theta/2$ direction is

$$B(\theta, \eta) = R(\theta) B(0, \eta) R(-\theta)$$

$$= \begin{bmatrix} \cosh\frac{\eta}{2} + \left(\sinh\frac{\eta}{2}\right)\cos\theta & \left(\sinh\frac{\eta}{2}\right)\sin\theta \\ \left(\sinh\frac{\eta}{2}\right)\sin\theta & \cosh\frac{\eta}{2} - \left(\sinh\frac{\eta}{2}\right)\cos\theta \end{bmatrix}. \quad (1.8)$$

With this point in mind, let us start with a circle of unit radius centered around the origin in the Cartesian coordinate system with the coordinate variables x^* and y^*, as is illustrated in Figure 1.1. The equation for this circle is

$$(x^*)^2 + (y^*)^2 = 1. \quad (1.9)$$

This equation is invariant under rotation of the x^*y^* coordinate system. The area of this circle is π.

We can next consider an ellipse whose equation in the xy coordinate system is

$$e^{-\eta}\left[x\cos\frac{\theta}{2} - y\sin\frac{\theta}{2}\right]^2 - e^{\eta}\left[x\sin\frac{\theta}{2} + y\cos\frac{\theta}{2}\right]^2 = 1. \quad (1.10)$$

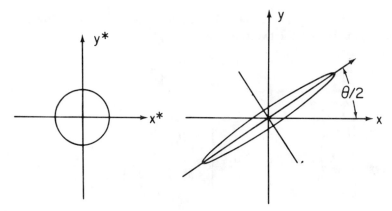

Fig. 1.1. Symplectic transformation of the circle in the x^*y^* coordinate system to the ellipse in the xy system. The area of the ellipse remains the same throughout the transformation process. In general, this elliptic deformation is preceded by a rotation in the x^*y^* coordinate system (Kim and Noz, 1983).

If η is positive, the major and minor axes of the ellipse are $e^{\eta/2}$ and $e^{-\eta/2}$, respectively. The major axis is along the $\theta/2$ direction. The area of this ellipse is π, and remains invariant as we change the values of η and θ. This is illustrated also in Figure 1.1.

It is then not difficult to imagine that we can obtain the ellipse of Equation (1.10) from the circle of Equation (1.9) by making a coordinate transformation from the x^*y^* to the xy system. Since the area of the ellipse is conserved, we can consider the following two "area-conserving" coordinate transformations:

(a) We can rotate the coordinate system without changing the area.

(b) We can perform the following scale transformation without changing the area element on the xy plane:

$$x \to x \exp(\eta/2), \quad y \to y \exp(-\eta/2). \tag{1.11}$$

Option (a) is well known. The scale transformation of Equation (1.11) is also very easy to visualize. However, not everybody is familiar with the combined effect of rotations and scale transformations. This is what the $Sp(2)$ group is about.

In fact, the coordinate transformation from the x^*y^* system for the circle of Equation (1.9) to the xy system for the ellipse of Equation (1.10) is achieved through the transformation matrix given in Equation (1.8).

In order to see the properties of successive scale transformations, let us consider the $x'y'$ coordinate system which is related to the xy system through the transformation

$$x' = x \exp(\xi/2), \quad y' = y \exp(-\xi/2). \tag{1.12}$$

Then the ellipse of Equation (1.10) is transformed into a new ellipse with the same area whose equation is

$$e^{-\eta'}[x'\cos(\theta'/2) + y'\sin(\theta'/2)]^2 +$$
$$+ e^{\eta'}[x'\sin(\theta'/2) - y'\cos(\theta'/2)]^2 = 1, \qquad (1.13)$$

where

$$\cosh\eta' = (\cosh\eta)\cosh\xi + (\sinh\eta)(\sinh\xi)\cos\theta, \qquad (1.14)$$

and

$$\tan\theta' = \frac{(\sin\theta)\sinh\eta}{(\sinh\xi)\cosh\eta + (\cosh\xi)(\sinh\eta)\cos\theta}. \qquad (1.15)$$

The above result shows that the $x'y'$ coordinate system is obtained from the x^*y^* system through the transformation matrix $B(\theta', \eta')$. Then the question is whether this matrix is a simple product of $B(0, \xi)$ and $B(\theta, \eta)$. The answer to this question is "No." An explicit calculation shows that

$$B(\theta', \eta') = B(0, \xi) B(\theta, \eta) R(\alpha). \qquad (1.16)$$

where the angle α is determined from θ and η and from the requirement that the $B(\theta', \eta')$ matrix be symmetric as indicated in Equation (1.8). From this calculation

$$\tan\frac{\alpha}{2} = \frac{\sin\theta}{\cos\theta + [\coth(\xi/2)]\coth(\eta/2)}. \qquad (1.17)$$

The existence of this rotation matrix in Equation (1.16) does not affect the geometry of circles and ellipses because the $R(\alpha)$ matrix in Equation (1.16) does not change the circle in the x^*y^* coordinate system.

This means that the two-parameter scale transformation matrices $B(\theta, \eta)$ alone cannot form a group, but they have to be supplemented by the rotation matrix $R(\alpha)$. Indeed, the $Sp(2)$ group consists of both the B and R matrices, and is therefore a three-parameter group.

If we use (x, y, t) to specify the coordinates in this $(2 + 1)$ space, then $SO(2, 1)$ consists of Lorentz transformations along the x and y directions and of the rotation on the xy plane around the t axis. The generators of this group are

$$K_1 = \begin{bmatrix} 0 & 0 & i \\ 0 & 0 & 0 \\ i & 0 & 0 \end{bmatrix}, \quad K_2 = \begin{bmatrix} 0 & 0 & 0 \\ 0 & 0 & i \\ 0 & i & 0 \end{bmatrix},$$

and

$$L_2 = \begin{bmatrix} 0 & -i & 0 \\ i & 0 & 0 \\ 0 & 0 & 0 \end{bmatrix}. \qquad (1.18)$$

They satisfy the following commutation relations:

$$[K_1, K_2] = iL_3, \quad [K_2, L_3] = iL_1, \quad [L_3, K_1] = iK_2. \tag{1.19}$$

These commutation relations are exactly like those for the generators of the $Sp(2)$ group given in Equation (1.5). This is precisely why we say that the $Sp(2)$ and $SO(2, 1)$ groups are locally isomorphic to each other.

In order to study the content of this isomorphism, let us write the transformation matrices generated by the above matrices:

$$B(0, \xi) = \exp(-i\xi K_1) = \begin{bmatrix} \cosh \xi & 0 & \sinh \xi \\ 0 & 1 & 0 \\ \sinh \xi & 0 & \cosh \xi \end{bmatrix},$$

$$B(\theta = \pi/2, \eta) = \exp[-i\eta K_2] = \begin{bmatrix} 1 & 0 & 0 \\ 0 & \cosh \eta & \sinh \eta \\ 0 & \sinh \eta & \cosh \eta \end{bmatrix}, \tag{1.20}$$

$$R(\theta) = \exp[-i\theta L_3] = \begin{bmatrix} \cos \theta & -\sin \theta & 0 \\ \sin \theta & \cos \theta & 0 \\ 0 & 0 & 1 \end{bmatrix}.$$

$B(0, \xi)$ and $B(\pi/2, \eta)$ are the Lorentz transformation or "boost" matrices along the x and y directions, respectively. The $R(\theta)$ matrix performs a rotation around the t axis. The boost along the direction which makes an angle θ with the x axis is therefore

$$B(\theta, \eta) = R(\theta) B(0, \eta) R(-\theta). \tag{1.21}$$

This relation is very similar to Equation (1.8) for the $Sp(2)$ group. The above matrix is symmetric, and its determinant is unity.

We can next study successive Lorentz transformations. Let us consider a boost along the θ direction followed by another boost along the x direction. Then we should end up with a boost in a new direction:

$$B(\theta', \eta') = B(0, \xi) B(\theta, \eta) R(\alpha). \tag{1.22}$$

The problem then is to determine the parameters θ', η', and α. An interesting point is that the algebras required to determine parameters are identical to those given in Equations (1.13) and (1.16) for the case of the $Sp(2)$ group. This is precisely the content of the isomorphism between the two groups.

While the $Sp(2)$ transformations are area-conserving transformations, the $SO(2, 1)$ transformation, whose determinant is unity, preserves the volume in the three-dimensional Euclidean space spanned by x, y, and t. In order to illustrate this point, let us start with a sphere of unit radius whose center is at the origin in the Lorentz frame whose coordinate variables are x^*, y^*, and t^*. The equation for this sphere takes the form

$$(x^*)^2 + (y^*)^2 + (t^*)^2 = 1. \tag{1.23}$$

If we perform a Lorentz transformation along the x direction:

$$x = x^* \cosh \eta + t^* \sinh \eta,$$
$$t = x^* \sinh \eta + t^* \cosh \eta, \tag{1.24}$$

the sphere of Equation (1.23) becomes an ellipsoid whose equation is

$$\tfrac{1}{2}\{e^{-2\eta}(x+t)^2 - e^{2\eta}(x-t)^2\} + y^2 = 1 \tag{1.25}$$

If we perform the same Lorentz transformation along the θ direction, the resulting ellipsoid becomes

$$\tfrac{1}{2}\{e^{-2\eta}(x\cos\theta + y\sin\theta + t)^2 + e^{2\eta}(x\cos\theta - y\sin\theta - t)^2\} +$$
$$+ (x\sin\theta - y\cos\theta)^2 = 1. \tag{1.26}$$

If we make another boost along the x direction:

$$x' = x\cosh\xi + t\sinh\xi, \quad t' = x\sinh\xi + t\cosh\xi, \tag{1.27}$$

the resulting equation will be that of another ellipsoid with the parameters θ' and η'. The algebra of determining these parameters is identical to that for the $Sp(2)$ ellipse given in Equation (1.13).

While we are accustomed to associate Lorentz transformations with hyperbolas and hyperbolic surfaces, it is interesting to note that ellipsoids can also play an effective role in illustrating the $SO(2, 1)$ group. This method of using ellipsoids is convenient in understanding and interpreting experimental data on Lorentz-deformed relativistic hadrons.

2. Finite-dimensional Representations of $SO(2, 1)$

We studied in Section 1 a simple geometrical picture of how transformations of $O(2, 1)$ and $SL(2, r)$ are different from the rotation group. This geometrical picture may be helpful in practical calculations. However, what we did in Section 1 is not necessarily the orthodox group theoretical approach of constructing irreducible representations. The standard approach is to construct first representations and representation spaces which are diagonal in the Casimir operators. The generators of $SO(2, 1)$ in differential form are

$$L_3 = -i\left(x\frac{\partial}{\partial y} - y\frac{\partial}{\partial x}\right),$$

$$K_1 = i\left(t\frac{\partial}{\partial x} + x\frac{\partial}{\partial t}\right), \qquad (2.1)$$

$$K_2 = i\left(t\frac{\partial}{\partial y} + y\frac{\partial}{\partial t}\right).$$

The Casimir operator in this case is

$$C = J_3^2 - K_1^2 - K_2^2. \qquad (2.2)$$

which commutes with the three generators of the group. We are interested in constructing representations and representation spaces which are diagonal in this operator. In addition to this Casimir operator, we can choose the representation to be diagonal in L_3 which commutes with C. We can then draw some immediate conclusions from what we know from the three-dimensional rotation group.

We can obtain the above Casimir operator from that of $SO(3)$ by replacing z by the imaginary time: (it). The representation space for the three-dimensional rotation group consists of tensors constructed from the coordinate variables x, y, z. For instance, the rotationally invariant combination in $SO(3)$ is $(x^2 + y^2 + z^2)$. The eigenvalue of $\mathbf{L}^2$ for this form is 0. For $\ell = 1$, there are three possible combinations. They are $(x \pm iy)$ and z. For $\ell = 2$ or higher, the representation space consists of the spherical harmonics: Y_ℓ^m (θ, ϕ). Let us see how these can be translated into the language of the $SO(2, 1)$ group.

When $t > r = (x^2 + y^2)^{1/2}$, we use the parametrization:

$$\tan\phi = \frac{y}{x}, \quad \cosh\eta = \frac{t}{\rho}, \quad \sinh\eta = \frac{r}{\rho}, \qquad (2.3)$$

where

$$\rho = (|t^2 - r^2|)^{1/2}.$$

In terms of the ϕ and η variables, the Casimir operator C takes the form

$$C = \frac{\partial}{\partial(\cosh\eta)}\left(\sinh^2\eta\,\frac{\partial}{\partial(\cosh\eta)}\right) + \left(\frac{1}{\sinh\eta}\right)^2\left(\frac{\partial}{\partial\phi}\right)^2. \qquad (2.4)$$

When $t < r$, we have to use the parametrization:

$$\tan \phi = \frac{y}{x}, \quad \sinh \eta = \frac{t}{\rho}, \quad \text{and} \quad \cosh \eta = \frac{r}{\rho}. \tag{2.5}$$

The C operator becomes

$$C = \frac{\partial}{\partial(\sinh \eta)}\left(\cosh^2 \eta \frac{\partial}{\partial(\sinh \eta)}\right) - \left(\frac{1}{\cosh \eta}\right)^2 \left(\frac{\partial}{\partial \phi}\right)^2. \tag{2.6}$$

We can obtain the above forms by replacing z with (it) in the $(\mathbf{L})^2$ operator of $SO(3)$:

$$(\mathbf{L})^2 = -\frac{\partial}{\partial(\cos \theta)}\left(\sin^2 \theta \frac{\partial}{\partial(\cos \theta)}\right) - \left(\frac{1}{\sin \theta}\right)^2 \left(\frac{\partial}{\partial \phi}\right)^2, \tag{2.7}$$

and the Casimir operator C of Equation (2.4) by substituting $i\eta$ for θ in Equation (2.7). C of Equation (2.6) is obtained from the same substitution after a 90° rotation $\theta \to \theta + 90°$. For this reason, solving the eigenvalue equation:

$$C \psi = \lambda \psi \tag{2.8}$$

is just like constructing the spherical harmonics when the eigenvalue λ is $\ell(\ell + 1)$, where ℓ takes integer values. The resulting solutions of the above differential equation are

$$\psi_\ell^m(\eta, \phi) = (\sinh \eta)^m \left[\left(\frac{d}{d(\cosh \eta)}\right)^m P_\ell(\cosh \eta)\right] e^{im\phi}, \quad \text{for } |t| > r, \tag{2.9}$$

$$\psi_\ell^m(\eta, \phi) = (\cosh \eta)^m \left[\left(\frac{d}{d(\sinh \eta)}\right)^m P_\ell(\sinh \eta)\right] e^{im\phi}, \quad \text{for } |t| < r,$$

with

$$m = -\ell, -\ell+1, \ldots, \ell-1, \ell. \tag{2.10}$$

If we multiply these solutions by ρ which is an invariant quantity, they become finite homogeneous polynomials of the ℓ-th degree in x, y, and t.

The above solutions are written in the form convenient to represent rotations around the origin on the xy plane. However, if we perform a Lorentz boost along the x direction, the resulting polynomial will be a linear transformation. If $\ell = 0$, the representation space is one-dimensional, and the polynomial is 1 which remains invariant under transformations. If $\ell = 1$, the polynomials are $(x \pm iy)$ for $m = \pm 1$ respectively, and t for $m = 0$.

If $\ell = 2$, the polynomials will be $(x \pm iy)^2$, $t(x \pm iy)$, and $(x^2 + y^2 + 2t^2)$ for $m = \pm 2$, $m = \pm 1$ and $m = 0$ respectively. Each polynomial satisfies the eigenvalue equation with the eigenvalue of $\ell(\ell + 1) = 6$. It is very easy to rotate these functions around the origin on the xy plane. If we boost the function $(x + iy)^2$ along the x direction through the transformation resulting in the replacement of x and t by x' and t' respectively, where

$$x' = x \cosh \alpha - t \sinh \alpha, \quad y' = y, \quad t' = x \sinh \alpha - t \cosh \alpha, \quad (2.11)$$

then

$$(x' + iy')^2 = \left(\frac{1 + \cosh \alpha}{2}\right)^2 (x + iy)^2$$

$$+ \left(\frac{\cosh \alpha - 1}{2}\right)^2 (x - iy)^2$$

$$- (\sinh \alpha)(\cosh \alpha + 1)(x + iy)t$$

$$- (\sinh \alpha)(\cosh \alpha - 1)(x - iy)t$$

$$+ \tfrac{1}{2} (\sinh \alpha)^2 (x^2 + y^2 + 2t^2). \quad (2.12)$$

This expression is still an eigenstate of the Casimir operator C. However, it is a linear combination of different m states. The number of terms is not more than $(2\ell + 1)$. For this reason, what we did above constitutes a finite-dimensional representation.

The procedure for obtaining these finite dimensional representations is identical to that for the three-dimensional rotational group. Let us start with the usual differential forms of L_1, L_2 and L_3 given in Section 1 of Chapter III. If we replace z by (it), then L_3 remains invariant, while L_1 and L_2 become:

$$L_1 \rightarrow iK_2, \quad \text{and} \quad L_2 \rightarrow iK_1. \quad (2.13)$$

Thus the Casimir operator C of Equation (2.2) is the same as that for the rotation group. The only difference is to replace the usual θ variable by $i\eta$. However, this replacement leads to the following non-trivial problems.

First, where can we use this representation in physics? Second, in the rotation group, the orthogonality relation is stated as

$$\int_{-1}^{1} d(\cos\theta) \int_{0}^{2\pi} d\phi \, (Y_{\ell'}^{m'}(\theta, \phi))^* \, Y_{\ell}^{m}(\theta, \phi) = \delta_{\ell\ell'} \delta_{mm'}. \quad (2.14)$$

In the case of $SU(1, 1)$, the corresponding integral is

$$\int_{0}^{\infty} (\sinh \eta) \, d\eta \int_{0}^{2\pi} d\phi \, (\psi_{\ell'}^{m'}(\eta, \phi))^* \, \psi_{\ell}^{m}(\eta, \phi), \quad (2.15)$$

where η ranges from 0 to infinity. The integral in this case does not converge for integer values of ℓ. For this reason, the generators K_1 and K_2 cannot be Hermitian, and the representation is not unitary. Is it possible to construct unitary representations as in the case of $SE(2)$?

In spite of these questions, the study of the finite-dimensional representations is a preliminary step for further developments.

3. Complex Angular Momentum

We have seen in Section 2 that the study of $SO(2, 1)$ can start from the Legendre functions. The study of $SO(3)$ is only a restricted case in which the appropriate boundary conditions are imposed on the solutions of the Legendre's differential equation. The boundary condition in this case is that it be analytic everywhere in the complex $\cos\theta$ plane. While $SO(3)$ is concerned with the region in which $|\cos\theta| \leq 1$, we are in $SO(2, 1)$ interested in the region where $|\cos\theta|$ is greater than one. The Legendre function does not have to analytic at $\cos\theta = \pm 1$. Indeed, the study of $SO(2, 1)$ is that of the Legendre function for $|\cos\theta| \geq 1$ (Serterio and Toller, 1964; Sciarrino and Toller, 1967).

The Legendre function is discussed extensively in standard textbooks (Whittaker and Watson, 1927; Morse and Feshbach, 1953). The fundamental definition of the Legendre function is given in an integral form:

$$P_\ell(z) = \frac{1}{2\pi i} \left(\frac{1}{2}\right)^\ell \int_C \frac{(t^2 - 1)^\ell}{(t - z)^{\ell+1}} \, dt, \tag{3.1}$$

where the counterclockwise contour C encloses $t = 1$ and $t = z$, but not $t = -1$ in the complex t plane.

The above integral representation satisfies the Legendre differential equation:

$$\frac{d}{dz}\left((1 - z^2)\frac{d}{dz}P_\ell(z)\right) + \ell(\ell + 1)P_\ell(z) = 0, \tag{3.2}$$

and the symmetry property:

$$P_\ell(z) = P_{-\ell-1}(z). \tag{3.3}$$

For integer values of ℓ, the integral representation of Equation (3.1) becomes a Legendre polynomial. Furthermore, the asymptotic behavior of the Legendre function for large values of z becomes

$$P_\ell(z) \to (z)^{\text{Re}(\ell)}. \tag{3.4}$$

In order that the representation be unitary, the solution of the differential equation given in Equation (3.2) should be an infinite series in $\cosh \eta$ and $\sinh \eta$ to make the function vanish fast enough for increasing values of η. Thus ℓ cannot take integer values. Because of the symmetry of Equation (3.3), the minimum value of $\text{Re}(\ell)$ expected from the Legendre function is $-1/2$. For this value, the orthogonality integral of Equation (2.15) converges. Indeed, the functions ψ_ℓ^m derived from the Legendre function with $\ell = -1/2$ through the differentiation in Equation (2.9) can form a representation space for infinite-dimensional unitary representations.

One important physical application of complex angular momentum is the Watson-Sommerfeld transformation of the nonrelativistic scattering amplitude (Regge, 1959), and its application to the Regge-pole model for high-energy scattering (Chew, 1962). Let us start with the scattering amplitude of the form:

$$f(k, \theta) = \frac{1}{2ik} \sum_{\ell=0}^{\infty} (2\ell + 1)(e^{2i\delta_\ell} - 1) P_\ell(\cos \theta), \tag{3.5}$$

where k and θ are the momentum and scattering angle respectively. We can write the above amplitude as

$$f(k, \theta) = \sum_{\ell} (2\ell + 1) a(\ell, k) P_\ell(\cos \theta). \tag{3.6}$$

If the potential is of the Yukawa type, $V(r) \sim (1/r) e^{-\mu r}$, $a(\ell, k)$ is analytic in the complex ℓ planc in the region $\text{Re}(\ell) > -1/2$, except for the poles in the first quadrant (Regge, 1959). $P_\ell(\cos \theta)$ is an entire function of ℓ. With this point in mind, we can write Equation (3.6) as the contour integral:

$$f(k, \theta) = \frac{1}{2\pi i} \int_C \frac{e^{i\pi \ell}}{\sin \pi \ell} a(\ell, k) P_\ell(\cos \theta) \, d\ell, \tag{3.7}$$

as is illustrated in Figure 3.1. The contour C can now be opened up and pushed to the vertical line along $\ell = -1/2 + is$, where $-\infty < s < \infty$, provided that the integral vanishes along the infinite circular section in the region right of $\text{Re}(\ell) = -1/2$ (Brown et al., 1963). During this process, the contour encloses the poles in the first quadrant. The result is

$$f(k, \theta) = \frac{1}{2\pi} \int_{-\infty}^{\infty} \frac{e^{i\pi(-1/2 + is)}}{\sin \pi(-1/2 + is)} a(-1/2 + is, k) P_{-1/2 + is}(\cos \theta) \, ds$$

$$+ \sum_i \frac{e^{i\pi \alpha_i}}{\sin \pi \alpha_i} a'(\alpha_i, k) P_{\alpha_i}(\cos \theta). \tag{3.8}$$

where the second term in the above expression consists of contributions from

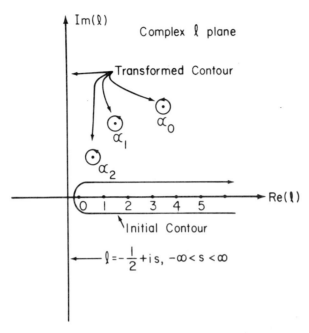

Fig. 3.1. Watson-Sommerfeld transformation of the scattering amplitude in the complex ℓ plane. For the Yukawa potential, the particle scattering amplitude is analytic in the region where $\operatorname{Re}(\ell)$ is greater than $-1/2$, except at the poles in the first quadrant.

the Regge poles. $a'(\alpha, k)$ is the derivative of $a(\alpha, k)$ with respect to α. The transformation from Equation (3.7) to Equation (3.8) is called the Watson-Sommerfeld transformation (Sommerfeld, 1949). Let α_0 be the one with the largest real part. Then the scattering amplitude, in the limit of large $\cos \theta$, becomes

$$f(k, \theta) \to (\cos \theta)^{\operatorname{Re}[\alpha_0(k)]}. \tag{3.9}$$

The position of the Regge pole depends on k.

This asymptotic behavior inspired many physicists in the 1960's, as it was able to predict the high-energy scattering cross section in the cross channel in which $\cos \theta$ is proportional to the total energy. This approach is commonly known as the Regge-pole model. One of the building blocks of this model is the Chew-Frautchi plot (Chew, 1962) in which the hadronic masses are plotted against their intrinsic angular momentum. As we noted in Chapter V, the intrinsic angular momentum is one of the Casimir operators of the Poincaré group. The Chew-Frautchi plot eventually became the hadronic mass spectrum in the harmonic oscillator model, and it is still not uncommon to call the oscillator spring constant "the slope of the Regge trajectory." As as noted before, the study of the Legendre function is the study of $O(2, 1)$ (Serterio and Toller, 1964; Sciarrino and Toller, 1967; Toller, 1968; Sollani and Toller, 1973). Indeed, the Regge-pole model has provided one of the stepping stones to the physicists' slow process of understanding the Poincaré group.

4. Unitary Representations of SU(1, 1)

Since $SU(1, 1)$ is one of the simplest non-compact groups, study in this area is extensive. Without a clear physical motivation, we are not going through all the papers published in the literature on this subject. The purpose of this section is to illustrate the scope of this subject by discussing the distribution of eigenvalues for infinite-dimensional unitary representations. Since the original work of Bargmann (1947), there have been many reformulations of the same problem. Among the many excellent papers on this subject, we choose to follow closely the lucid presentation of Holman and Biedenharn (1966) who emphasize the connection between $SU(2)$ and $SU(1, 1)$.

Since $SU(1, 1)$ is noncompact and contains no invariant subgroups, the dimension of its unitary representation is necessarily infinite. As before, the generators of the group are K_1, K_2 and J_3. This time, these operators are Hermitian. The Casimir operator C takes the form of Equation (2.2). As in the case of Section 3, we shall consider representations diagonal in J_3. The Hilbert space consists of normalized eigenstates of J_3:

$$J_3|m\rangle = m|m\rangle, \tag{4.1}$$

where m has to be an integer or half integer. The $SU(1, 1)$ transformation on $|m\rangle$ will result in an infinite sum:

$$\sum_{n=-\infty}^{\infty} A_{mn}|n\rangle, \quad \text{with} \quad \sum_{k} (A_{mk})^*(A_{kn}) = \delta_{mn}, \tag{4.2}$$

unlike the case considered in Section 3 where the sum was finite. We are considering here a unitary infinite-dimensional representation.

For the rotation group, we know how eigenvalues of J_3 are distributed for a given value of the total angular momentum. The purpose of this section is to see how they are distributed in the case of $SU(1, 1)$. As in the case of the rotation group, let us consider the step-up and step-down operators:

$$K_\pm = K_1 \pm iK_2. \tag{4.3}$$

then from the commutation relations $[J_3, K_\pm] = \pm J_3$,

$$J_3 K_- = (K_- - 1)J_3,$$
$$J_3 K_+ = (K_+ + 1)J_3. \tag{4.4}$$

The application of $K_\pm$ on $|m\rangle$ results in $|m \pm 1\rangle$. Furthermore, from the definition of C in Equation (2.2),

$$K_+ K_- = -C + J_3(J_3 - 1),$$
$$K_- K_+ = -C + J_3(J_3 + 1), \qquad (4.5)$$

so that

$$K_- K_+ - K_+ K_- = 2J_3. \qquad (4.6)$$

Since K_1 and K_2 are Hermitian operators:

$$(K_\pm)^\dagger = K_\mp, \qquad (4.7)$$

$K_+ K_-$ and $K_- K_+$ must be positive and Hermitian:

$$\langle m | K_\pm K_\mp | m \rangle = |K_\mp | m \rangle|^2 \geq 0. \qquad (4.8)$$

Hence the eigenvalues of

$$-C + J_3(J_3 - 1),$$
$$-C + J_3(J_3 + 1). \qquad (4.9)$$

must be positive or zero. By applying the above operators on $|m\rangle$, we derive the following inequalities:

$$-C + m(m-1) \geq 0,$$
$$-C + m(m+1) \geq 0, \qquad (4.10)$$

for sufficiently large $|m|$. The letter C is used also for the eigenvalue of the Casimir operator. For a given positive m, we can construct a chain of states:

$$|m\rangle, \quad |m+1\rangle, \quad |m+2\rangle, \ldots, \qquad (4.11)$$

by applying repeatedly K_+ on $|m\rangle$. From Equation (4.6), we obtain

$$|K_+|m\rangle|^2 = \langle m|(2J_3 + K_+ K_-)|m\rangle$$
$$= 2m + |K_-|m\rangle|^2. \qquad (4.12)$$

The chain does not terminate. If we apply the step-down operator K_- repeatedly to $|m\rangle$ to obtain

$$|m\rangle, \quad |m-1\rangle, \quad |m-2\rangle, \ldots, \qquad (4.13)$$

one of the two alternatives occurs:

(a)$^+$ The chain never terminates.

(b)$^+$ The chain terminates.

Let us first consider the case (a)$^+$. The positivity conditions of Equation (4.10) in terms of the eigenvalues m imply that

$$C < 0 \text{ if } m \text{ is an integer,} \tag{4.14}$$

$$C < -1/4 \text{ if } m \text{ is a half-integer.} \tag{4.15}$$

These conditions can be restated as

$$C < -1/4 \text{ for both integer and half-integer values of } m, \tag{4.16}$$

$$-1/4 < C < 0 \text{ for integer values of } m. \tag{4.17}$$

If we write C as

$$C = \ell(\ell + 1), \tag{4.18}$$

then the condition of Equation (4.15) is translated into

$$\ell = -1/2 + is, \tag{4.19}$$

where s is a real parameter. We have already seen one example for this case in Section 3.

In the case of (b)$^+$, let $|k\rangle$ be the last non-vanishing state in the descending chain. This means

$$K_-|k\rangle = 0, \tag{4.20}$$

and therefore

$$K_+K_-|k\rangle = [-C + J_3(J_3 - 1)]|k\rangle = 0,$$

or

$$C = k(k - 1). \tag{4.21}$$

Equation (4.12) applied to $|k\rangle$ leads to $k = |K_+|k\rangle|^2/2$, which means that k must be positive or zero. The case $k = 0$ occurs only in the identity representation for which

$$K_+|0\rangle = K_-|0\rangle = 0, \quad \text{and} \quad C = 0. \tag{4.22}$$

Let us next consider the case where the starting value of m is negative. Then the infinite chain of states

$$|m\rangle, |m - 1\rangle, \ldots \tag{4.23}$$

is obtained by the repeated application of K_- on $|m\rangle$. Then, as we apply the step-up operator K_+ repeatedly to $|m\rangle$ we again meet one of the two alternatives (a) and (b) above, which in this case we shall label (a)$^-$ and (b)$^-$:

(a)$^-$ the chain never terminates;

(b)$^-$ the chain terminates.

For case (a)⁻ the positivity conditions on Equation (4.10) again imply one of the conditions of Equations (4.16) and (4.17). In the case of (b)⁻, let $|k\rangle$ be the last non-vanishing state in the ascending chain, i.e.,

$$K_+|k\rangle = 0, \tag{4.24}$$

and

$$K_-K_+|k\rangle = [-C + J_3(J_3 + 1)]|k\rangle = 0,$$

or

$$C = k(k+1). \tag{4.25}$$

Now, Equation (4.12) shows that k must be negative and non-zero. Again, the case $k = 0$ occurs only in the identity representation described by Equation (4.22).

We have obtained a system of classification of the representations of $SU(1, 1)$ in terms of the case (a)$^\pm$ and (b)$^\pm$. We shall call the representations obtained under cases (a)$^\pm$, for which all integer (or half-integral) states $|m\rangle$ exist, the continuous series of irreducible representations. We shall call those obtained under the cases (b)⁺ and (b)⁻, respectively, the positive and negative discrete series representations. In addition to these, of course, there exists the unique, one-dimensional identity representation.

Let us summarize what we did above for both positive and negative values of m. In the case of continuous series,

$$-1/4 < C < 0 \text{ for integer values of } m, \tag{4.26}$$

$$C < -1/4 \text{ for both integer and half-integer values of } m. \tag{4.27}$$

For discrete series, there is a positive lower bound for positive values of m, while the upper bound on negative values of m is negative. The absolute value of m has to be equal to or greater than a positive number $|k|$ which may be integer or half integer. C in this case is $|k|(|k| - 1)$. Figure 4.1 shows the distribution of allowable values of m.

With this preparation, the reader can go to the papers of Holman and Biedenharn (1966, 1968), and the paper of Basu and Wolf (1982), where the subject was completely and thoroughly discussed. As for the mathematical methods, Bargmann in his original paper (1947) uses the *multiplier representation* based on the conformal representation of $SL(2, c)$. This method has been explained thoroughly in Miller's book on Lie theory (1968).

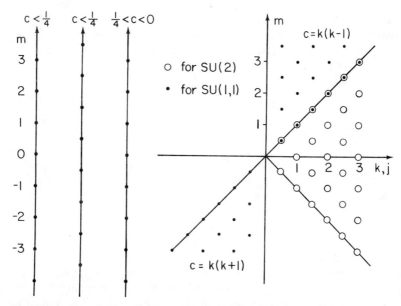

Fig. 4.1. Distribution of allowable eigenvalues of J_3 for unitary representations of $SU(1, 1)$. The eigenvalues for $SU(2)$ are included for comparison.

5. Exercises and Problems

Exercise 1. Special relativity deals with two coordinate frames which move with uniform velocity relative to each other. Therefore, for a particle in a circular orbit, we have to bring the particle back to the rest frame first, and then boost it by the same amount in a different direction. The net result is a pure boost which changes the initial velocity to the final velocity preceded by a rotation. The time rate of change of this rotation angle is called the Thomas precession. Calculate the Thomas procession from the expression given in Equation (1.17).

In the calculation of Section 1, since the magnitude of velocity is the same, we can let

$$\eta = \xi = \frac{1}{2} \ln \frac{1-\beta}{1+\beta}, \tag{5.1}$$

where β is the magnitude of velocity. The final angle is 0, and we can let the initial angle be

$$\theta = \pi - \delta\theta.$$

Thus

$$\sin\theta = \delta\theta, \tag{5.2}$$

for small $\delta\theta$. In this limit, Equation (1.17) becomes

$$\delta\alpha = 2\left(\sinh\frac{\eta}{2}\right)^2 \delta\theta. \tag{5.3}$$

This means that the time rate of the Thomas precession is proportional to the time rate of the rotation of the velocity:

$$\omega_T = (\cosh\eta - 1)\frac{d\theta}{dt}. \tag{5.4}$$

For the exact kinematics of the Thomas precession in terms of the $O(3)$-like little group, see Han *et al.* (1985).

Exercise 2. If a nucleus consists of two nucleons, it is very easy to study its structure. However, if there are many nucleons in the nucleus, then the effective force on a single nucleon is not necessarily isotropic. In fact, the surface of the nucleus can be described by the equation:

$$(x^2 + y^2)e^b + z^2 e^{-2b} = R^2. \tag{5.5}$$

This is an equation for a spheroid whose volume remains unchanged as the parameter b takes different values. This equation indeed describes a nucleus undergoing pancake/football-like deformations. Is this deformation describable in terms of $Sp(2)$?

The deformation of Equation (5.5) is achieved by an elongation/deformation on the xz plane followed by the same deformation on the yz plane. The elongation/deformation is achieved by the matrix

$$\begin{bmatrix} e^{b/2} & 0 \\ 0 & e^{-b/2} \end{bmatrix} \tag{5.6}$$

If we allow full $Sp(2)$ transformations on both planes, the nucleus can also perform rotations around the y axis and x axis respectively. In this case, the transformation becomes the subset of volume preserving transformation of $SL(3, r)$. This is called the collective model of nuclei (Bohr and Mottelson, 1969). For application of the symplectic group in nuclear deformation, see the papers by Moshinsky and Winternitz (1980), Aricks *et al.* (1979), Rosensteel and Rowe (1980), Deenen and Quesne (1982).

Exercise 3. Let us next discuss another example of elliptic deformation. The concept of phase space plays an important role in statistical mechanics, semi-classical problems, and foundations of quantum mechanics. In these areas of physics, the Wigner function defined as

$$W(x, p) = (1/\pi)\int \psi^*(x + x')\psi(x - x')e^{2ipx'} dx' \tag{5.7}$$

serves many useful purposes (Wigner, 1932). How does the symplectic group serve useful purposes for studying the Wigner function?

The simplest way to illustrate the Wigner function is to evaluate the integral of Equation (5.7) for the one-dimensional harmonic oscillator, whose Hamiltonian tkaes the form

$$H = \tfrac{1}{2} mp^2 + \tfrac{1}{2} m\omega^2 x^2. \tag{5.8}$$

The solutions of the Schrödinger equation with this Hamiltonian are well known. If we use the generating function of the Hermite polynomials, it is easy to carry out the integration in Equation (5.7). In terms of the simplified variables:

$$u = (m\omega)^{1/2} x, \quad v = (1/m\omega)^{1/2} p, \tag{5.9}$$

and

$$r^2 = 2(u^2 + v^2), \tag{5.10}$$

the Wigner function for the n-th excited-state harmonic oscillator takes the form

$$W_n(u, v) = (n!/\pi) [\exp(-r^2/2)] \sum_{k=0}^{n} \frac{(-1)^{n-k} r^{2k}}{[(n-k)!(k!)^2]}. \tag{5.11}$$

The above expression is rotationally invariant in the uv plane, and the function takes the same value on a circle centered at the origin.

From Equation (5.9), it is clear that the story on the xp plane is a symplectic deformation of the uv plane. If we combine this with the rotation in the xp plane, the net effect is the general linear canonical transformation. The basic symmetry of this Wigner function in phase space is therefore that of the $Sp(2)$ group.

The study of the Wigner function is so extensive in the literature that it is by now an almost independent branch of physics. The papers on this subject include those of Wigner (1932, 1971), Moyal (1949), Bartlett and Moyal (1949), Nix (1969), Klauder and Sudarshan (1968), Balasz and G. G. Zippel (1973), Davies and Davies (1975), O'Connell and Rajagopal (1982), Carruthers and Zachariasen (1983), and the references quoted in these papers.

Problem 1. Show that $W_n(r)$ of Equation (5.11) satisfies the two-dimensional harmonic oscillator Schrödinger equation:

$$-\frac{1}{2r} \left(\frac{d}{dr} \right)^2 (rW_n(r)) + \frac{1}{2} r^2 W_n(r) = (2n+1) W_n(r). \tag{5.12}$$

W_n can naturally be normalized as

$$2\pi \int_0^\infty W_n(r) \, W_m(r) \, r \, dr = \delta_{nm}.$$

Show then that the following relation is satisfied for two different Wigner functions:

$$\frac{1}{2\pi} \left| \int \psi^*(x) \phi(x) \, dx \right|^2 = \int W_\phi(x, p) \, W_\psi(x, p) \, dx \, dp. \tag{5.13}$$

Problem 2. In Section 3 of Chapter 4, we discussed the irreducible representations of $SU(2)$ and the spinors and their symmetric combinations, particularly those for total spin 1/2, 1, and 3/2. Discuss the transformation properties of those spinors with all possible combinations of dotted and undotted α and β.

Problem 3. Show that it is possible to construct the ψ_ℓ^m of Equation (2.9) starting from $m = 0$ for an arbitrary complex value of ℓ.

Problem 4. Prove that the Legendre function $P_{-1/2+is}(\cosh \eta)$ forms an orthonormal set with respect to the integral of Equation (2.15).

Problem 5. Show that the asymptotic behavior of the Legendre function for large values of z is given by Equation (3.4). See Morse and Feshbach (1953).

Problem 6. Show that the integral representation given in Equation (3.1) satisfies the Legendre differential equation of Equation (3.2) for complex ℓ.

Problem 7. In Section 4 of Chapter I, we noted that the transformation of Equation (1.1) corresponds to the conformal mapping:

$$w = \frac{az + b}{b^*z + c^*}$$

Find the little group which leaves i invariant: $w = z = i$. See Lang (1974).

Problem 8. Explain why the Chew-Frautchi plot (Chew, 1962) is basically the eigenvalue distribution of the three-dimensional harmonic oscillator.

Chapter X

Homogeneous Lorentz Group

We have so far discussed subgroups of the Lorentz group which can describe the internal space-time symmetries of relativistic particles. Transformations of these subgroups leave the four-momentum of a particle invariant. The four-momentum which remains invariant under transformations of the homogeneous Lorentz group has to vanish. Since there are no particles with vanishing four-momentum, this group does not serve any useful purpose as a little group. However, this fact alone cannot eliminate the homogeneous Lorentz group from our consideration.

The construction of representations of the Poincaré group consists of two steps. The first step is to construct representations of the little group for a given four-momentum. The second step is to generalize the representation of the little group for all possible values of the four-momentum belonging to the same orbit. Lorentz transformations play the essential role during the process of orbit completion.

There are two different approaches to this orbit completion process. The first approach is to construct the representation and/or representation space of the little group in a given Lorentz frame, followed by the Lorentz generalization of the representation by performing necessary transformations on the representation space. The second approach is to construct the representation of the little group in a covariant manner. We have been using the first method throughout Chapters III–VIII. The purpose of this chapter is to discuss the second method.

In order to proceed in a covariant manner, we have to know what quantities should remain Lorentz-invariant, in addition to the Casimir operators. For instance, in the rotation group, the total angular momentum is the sum of the orbital and spin angular momenta. The Casimir operator for the rotation group is (total angular momentum)^2. However, this is not the only quantity which remains invariant under rotations. The (total orbital angular momentum)^2 and the (total spin)^2 each are invariant quantities.

In the case of the Poincaré group, the six generators of spinor transformations form a Lie algebra of the homogeneous Lorentz group. There are two Casimir operators for this subgroup, and they commute not only with the six

Homogeneous Lorentz Group

generators applicable to spinors, but also with the ten generators of the Poincaré group. The purpose of this chapter is to exploit this property.

In Section 1, we spell out the proposed procedure. Section 2 deals with a method of constructing all finite-dimensional representations of $SL(2, c)$. In Section 3, the transformation property of Maxwell's equations is discussed as a further illustrative example of spinor representations. In Section 4, we discuss the metric and the normalization useful for unitary representation of the Dirac spinors.

In Section 5, we go back to the harmonic oscillator case, and discuss in detail when the method of homogeneous Lorentz group works and when it does not. It is pointed out in Section 6 that, as in the case of $SO(3)$ and $SL(2, c)$, the step-up and step-down operators may be useful in constructing representations of the homogeneous Lorentz group. We summarize in Section 7 the methods of constructing representations of the Poincaré group.

1. Statement of the Problem

As we stated in Chapter III, the Poincaré group is generated by four translation generators and six generators of Lorentz transformations. We may again write these as

$$P_\mu \quad \text{and} \quad M_{\mu\nu}. \tag{1.1}$$

The operator $M_{\mu\nu}$ which generates Lorentz transformations is the sum of that applicable to the space-time coordinates and that for internal space-time or spin coordinates:

$$M_{\mu\nu} = L^*_{\mu\nu} + S_{\mu\nu}. \tag{1.2}$$

As in Chapter V, we shall use $L^*_{\mu\nu}$ for the particle (hadron or electron) coordinate. $S_{\mu\nu}$ is for internal degrees of freedom.

We observed in Chapter III that the Poincaré group is a semi-direct product of the translation group generated by P_μ and the group of the homogeneous Lorentz transformations generated by $M_{\mu\nu}$. The translation subgroup is an Abelian invariant subgroup, and, therefore, the above-mentioned homogeneous Lorentz group is a quotient group consisting of cosets of the original Poincaré group over the translation group. The purpose of this section is to discuss *another* homogeneous Lorentz group which is a subgroup of the Poincaré group.

The six $L^*_{\mu\nu}$ operators commute with $S_{\mu\nu}$, but they do not commute with P_μ. The $S_{\mu\nu}$ operators commute with both P_μ and $L^*_{\mu\nu}$. For this reason, we can regard the Poincaré group as a *direct product* of the group generated by P_μ and $L^*_{\mu\nu}$ and the homogeneous Lorentz group generated by $S_{\mu\nu}$. This is precisely why we are interested in representations of the homogeneous Lorentz group. After constructing a proposed representation, we should check whether the direct product should be diagonal in the Casimir operators of the Poincaré group (Michel, 1962). As for the new Poincaré group generated by P_μ and $L^*_{\mu\nu}$, its representation is the Klein-Gordon equation.

The Casimir operators of the homogeneous Lorentz group are

$$C_1 = \tfrac{1}{2} S^{\mu\nu} S_{\mu\nu} \quad \text{and} \quad C_2 = \tfrac{1}{4} \varepsilon_{\mu\nu\alpha\beta} S^{\mu\nu} S_{\alpha\beta}. \tag{1.3}$$

These operators commute with all ten generators of the Poincaré group, but are quite different from the Casimir operators of the Poincaré group. Let us see how useful this Lorentz group is in studying representations of the original Poincaré group and in understanding internal space-time symmetry of relativistic particles.

2. Finite-dimensional Representations of the Homogeneous Lorentz Group

By now, we are quite familiar with the fact that the homogeneous Lorentz group is generated by the S_i and K_i which generate rotations and boosts respectively, and also with the commutation relations these generators satisfy. The commutation relations for these operators remain invariant under the sign change of the boost operators.

As we noted in Chapter III, the Casimir operators for this group are

$$\begin{aligned} C_1 &= (\mathbf{S})^2 - (\mathbf{K})^2, \\ C_2 &= \mathbf{S} \cdot \mathbf{K}. \end{aligned} \tag{2.1}$$

It is often convenient to define new operators:

$$\begin{aligned} F_i &= \tfrac{1}{2}(S_i - iK_i), \\ G_i &= \tfrac{1}{2}(S_i + iK_i). \end{aligned} \tag{2.2}$$

Then these new operators satisfy the following commutation relations:

$$\begin{aligned} [F_i, G_j] &= 0, \\ [F_i, F_j] &= i\varepsilon_{ijk} F_k, \\ [G_i, G_j] &= i\varepsilon_{ijk} G_k. \end{aligned} \tag{2.3}$$

It is clear from the above commutation relations that F_i and G_i generate their own respective rotation groups. The Lorentz group can be regarded as a direct product of these two rotation groups. The Casimir operators for these groups are F^2 and G^2 respectively. Their eigenvalues are $f(f+1)$ and $g(g+1)$. The representation of the Lorentz group is $(2f+1)(2g+1)$-

dimensional. In terms of F^2 and G^2, the Casimir operators of Equation (2.1) are

$$C_1 = \frac{1}{2}(F^2 + G^2),$$

$$C_2 = -\frac{i}{4}(F^2 - G^2). \qquad (2.4)$$

This means that the representations with eigenvalues of F^2 and G^2 give fixed values of C_1 and C_2. We can therefore label the representation by two numbers (f, g). Representations constructed in this way are finite-dimensional and non-unitary.

In addition to the Casimir operators, we can choose F_3 and G_3 to be diagonal. Since

$$J_i = F_i + G_i, \qquad (2.5)$$

J_3 can be diagonal for the Lorentz group. Because the Lorentz group in this case is regarded as a direct product of the two rotation groups, the eigenvalues of J^2 is $j(j+1)$, with

$$j = (f+g), (f+g-1), \ldots |f-g|. \qquad (2.6)$$

J^2, F^2, and G^2 can be simultaneously diagonalized, but J^2 does not commute with F_3 or G_3.

As in the case of the three-dimensional rotation group, we should start constructing representations by making tensor products of spinors. In the $SL(2, c)$ case, as we studied in Chapter IV, the boost operators take two different signs. Indeed, this is why we are able to separate the generators into F_i and G_i.

Let us first study a Dirac particle with non-zero mass. When $S_i = \frac{1}{2}\sigma_i$ and $K_i = (i/2)\sigma_i$, $F_i = \frac{1}{2}\sigma_i$ and $G_i = 0$. On the other hand, when $K_i = -(i/2)\sigma_i$, $F_i = 0$ and $G_i = \frac{1}{2}\sigma_i$. Since the boost matrices for the upper and lower components of the Dirac wave functions have opposite signs, the Dirac particle is represented by the direct sum of $(f = 1/2, g = 0)$ and $(f = 0, g = 1/2)$. Indeed the Dirac equation is a finite-dimensional non-unitary representation of the homogeneous Lorentz group. This interpretation of the Dirac equation is quite consistent with the analysis given in Chapter IV.

Let us next consider four-vectors. According to Section 4 of Chapter IV, a four-vector can be constructed from a direct product of a spinor with one sign of the boost operator and another spinor with the opposite sign. Indeed, this direct product constitutes the representation $(1/2, 1/2)$. This is true for both massive and massless fields. Massless fields contain gauge degrees of freedom.

For the general case of (f, g) we know how to construct representations of each of the two rotation groups for higher values of f and g. The eigenvalues of J_3 will run from $(f + g)$ to $-(f + g)$. It is possible to construct a general theory starting from these observations (Weinberg, 1964a). However, we shall continue the practice of studying each case in detail as an illustrative example.

As for massless particles, in a series of papers published in 1964, Weinberg was interested in constructing irreducible representations which are totally independent of gauge degrees of freedom and contain only the minimal number of variables. Gauge-dependent four-potentials are not included in this construction scheme. In order that the massless fields be gauge-independent, the state vectors should satisfy the conditions:

$$(F_1 - iF_2)|m\rangle = 0,$$
$$(G_1 - iG_2)|m\rangle = 0, \qquad (2.7)$$

which are equivalent to

$$N_1^{(\pm)}|m\rangle = N_2^{(\pm)}|m\rangle = 0, \qquad (2.8)$$

where the N operators defined in Section 5 of Chapter VII are the generators of gauge transformations. These operators are the translation-like generators of the $E(2)$-like little group. As we noted in Chapter VII, the above conditions lead to the polarization of the spin-1/2 particle. Indeed,

$$F_3|m\rangle = f|m\rangle,$$
$$G_3|m\rangle = -g|m\rangle, \qquad (2.9)$$

where m is the eigenvalue of J_3. Since $J_3 = F_3 + G_3$,

$$m = f - g. \qquad (2.10)$$

For a right-handed particle with $m = j$, the field can be one of the following representations:

$$(j, 0), \quad (j + 1/2, 1/2), \quad (j + 1, 1), \ldots.$$

The left-handed particles can be associated with one of the representations:

$$(0, j), \quad (1/2, j + 1/2), \quad (1, j + 1), \ldots. \qquad (2.12)$$

According to this scheme, the simplest representation for photons is the direct sum of $(1, 0)$ and $(0, 1)$. This representation corresponds to the solutions of Maxwell's equations for plane waves. The neutrino and antineutrino are represented by $(1/2, 0)$ and $(0, 1/2)$ respectively (Weinberg, 1964a).

The neutrino problem has been discussed in detail in Chapter VII. As for photons, we discussed in detail the four-potential representation also in Chapter VII. In the above-mentioned representation of Weinberg, the photon state vectors can be constructed as a product of two spinors. The positive and negative helicity states are

$$|+\rangle = a_1 a_2, \quad |-\rangle = \beta_1 \beta_2, \tag{2.13}$$

respectively. Under rotations these state vectors can be identified as

$$|+\rangle = -(x + iy), \quad |-\rangle = (x - iy). \tag{2.14}$$

We shall study Lorentz transformation properties of the above spinor combinations in Section 3.

3. Transformation Properties of Electric and Magnetic Fields

We stated in Section 1 that the spinor representations $a_1 a_2$ and $\beta_1 \beta_2$ describe electric and magnetic fields. Since they are invariant under gauge transformations, their little group is simply $O(2)$ and unitary. We are now interested in their Lorentz transformation properties and in whether they are really like those of electric and magnetic fields.

Let us first discuss the transformation properties of electromagnetic fields. If a plane wave propagates along the z direction, the electric and magnetic fields are along the x and y axes respectively. In the case of plane waves or free photons, the magnitude of the magnetic field is the same as that of the electric field.

Under the Lorentz boost by velocity β, the longitudinal components of the eletric and magnetic fields remain unchanged. The transverse component become

$$\begin{aligned} \mathbf{E}'_T &= (E_T - \boldsymbol{\beta} \times \mathbf{B}_T)/(1 - \beta^2)^{1/2}, \\ \mathbf{B}'_T &= (\mathbf{B}_T + \boldsymbol{\beta} \times \mathbf{E}_T)/(1 - \beta^2)^{1/2}, \end{aligned} \tag{3.1}$$

where $\mathbf{E}$ and $\mathbf{B}$ are the electric and magnetic fields respectively.

If the photon is boosted along the z axis so that the momentum or energy is increased by

$$\begin{aligned} p'_0 &= (p_0 + \beta p_0)/(1 - \beta^2)^{1/2} \\ &= \left(\frac{1+\beta}{1-\beta}\right)^{1/2} p_0, \end{aligned} \tag{3.2}$$

then according to Equation (3.1), the transverse electric field which stays in the x direction is also increased by

$$E'_x = \left(\frac{1+\beta}{1-\beta}\right)^{1/2} E_x. \tag{3.3}$$

The magnetic field remains in the y direction and is increased by the same amount.

Let us next boost along the x direction, so that the momentum along the z direction remains the same, but the new momentum has an x component:

$$p'_z = p_0, \quad \text{and} \quad p'_x = p_0 \sinh \eta. \tag{3.4}$$

then the magnitude of the new momentum or energy is

$$p'_0 = (\cosh \eta) \, p_0. \tag{3.5}$$

The magnetic field stays in the y direction, but becomes increased by

$$B'_y = (\cosh \eta) \, B_y. \tag{3.6}$$

The x component of the electric field remains unchanged: $E'_x = E_x$, but the new electric field has a z component:

$$E'_z = -(\sinh \eta) \, B_y = -(\sinh \eta) \, E_x, \tag{3.7}$$

so that it will be perpendicular to the momentum and the magnetic field. Its magnitude is also increased by the same amount as the magnetic field.

Let us identify the spinor combinations of Equation (2.13) as

$$\alpha_1 \alpha_2 = E_x + iB_y, \quad \text{and} \quad \dot{\beta}_1 \dot{\beta}_2 = E_x - iB_y. \tag{3.8}$$

Then, when we boost along the z direction, each of the above spinors becomes increased by $\exp(\xi/2)$, where

$$e^\xi = \left(\frac{1+\beta}{1-\beta} \right)^{1/2}. \tag{3.9}$$

Therefore both the electric and magnetic fields become increased by the above factor. This is consistent with the transformation law given in Equation (3.3).

As we calculated in Section 7 (Exercise 3) of Chapter VII, when we boost along the x direction, the spinor transformation matrix for each of the above α's is

$$\exp[-i(\eta/2)K_1] = \begin{bmatrix} \cosh \dfrac{\eta}{2} & \sinh \dfrac{\eta}{2} \\ \sinh \dfrac{\eta}{2} & \cosh \dfrac{\eta}{2} \end{bmatrix}, \tag{3.10}$$

and the matrix for $\dot{\beta}$ is the inverse of the above form. Then the new spinors become

$$\alpha' = \begin{bmatrix} \cosh \dfrac{\eta}{2} \\ \sinh \dfrac{\eta}{2} \end{bmatrix} = (\cosh \eta)^{1/2} \begin{bmatrix} \cos \dfrac{\theta}{2} \\ \sin \dfrac{\theta}{2} \end{bmatrix}, \qquad (3.11)$$

$$\beta' = \begin{bmatrix} -\sinh \dfrac{\eta}{2} \\ \cosh \dfrac{\eta}{2} \end{bmatrix} = (\cosh \eta)^{1/2} \begin{bmatrix} -\sin \dfrac{\theta}{2} \\ \cos \dfrac{\theta}{2} \end{bmatrix}. \qquad (3.12)$$

where

$\tan \theta = \sinh \eta$.

New spinors α' and β' specify the positive and negative helicities with respect to the new momentum respectively. Each of the above spinors gained the normalization constant $(\cosh \eta)^{1/2}$. Thus both $\alpha_1 \alpha_2$ and $\beta_1 \beta_2$ became bigger by $(\cosh \eta)$. This is exactly the transformation property of the electric and magnetic fields.

With this result, we can now consider the spinor combinations:

$$\begin{aligned} \phi_{11} &= \alpha_1 \alpha_2, \quad \phi_{1-1} = \beta_1 \beta_2, \\ \phi_{10} &= (\alpha_1 \beta_2 + \beta_1 \alpha_2)/\sqrt{2}, \\ \phi_{00} &= (\alpha_1 \beta_2 - \beta_1 \alpha_2)/\sqrt{2}, \end{aligned} \qquad (3.13)$$

and

$$\begin{aligned} \dot{\phi}_{11} &= \dot{\alpha}_1 \dot{\alpha}_2, \quad \dot{\phi}_{1-1} = \dot{\beta}_1 \dot{\beta}_2, \\ \dot{\phi}_{10} &= (\dot{\alpha}_1 \dot{\beta}_2 + \dot{\beta}_1 \dot{\alpha}_2)/\sqrt{2}, \\ \dot{\phi}_{00} &= (\dot{\alpha}_1 \dot{\beta}_2 - \dot{\beta}_1 \dot{\alpha}_2)/\sqrt{2}. \end{aligned} \qquad (3.14)$$

Both ϕ_{00} and $\dot{\phi}_{00}$ are singlets and they are invariant both under rotations and boosts. Among the remaining six, we now know the transformation properties of $\alpha_1 \alpha_2$ and $\beta_1 \beta_2$, and we can obtain others by rotating either of them. Therefore, it is possible to construct from these six spinor combinations a six-component antisymmetric tensor whose transformation property is identical to that of the electromagnetic field tensor.

From the $SL(2, c)$ standpoint, there is a non-trivial difference between the four-potential and electromagnetic field. As we studied in Chapter VII, the four potential is represented by a direct product of one dotted and one undotted spinors. The most crucial difference between the two different representations is that the transformations on the four-potential can be made unitary. Transformations on the electric and magnetic fields are non-unitary as in the case of the Dirac spinors. The amplitudes of the Dirac and Maxwell fields are proportional to $(p_0)^{1/2}$ and p_0 respectively.

4. Pseudo-unitary Representations for Dirac Spinors

We have seen in Sections 2 and 3, and also in Section 5 of Chapter IV that Lorentz boosts change the normalization of the $SL(2, c)$ spinors. This is a manifestation of the fact that finite-dimensional representations of the non-compact groups are not unitary, and that the Lorentz group is a non-compact group.

However, the boost generators of $SL(2, c)$ are double-valued. This property assures us that if a given boost changes the normalization by a number $\exp(\eta/2)$, there is another boost in the same representation which will change the normalization by $\exp(-\eta/2)$. Indeed, the normalization of the dotted spinor is the inverse of that of the undotted spinor. Therefore, if we define the inner product of two spinors in such a way that one normalization factor will cancel the other, then the unitarity will be maintained.

Let us see whether this is possible for Dirac particles in the Weyl representation. In this representation, the undotted and dotted spinors constitute the upper and lower components of the four-component Dirac spinor. If the inner product is defined in such a way that the upper component is multiplied by the lower component, the unitarity will be guaranteed during the process of Lorentz boost. In the Weyl representation, γ_0 takes the form

$$\gamma_0 = \begin{bmatrix} 0 & 1 \\ 1 & 0 \end{bmatrix}. \tag{4.1}$$

The boost operator is

$$B(\boldsymbol{\eta}) = \begin{bmatrix} \exp(\boldsymbol{\eta} \cdot \boldsymbol{\sigma}/2) & 0 \\ 0 & \exp(-\boldsymbol{\eta} \cdot \boldsymbol{\sigma}/2) \end{bmatrix}. \tag{4.2}$$

This boost matrix is not unitary, but is clearly Hermitian. γ_0 is therefore Lorentz-invariant:

$$[B(\boldsymbol{\eta})]^\dagger \gamma_0 B(\boldsymbol{\eta}) = \gamma_0. \tag{4.3}$$

Indeed, the Dirac spinors are unitary if γ_0 is placed between the two spinors when we take their inner product.

In the Dirac representation, γ_0 takes the form

$$\gamma_0 = \begin{bmatrix} 1 & 0 \\ 0 & -1 \end{bmatrix}. \tag{4-4}$$

Instead of taking the Hermitian conjugate of the spinor U, we often take the quantity $\bar{U} = U^\dagger \gamma_0$. Then the product of two spinors $\bar{U}V$ is known to be a Lorentz invariant quantity (Bjorken and Drell, 1964).

As we discussed in Section 4 of Chapter I, the inner product with the metric of the form of Equation (4.4) is called a pseudo-unitary representation. As was pointed out in Section 5 of Chapter IV, the Weyl representation is unitarily equivalent to the Dirac representation. Therefore, in this section, we have explained the physics of the above pseudo-unitary representation using the Weyl representation.

The pseudo-unitary representation as described above is useful only if there exist non-vanishing inner products between dotted and undotted spinors. There are non-vanishing spinor products in this pseudo-unitary space for massive Dirac particles. However, in the case of massless particles, the spin of the dotted spinor has to be opposite to that of the undotted spinor. For this reason, there is no non-zero pseudo-unitary inner product. The $SL(2, c)$ representations remain non-unitary for massless particles.

A similar explanation of the pseudo-unitary representation can be given in the Foldy-Wouthuysen picture of the Dirac equation in which the energy matrix takes a diagonal form, as was discussed in the literature (Streater and Wightman, 1964; Bogoliubov et al., 1975; Barut and Raczka, 1977; and S. T. Ali, 1981).

5. Harmonic Oscillator Wave Functions in the Lorentz Coordinate System

We study in this section finite-dimensional representations of the homogeneous Lorentz group using covariant harmonic oscillators. In the covariant harmonic oscillator regime in which two quarks are bound together within a hadron, as was noticed before, there are two space-time coordinates. One is for the hadron, and the other is the internal space-time separation between the quarks. This internal space-time coordinate acts like the spin coordinate in the present case. Thus, if we use x_μ as the internal coordinate and use the notation $L_{\mu\nu}$ for $S_{\mu\nu}$, then

$$L_{\mu\nu} = i\left(x_\mu \frac{\partial}{\partial x_\nu} - x_\nu \frac{\partial}{\partial x_\mu}\right),$$

and

$$L_i = -i\varepsilon_{ijk} x_j \frac{\partial}{\partial x_k},$$

$$K_i = i\left(x_i \frac{\partial}{\partial t} + t \frac{\partial}{\partial x_i}\right). \tag{5.1}$$

In Chapter V, we were primarily interested in the normalizable solutions of the harmonic oscillator equation whose forms depend on the four-momentum of hadrons. They constitute the representation space for the $O(3)$-like little group for massive particles. In order to obtain them, we used the coordinate system in which the t' variable is separated from the three space-like variables which span the three-dimensional Euclidean space. There are many other coordinate systems in which the oscillator equation is separable, and the coordinate system determines which quantities are diagonal (Kalnins and Miller, 1977). We are interested in the coordinate system in which the sub-Casimir operators C_1 and C_2 of Equation (2.1) which we would construct using the above generators are diagonal.

If we evaluate C_2 using the explicit expression for S_i and K_i given in Equation (5.1), this operator vanishes for the present case of spinless quarks. In order to obtain solutions diagonal in C_1, let us consider the hyperbolic coordinate system with the coordinate variables: ρ, α, θ and ϕ. They are related to the Cartesian variables by

$$\rho = [|t^2 - r^2|]^{1/2}, \tag{5.2}$$

where

$$r = (x^2 + y^2 + z^2)^{1/2},$$
$$x = r\sin\theta\cos\phi,$$
$$y = r\sin\theta\sin\phi, \tag{5.3}$$
$$z = r\cos\theta,$$
$$t = \pm\rho\cosh\alpha, \quad r = |\rho\sinh\alpha| \text{ when } |t| > r,$$

and

$$t = \rho\sinh\alpha, \quad r = \rho\cosh\alpha \text{ when } |t| < r.$$

Among the above four variables, ρ is Lorentz invariant. For convenience, we shall call this coordinate system the *Lorentz coordinate system*.

In the Lorentz coordinate system, the sub-Casimir operator C_1 takes the form (Kim, et al., 1979b)

$$C_1 = \left(\frac{1}{\sinh\alpha}\right)^2 \frac{\partial}{\partial\alpha}\left(\sinh^2\alpha \frac{\partial}{\partial\alpha}\right) - \left(\frac{1}{\sinh\alpha}\right)^2 (\mathbf{L})^2, \tag{5.4}$$

in the time-like region. In the space-like region,

$$C_1 = \left(\frac{1}{\cosh \alpha}\right)^2 \left(\frac{\partial}{\partial \alpha} \cosh^2 \alpha \frac{\partial}{\partial \alpha}\right) + \left(\frac{1}{\cosh \alpha}\right)^2 (\mathbf{L})^2, \qquad (5.5)$$

where

$$(\mathbf{L})^2 = -\left(\frac{1}{\sin \theta}\right)^2 \frac{\partial}{\partial \theta}\left(\sin^2 \theta \frac{\partial}{\partial \theta}\right) - \left(\frac{1}{\sin \theta}\right)^2 \left(\frac{\partial}{\partial \phi}\right)^2. \qquad (5.6)$$

The harmonic oscillator differential equation of Equation (V.1.6) then becomes

$$\left(\frac{1}{\rho}\right)^3 \frac{\partial}{\partial \rho}\left(\rho^3 \frac{\partial V}{\partial \rho}\right) + \left[\left(\frac{1}{\rho}\right)^2 C_1 - \rho^2\right] V = \varepsilon V. \qquad (5.7)$$

We can now consider the form of the solution

$$V(x) = R(\rho) B(\alpha, \theta, \phi). \qquad (5.8)$$

In terms of the R and B functions, the differential equation of Equation (5.4) is separated into

$$\frac{1}{2}\left[-\left(\frac{1}{\rho}\right)^3 \frac{\partial}{\partial \rho}\left(\rho^3 \frac{\partial}{\partial \rho}\right) + \eta \left(\frac{1}{\rho}\right)^2 + \rho^2\right] R(\rho) = \varepsilon R(\rho). \qquad (5.9)$$

and

$$C_1 B(\alpha, \theta, \phi) = \eta B(\alpha, \theta, \phi).$$

In order that the radial equation have regular solutions,

$$\eta = n(n+2), \qquad (5.10)$$

where n takes integer values. The radial function in this case becomes

$$R_{\mu n}^{\ell}(\rho) = (2(\mu!)/[\Gamma(\mu + n + 2)]^3)^{1/2} \rho^n L_\mu^{n+1}(\rho^2) \exp(-\rho^2/2), \qquad (5.11)$$

with

$$\varepsilon = \pm(2\mu + n), \quad \mu = 0, 1, 2, \ldots,$$

where $L_\mu^{n+1}(\rho^2)$ is again the associated Laguerre function. This form of the radial function is the same as the one for the $O(4)$ case given in Section 5 of Chapter V. The eigenvalue ε is positive in the space-like region, while it is negative in the time-like region.

From the form of C_1 given in Equations (5.4) and (5.5), it is clear that we can separate the B function as

$$B_n^{\ell m}(\alpha, \theta, \phi) = A_n^\ell(\alpha) Y_\ell^m(\theta, \phi). \qquad (5.12)$$

As is indicated in Equations (5.4) and (5.5), the A function will take different forms for space-like and time-like regions. For the time-like region, we use the notation

$$A_n^\ell(\alpha) = T_n^\ell(\alpha), \tag{5.13}$$

and for the space-like region

$$A_n^\ell(\alpha) = S_n^\ell(\alpha). \tag{5.14}$$

Then T and S satisfy the following differential equations.

$$\frac{d^2}{d\alpha^2}(\sinh^2 \alpha\, T_n^\ell(\alpha)) - [n(n+2) + \ell(\ell+1)]\, T_n^\ell(\alpha) = 0,$$

$$\frac{d^2}{d\alpha^2}(\cosh^2 \alpha\, S_n^\ell(\alpha)) - [n(n+2) - \ell(\ell+1)]\, S_n^\ell(\alpha) = 0. \tag{5.15}$$

If $\ell = 0$, the solutions to the above equations take the form

$$T_n^0(\alpha) = \sinh[(n+1)\alpha]/\sinh \alpha,$$
$$S_n^0(\alpha) = \cosh[(n+1)\alpha]/\cosh \alpha. \tag{5.16}$$

For non-vanishing values of ℓ,

$$T_n^\ell(\alpha) = (\sinh \alpha)^\ell \left(\frac{1}{\sinh \alpha}\frac{d}{d\alpha}\right)^\ell T_n^0(\alpha),$$

$$S_n^\ell(\alpha) = (\cosh \alpha)^\ell \left(\frac{1}{\cosh \alpha}\frac{d}{d\alpha}\right)^\ell S_n^0(\alpha), \tag{5.17}$$

We have discussed above the solutions of the oscillator differential equation which are diagonal in the sub-Casimir operators of the $SO(3, 1)$ group. These solutions are quite different from the normalizable wave functions discussed in Chapter V. As we can see in Figure 5.1, the normalizable wave functions diagonal in the Casimir operators of the Poincaré group are localized within an elliptic region around the origin. However, the solutions of the oscillator equation discussed in this section are not normalizable and are localized hyperbolically around the light cones.

The question is then whether a solution with this hyperbolic distribution can be converted into a normalizable wave function with an elliptic distribution. If we replace t by (it), then the solutions given in this section which are diagonal in the sub-Casimir operators of Equation (5.4) or (5.5) become the $O(4)$-based solutions given in Chapter V. The $O(4)$-based solutions correspond to the circle in Figure 5.1. The ellipse in the same figure corresponds to a Lorentz-deformed wave function which can be written as a linear

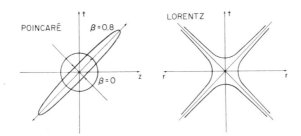

Fig. 5.1. Region of localization in the $O(4)$ and the $O(3, 1)$ coordinate systems. The normalizable $O(4)$ wave functions can be made diagonal in the Casimir operators of the Poincaré group after suitable linear combination. The wave functions in the $O(3, 1)$ coordinate system are not normalizable and cannot be made diagonal in the Casimir operators of the Poincaré group.

expansion of orthonormal $O(4)$ wave functions. It in now clear that the connection between the solutions of the present section and the normalizable wave functions require more than a linear transformation.

Because of the Lorentz invariance of the Casimir operators, it is mathematically appealing to study solutions of the oscillator equation diagonal in the operators given in equation (5.1), instead of those of the original Poincaré group (Michel, 1962). In fact, it is not difficult to find in the literature hadronic models based on this representation (Feynman et al., 1971; Leutwyler and Stern, 1978). However, the crucial point is that it is very difficult to construct normalizable wave functions in these models.

It is also appealing to study unitary representations diagonal in C_1 and C_2. In fact, the analysis only for the part which does not depend on the ρ variable has been carried out by Gel'fand et al. (1966). In the case of harmonic oscillators, the study will also require that of the radial functions. The question is whether it is possible to construct normalizable solutions of the harmonic oscillator equation. This analysis would be very similar to the $SO(2, 1)$ case where complex angular momentum plays a predominant role. In the present $SO(3, 1)$ case, we still need a physical motivation for carrying out this analysis, even though it would produce beautiful mathematics.

6. Further Properties of the Homogeneous Lorentz Group

As we noted in Sections 2 and 5, the study of the homogeneous Lorentz group sometimes gives the state vectors which correspond to observable particles. It was Dirac (1945) who observed that finite-dimensional representations of the Lorentz group are non-unitary and who suggested the study of unitary, though infinite-dimensional, representations of this group. For this purpose, Dirac suggested further the use of the harmonic oscillator wave functions normalizable in four-dimensional space-time. We discussed these wave functions in Chapters V and VI.

Harish-Chandra (1947) started the investigation of infinite-dimensional unitary representations of the homogeneous Lorentz group. Harish-Chandra's method is based on the step-up and step-down operators which we use often in the rotation group and which serves a useful purpose in studying unitary representations of $SU(1, 1)$. This method was developed further by Naimark (1957), and has now been polished enough for a text book on quantum mechanics (Bohm, 1979), as well as on mathematical physics (Miller, 1972).

Since it is customary to use the notation J_i and K_i for the rotation and boost generators in the literature on this subject, we shall change the notation from S_i which we used in Sections 1 and 2 to J_i. In terms of $J_\pm = J_1 \pm iJ_2$ and $K_\pm = K_1 \pm iK_2$, the commutation relations for the Lorentz group can be written as

$$[J_3, J_\pm] = \pm J_\pm, \quad [J_+, J_-] = 2J_3,$$
$$[J_3, K_3] = [J_\pm, K_\pm] = 0,$$
$$[K_3, K_\pm] = \mp J_\pm, \quad [K_+, K_-] = -2J_3, \qquad (6.1)$$
$$[J_\pm, K_3] = -K_\pm, \quad [J_\pm, J_\mp] = \pm 2K_3,$$
$$[J_3, K_\pm] = \pm K_\pm.$$

Because we are familiar with the rotation group, we construct the representation space in terms of the wave functions diagonal in J^2 and J_3. These wave functions have the well-known property:

$$\langle j', m'|J_3|j, m\rangle = m\, \delta_{mm'}\, \delta_{jj'},$$
$$\langle j', m'|J_\pm|j, m\rangle = [(j \pm m + 1)(j \mp m)]^{1/2}\, \delta_{jj'}\, \delta_{m', m \pm 1}. \qquad (6.2)$$

Next, strictly in terms of the rotation group, the boost operators K_i form a three-dimensional vector, or a tensor operator of rank 1. The operators $Q_0 = \sqrt{2}\,K_3$ and $Q_\pm = \mp K_\pm$ constitute a spherical vector. We can thus apply the Wigner-Eckart theorem to write

$$\langle j', m'|K_3|j, m\rangle = (1/\sqrt{2})\, N(j, j')\, C(j', m'|1, 0; j, m), \qquad (6.3)$$

and

$$\langle j', m'|K_\pm|j, m\rangle = \mp N(j, j')\, C(j', m'|1, \pm 1; j, m). \qquad (6.4)$$

where $N(j, j')$ is independent of m and vanishes unless $j' = j$ or $j' = j \pm 1$. $C(j', m'|\ldots)$ are the Clebsch-Gordan coefficients. These coefficients can be calculated or can be found in a standard table. The Wigner-Eckart theorem is discussed in textbooks on quantum mechanics, as well as standard referencce books on the rotation group (Edmonds, 1957). It is then possible to write the effect of the boost generators $|j, m\rangle$ as:

$$K_3 |j, m\rangle = [(j - m)(j + m)]^{1/2} A_j |j-1, m\rangle - m B_j |j, m\rangle$$
$$- [(j + m + 1)(j - m + 1)]^{1/2} C_j |j+1, m\rangle, \qquad (6.5)$$

$$K_\pm |j, m\rangle = \pm [(j \mp m)(j \mp m - 1)]^{1/2} A_j |j+1, m \pm 1\rangle$$
$$- [(j \pm m + 1)(j \mp m)]^{1/2} B_j |j, m \pm 1\rangle$$
$$\pm [(j \pm m + 1)(j \pm m + 2)]^{1/2} C_j |j+1, m \pm 1\rangle. \qquad (6.6)$$

where A_j and B_j are coefficients which depend only on j. The remaining task is then to determine these coefficients.

After reducing the problem into that of determining the coefficients A_j, B_j, and C_j, we need a definite model to continue the study. For instance, these coefficients may be operators acting on the variables not depending on angles. In the case of the covariant oscillator model discussed in Section 4 of Chapter V, the coefficients are differential operators of t and r. As we stated in Chapter V, the complete study of this problem remains as a future research problem.

As for other models discussed in the literature, Naimark assumes that coefficients A_j, B_j and C_j are constants, and considers the case in which $C_j = A_{j+1}$. The commutation relation $[K_\pm, K_3] = \pm J_\pm$ applied to $|j, m\rangle$ leads to:

$$\pm J_\pm |j, m\rangle = K_\pm(K_3|j, m\rangle) - K_3(K_\pm |j, m\rangle). \qquad (6.7)$$

This means that the left-hand side can be calculated from Equation (6.6), while the right-hand side, after repeated applications of the operations defined in Equation (6.5) and (6.6), becomes quadratic in A and B coefficients. After simplification, the equations for these coefficients become independent of m. They are:

$$[(j + 1)B_j - (j - 1)B_{j-1}] A_j = 0,$$
$$[(j + 2)B_{j+1} - (j)B_j] A_{j+1} = 0, \qquad (6.8)$$
$$(2j - 1)(A_j)^2 - (2j + 3)(A_{j+1})^2 - (B_j)^2 = 0.$$

The solution is

$$B_j = i(j_0 j_1)/[j(j + 1)], \qquad (6.9)$$
$$A_j = i[(j^2 - j_0^2)(j^2 - j_1^2)/(4j^2 - 1)]^{1/2}/j. \qquad (6.10)$$

The above coefficients depend on the parameters j_0 and j_1 which specify the representation.

The concept of constructing models of elementary particles starting from representations of the homogeneous Lorentz group still remains as an attractive proposition, and there are many papers in the literature on this subject (Hanson and Regge, 1974; Roman and Haavisto, 1981; Biedenharn and van Dam, 1974; Aldinger *et al.*, 1983; Bohm *et al.*, 1983).

7. Concluding Remarks

The purpose of this book was to study the representations of the Poincaré group for relativistic particles. The study consists of two steps. The first step is to work out the little groups for a given four-momentum. The second step is to complete the orbit by boosting and rotating the system. As we did in this chapter, we can approach the problem by studying the representations of the homogeneous Lorentz group for internal degrees of freedom. Table 7.1 summarizes the major conclusions.

For particles with non-zero mass, the little group is like $O(3)$. The little group applicable to electrons is $SU(2)$ which is locally isomorphic to $SO(3)$. The little group for massive vector mesons and also for the covariant harmonic oscillator is $SO(3)$. For massive particles, it is convenient to study the little group in the Lorentz frame in which the particle is at rest.

TABLE 7.1
Representations of the Poincaré group for relativistic particles. When we use the word "unitary," it is important to specify where it applies. For instance, if we restrict ourselves to the little groups only, then all the particles have their own unitary representations. If we talk about the overall transformation properties, not many particles have unitary representations.

	Little groups	Orbit completion	Overall transformation
Massive particles:			
Dirac particles	Unitary	Pseudo-unitary*	Pseudo-unitary*
Vector mesons	Unitary	Non-unitary	Non-unitary
Cov. Harmonic Oscillators	Unitary	Unitary	Unitary
Massless particles:			
Neutrinos	Unitary	Non-unitary	Non-unitary
Photons: A_μ	Non-unitary	Non-unitary	Non-unitary
		Unitary up to gauge transformation	
Photons: A_μ with helicity gauge	Unitary	Unitary	Unitary
Photons: electric and magnetic fields	Unitary	Non-unitary	Non-unitary

* The pseudo-unitarity of massive Dirac particles has been discussed in Section 4.

There is no Lorentz frame in which the massless particle is at rest. For this reason, we align the coordinate system in such a way that the momentum is along the z direction. The little group in this case is a three-parameter subgroup of the Lorentz group whose generators satisfy the Lie algebra of the two-dimensional Euclidean group. It was noted that a massless particle has three internal degrees of freedom. One of them is the rotation around the momentum associated with the helicity. The remaining two are the gauge degrees of freedom. The requirement of gauge invariance allows both positive and negative helicity states for massless particles with spin 1. However, for massless particles with spin 1/2, the gauge invariance leads to only one helicity (either positive or negative, not both) state, as we observe in the polarization of neutrinos.

We noted that the little group for massless particles can be obtained from the large-momentum/small-mass limit of the massive case. We observed also that the four-vectors and Maxwell tensors can be constructed from direct products of two spinors for both massive and massless cases.

The little group for negative (mass)^2 does not appear to have direct application in physics, because there is no reason to expect that particles can move faster than light. However, $SO(2, 1)$ and $SU(1, 1)$ have other physical applications. Perhaps the most important application is in the form of the symplectic group $Sp(2)$ which is becoming increasingly important in both new and traditional branches of physics. It is also an important fact that most of the practical calculations in special relativity are done in the $SO(2, 1)$ or $SU(1, 1)$ regime, because we deal most often with motions of particles in a plane. $SU(1, 1)$ is the simplest non-trivial non-compact group while $SU(2)$ is the simplest non-trivial compact group. Therefore $SU(1, 1)$ occupies its unique place in the study of non-compact groups which are becoming more popular among physicists. We also take advantage of the fact that $SU(1, 1)$ is unitarily equivalent to $SL(2, r)$ so that two-dimensional geometry can be used for computational purposes.

The homogeneous Lorentz group plays three different roles in the study of the Poincaré group. The first role is to be the little group for particles with zero four-momentum. Since we do not expect to observe such particles, $SO(3, 1)$ or $SL(2, c)$ is not useful as a little group. Second, the homogeneous Lorentz group is a factor group of the Poincaré group over the translation subgroup, and is the transformation group for the process of orbit completion. Third, as we discussed in this chapter, this group is useful in studying the Lorentz-invariance properties of the internal degrees of freedom.

We have to be careful when we use the word "unitary." As is indicated in Table 7.1, this word can apply to the representations of the little groups only, or to all Lorentz transformations including orbit completion. Some of the early original articles (Wigner, 1939; Dirac, 1945; Bargmann and Wigner, 1946) appear to apply this word to the overall transformation, based on the probabilistic interpretation of quantum mechanics. However, the development of covariant quantum field theory does not require this overall unitarity. For example, the free-particle Dirac equation we use in quantum electrodynamics is not a unitary representation, but is a pseudo-unitary representation as is explained in Section 4.

In spite of this new development, we emphasize the traditional approach to the unitarity problem. Indeed, we studied in detail the transformation properties of photon polarization vectors. The representation is finite-dimensional and unitary in spite of the fact that the Lorentz group is non-compact. This is due to the fact that the $E(2)$-like litle group contains an Abelian invariant subgroup which can be factored out, and the little group contains only one degree of freedom which is rotation around the momentum. This representation is often called Wigner's trivial representation (Wigner, 1939, 1962b; Mackey, 1963, 1968; Lipsman, 1974). It was observed that the electromagnetic four-potential in the helicity gauge is the correct description of the trivial representation.

Another unitary representation listed in Table 7.1 is the covariant harmonic oscillator model. This model was originally proposed by Dirac (1945) for studying the covariant Fock space needed in quantum field theory. However, the model became more useful after Yukawa's proposal (1953) to describe relativistic extended particles and its application to the quark model (Fujimura et al., 1970; Feynman et al., 1971). It was shown in Chapters V and VI that this harmonic oscillator model has far reaching applications in physics and mathematics. It was shown in Chapter V that the solutions of the oscillator equation form a representation space for unitary irreducible representations of the Poincaré group. In Chapter VI, it was shown that this harmonic oscillator model meets Dirac's requirement for his "instant form" quantum mechanics (Dirac, 1949). Furthermore, the model has a very simple Lorentz deformation property in Dirac's light-cone coordinate system (Dirac, 1949). The next question about this harmonic oscillator model is whether it can explain observable features in the real world. We shall discuss these features in Chapters XI and XII.

Chapter XI

Hadronic Mass Spectra

Historically quantum mechanics was developed for the purpose of explaining discrete spectra. The first discrete spectrum was that of the hydrogen atom. The second discrete spectrum was that of nuclei which are quantum bound states of nucleons. Quantum mechanics has been very successful in explaining these two traditional spectra.

The third discrete spectrum is the mass spectrum of strongly interacting "elementary particles" commonly called "hadrons." It is by now firmly established that hadrons are quantum bound states of quarks. One of the wonders of modern high-energy physics is that the use of nonrelativistic quantum mechanics leads to a reasonable understanding of hadronic mass spectra. Then what is the theoretical basis for using nonrelativistic bound-state picture for this highly relativistic situation?

The answer to the above question is very simple. When we study hadronic mass spectra, we often employ the language of the direct product of $O(3)$, $SU(2)$ and $SU(3)$ (Greenberg and Resnikoff, 1967; Feynman et al., 1971), where $SU(2)$ describes quark spins, and $SU(3)$ is used for other internal quantum numbers such as "flavor" and "color." $O(3)$ means that nonrelativistic quantum mechanics with rotational symmetry is used. Because $O(3)$ is not Lorentz-invariant, it is very easy to make a hasty conclusion that the quark model is inherently nonrelativistic. This is not correct logic. As we studied in Chapters III and V, the little group for massive particle is locally isomorphc to $SO(3)$. Indeed, $O(3)$ of the quark model is a manifestly relativistic concept.

With this point in mind, we can use nonrelativistic models in the Lorentz frame in which the hadron is at rest. Among those models, the three-dimensional isotropic harmonic oscillator serves many useful purposes. As we discussed in Chapter V, the oscillator formalism is mathematically simple and can be made relativistic. In addition, the formalism readily accommodates three-particle kinematics and symmetry properties. The formalism of course contains the above-mentioned $O(3)$ symmetry.

In this chapter, we are interested in seeing whether there is evidence in the mass spectrum data for the existence of the harmonic oscillator excitations.

Specifically, we should study the (mass)^2 spectrum derived in Chapter V and VI. The oscillator excitations should create an equal-spaced spectrum. In order to examine this linearity, we need at least three levels, including the ground state. This is at present possible for non-strange baryons which are bound states of three quarks. For this reason, we have to go through the process of three-particle symmetry classification.

In Section 1, we briefly explain what the quark model is. In Section 2, we carry out explicit calculation for three-particle symmetry classification scheme following the procedure outlined by Dirac for the general case. The permutation operators which commute with the Hamiltonian and with one another are constructed and their eigenvalues are calculated. In Section 3, we construct wave functions which are diagonal in the permutation operators defined by Dirac. It is shown that this method leads to the three-quark wave functions (Feynman *et al.*, 1971). Section 4 deals with the problem of combining two symmetrized wavefunctions.

In Section 5, the three-body spin wave functions are considered. It is pointed out that the three-quark spin wave functions are identical to those of the three-electron system. In Section 6, the unitary spin wave functions are considered for the three-quark system. It is shown that these wave functions specify the well-known $SU(3)$ multiplets. The $SU(3)$ states are then combined with the spin wave functions of Section 5 to generate the $SU(6)$ states.

In Section 7, we discuss three-body spatial wave functions using harmonic oscillators. The oscillator scheme produces reasonably accurate mass spectra and is mathematically simple for representing the symmetry properties. In Section 8, these oscillator wave functions are combined with the $SU(6)$ states to generate totally symmetric baryonic wave functions.

In Section 9, we note that mesons present to us a relatively simple problem and discuss mesonic multiplets in the quark model. Section 10 consists of a brief review of the experimental efforts to identify the hadronic states in the quark model. Section 11 contains exercises and problems which may be helpful in studying the $SU(3)$ group.

1. Quark Model

In the quark model, hadrons are quantum bound states of quarks and/or antiquarks. Baryons such as the proton and neutron are bound states of three quarks, and mesons such as the π and K mesons are bound states of a quark and antiquark. Like electrons, quarks are spin-1/2 fermions and have negligible size. Unlike electrons quarks carry two additional quantum numbers commonly called *flavor* and *color*. The traditional name for flavor is *unitary spin*.

In spite of the importance of understanding relativistic bound-state problems, the quark model was originally developed to explain selection rules in hadronic processes. The idea started from the concept of isotopic spins in the system of π mesons and nucleons. Let us look at the nucleons. The proton and neutron have approximately the same mass, and further study of their properties had led us to believe that they belong to the same isotopic multiplet, and that the only difference between them is the electromagnetic property. For example, it is by now firmly believed that the neutron and the proton will have the same mass once the electromagnetic interaction is turned off. This isotopic spin is often called *isospin* for simplicity. What is then isospin?

The symmetry of electron spin is governed by the $SU(2)$ group. The spin can be up or down. Likewise, we can consider a Hilbert space of nucleonic states, and use I and I_3 to specify the total isospin and its third component. I for the nucleonic system is $1/2$. If the nucleon is proton, its I_3 is $1/2$. The neutron's I_3 is $-1/2$. There are three π mesons with approximately the same mass separated only by electromagnetic interaction. The total isospin for this mesonic system is 1, and the eigenvalues of I_3 for π^+, π^0, and π^- are 1, 0 and -1 respectively. The nucleonic multiplet and mesonic multiplets are represented by a spinor and a vector in isospin space resepctively. From these vectors and spinors, we can construct scalar quantities which are invariant under rotations in isospin space. Indeed, the observed strong-interaction symmetries are consistent with the rotational symmetry in isospin space. This aspect of strong interaction is widely discussed in textbooks (Frazer, 1966).

It was observed that there are, in addition to the nucleons, six more particles which may be put into the same multiplet, and that not all hadronic transitions are due to strong interactions. It was absolutely necessary to add another dimension to the internal symmetry space, called the hyperchange, to explain the selection rules for these eight particles which are similar to the nucleon. If we add another dimension to the nucleonic multiplet, the symmetry group has to be enlarged from $SU(2)$ to $SU(3)$. However, since there are eight particles in the multiplet, it is not possible to describe this in terms of the fundamental representation of the $SU(3)$ group whose dimension is three. This line of reasoning had led to the concept of quarks (Gell-Mann, 1964; Zweig, 1964).

Quarks have fractional charges and fractional baryon numbers as are listed in Table 1.1. Three quarks or a quark/antiquark pair can form an integer charge and an integer baryon number to become an observable hadron. For example, the proton consists of two u quarks and one d quark. The way in which the quarks form observable baryonic multiplets is discussed in Section 6. It is not necessary to present here a full-fledged discussion of the quark model, because the model has been extensively discussed in the physics literature (Kokkedee, 1969; Lichtenberg, 1970; Greenberg, 1982; Huang, 1982).

TABLE 1.1

Quantum numbers for quarks and antiquarks, B, Q, S, C, B, T are baryon number, charge, strangeness, charm, bottom and top respectively. I is the total isospin quantum number, and I_3 is its third component. The u and d quarks form a doublet in isospin space, while the s, c, b, and t quarks are isospin singlets.

Quantum nos.		B	Q	I	I_3	S	C	B	T
Quarks	u	1/3	2/3	1/2	1/2	0	0	0	0
	d	1/3	−1/3	1/2	−1/2	0	0	0	0
	s	1/3	−1/3	0	0	−1	0	0	0
	c	1/3	2/3	0	0	0	1	0	0
	b	1/3	−1/3	0	0	0	0	−1	0
	t	1/3	2/3	0	0	0	0	0	1
Antiquarks	$\bar{u}$	−1/3	−2/3	1/2	−1/2	0	0	0	0
	$\bar{d}$	−1/3	1/3	1/2	1/2	0	0	0	0
	$\bar{s}$	−1/3	1/3	0	0	1	0	0	0
	$\bar{c}$	−1/3	−2/3	0	0	0	−1	0	0
	$\bar{b}$	−1/3	1/3	0	0	0	0	1	0
	$\bar{t}$	−1/3	−2/3	0	0	0	0	0	−1

We are interested only in those aspects of the quark model which are needed in studying the bound-state property of hadrons. The key question is whether the quarks can be regarded as constituent particles within quantum bound states. For instance, the proton and electrons are clearly constituent particles in the hydrogen atom. The consequence of this bound-state picture is that the hydrogen atom has a localized probability of electron around the proton with the radius determined by the electron mass and the strength of interaction between the proton and electron. The localization condition imposed on the hydrogen wave function is responsible for the discreteness of the energy spectrum. The electron can sometimes be separated from the proton to become a free particle.

If the hadron is a bound state of quarks, it should have a non-zero radius determined by the interaction between the quarks, and should show evidence for the existence of a discrete mass spectra due to the localization condition. Since quarks have fractional charges, it is not advisable to consider unbound or free quarks. Therefore, unlike the case of the hydrogen atom, quarks in hadrons cannot be separated.

We do not yet know the exact form of force between the quarks. However, from both phenomenological and field theoretic approaches, the present indications are that the force is very weak at short distances and becomes very strong at large distances. The potential governing the mass is expected to be linear in distance between the quarks. It is therefore reasonable to study the harmonic oscillator potential for the (mass)^2 spectrum (Critchfield, 1976). This means that we should see only equal-spaced (mass)^2 spectra in high-energy laboratories. This is not the case.

What we see in the real world is the perturbed mass spectra. Quarks have spins and unitary spins. In addition to the approximate harmonic oscillator force, their spins and unitary spins can remove the degeneracy of the harmonic oscillator system. Precisely for this reason, we have to know how to construct unperturbed symmetric wave functions and then perform perturbations on them.

Since mesons are two-body states of two different particles, there are no problems connected with identical particles, and the two-body problem should not cause any mathematical complications. However, baryons are bound states of three quarks. It is by now firmly established *experimentally* that the baryonic wave functions be totally symmetric under the exchange of quarks. This exchange degeneracy is analogous to the case of many electrons. As Dirac pointed out in Sections 55 and 56 of his classic book on quantum mechanics (Dirac, 1958), the Hamiltonian should be invariant under the exchange of quarks, and physical states should therefore be eigenstates of the permutation operators whose eigenvalues correspond to constants of motion.

When we attempt to construct a totally symmetric wave function from spin 1/2 particles, we are led to the question of the Pauli exclusion principle. This is in fact a very serious question, and there have been many attempts to rectify the situation. At present, the prevailing view is that there is an additional quark quantum number called "color" (Greenberg, 1964). There are three colors forming a unitary multiplet. The observable three-quark system always manifests itself in a color singlet or a totally antisymmetric state of this quantum number. For this reason, the hadronic wave functions are totally symmetric in all other quantum numbers.

The hypothesis of the color space leads to very rich experimental and theoretical consequences (Greenberg and Nelson, 1977; Marciano and Pagels, 1978). However, since it does not have a direct relation to what we plan to establish in this chapter, we shall not discuss this subject further.

Once the quarks are identified, it is not difficult to add their quantum numbers to calculate the resulting quantum number for the hadron. The nontrivial aspect of constructing wave functions is to make symmetric combinations to construct irreducible representations. Indeed, this is also a widely discussed subject (Greenberg and Resnikoff, 1967; Shapiro, 1968). The remarkable fact is that this symmetry problem was considered in depth by Dirac in his book on quantum mechanics (1958),* many years before the invention of the quark model in 1964. The method used by Feynman *et al.* (1971) is along the line suggested by Dirac. In this chapter, we shall give a full discussion of Dirac's symmetry classification method for the three-particle system.

---

* The first edition of Dirac's book entitled *Principles of Quantum Mechanics* was published in 1930.

2. Three-particle Symmetry Classifications According to the Method of Dirac

In his Sections 55 and 56 entitled 'Permutations as Dynamical Variables' and 'Permutations as Constants of Motion,' respectively, Dirac (1958) clearly spelled out his original ideas about the dynamical roles permutations play in quantum mechanics. The purpose of the present section is to work out a concrete illustrative example which might be helpful in understanding Dirac's original treatment. We shall carry out explicit calculations for the three-particle system.

Let us consider three similar objects labeled as 1, 2, 3 respectively. As we studied in Section 7 of Chapter I, we can perform six different permutations on these three objects. First, there are three odd permutations of the form

$$(12), \quad (23), \quad (31), \tag{2.1}$$

where each number is replaced by the succeeding number in parentheses, while the first one goes to the last position. In addition, there are two even permutations of the form

$$(123), \quad (132), \tag{2.2}$$

The above five permutations together with the identity form the six permutations which can be performed on the three objects. The identity is an even permutation.

As we noted in Section 7 of Chapter I, there are three operators which commute with all of the above permutations:

$$X_1 = I,$$
$$X_2 = [(12) + (23) + (31)]/3, \tag{2.3}$$
$$X_3 = [(123) + (132)]/2,$$

where I is the identity operator. X_1 can also be written as

$$X_1 = [I + (12) + (23) + (31) + (123) + (132)]/6. \tag{2.4}$$

X_1, X_2, and X_3 are the Casimir operators of the permutation group of three objects.

If the Hamiltonian is invariant under permutations, the above three X_i's can be simultaneously diagonalized. The next question is how to find eigenvalues for these operators. Here again, we follow the steps outlined by Dirac in his Equation (14) of Section 56. By explicit calculation, we derive

$$X_1^2 = X_1, \quad X_1 X_2 = X_2, \quad X_1 X_3 = X_3,$$
$$X_2^2 = (X_1 + 2X_3)/3, \quad X_2 X_3 = X_2. \tag{2.5}$$
$$X_3^2 = (X_1 + X_3)/2.$$

Following Dirac's Equation (15) of Section 56 for the general case, we consider the following arbitrary function of the X operators:

$$B = X_1 + X_2 + X_3. \tag{2.6}$$

Then

$$B^2 = (11/6)X_1 + 4X_2 + (19/6)X_3, \tag{2.7}$$

$$B^3 = (10/4)X_1 + 13X_2 + (37/4)X_3. \tag{2.8}$$

Because X_1 is the identity operator, its eigenvalue is always 1. By eliminating X_2 and X_3 from Equations (2.6)–(2.8), we arrive at

$$B^3 = (9/2)B^2 + 5B - (3/2) = 0. \tag{2.9}$$

which is Dirac's Equation (16) of Section 56 for the "arbitrary" B given in Equation (2.6). The above cubic equation has three roots:

$$B_1 = 3, \quad B_2 = 1, \quad B_3 = 1/2. \tag{2.10}$$

We can use each of these three numbers to calculate the left-hand side of Equations (2.6)–(2.8), which then become three simultaneous linear equations. The solutions to these linear equations will indeed be the eigenvalues for the X operators. They are given in Table 2.1.

The choice of the B function in Equation (2.6) was arbitrary. However, the eigenvalues of the operators $X_1, \ldots, P$ are independent of the form of B. For instance, even if we choose $B = X_1 + 2X_2 + 3X_3$, the eigenvalue distribution would be the same as the one given in Table 2.1 (Problem 5 in Section 11).

If the Hamiltonian is invariant under exchange of particles, one of the five non-trivial permutations can also be simultaneously diagonalized, because it commutes with X_1, X_2, and X_3. Let us choose this particular permutation to be

$$P = (23). \tag{2.11}$$

Other permutations which do not commute with the above P cannot be simultaneously diagonalized. Since $P^2 = I$, the eigenvalue of this operator is either $+1$ or -1. We are interesed here in how these eigenvalues are distributed. For this purpose, we introduce the operator P' defined as

$$P' = 3X_2 - P = (12) + (31), \tag{2.12}$$

or

$$P + P' = 3X_2. \tag{2.13}$$

P and P' satisfy also the relation

$$P'P = 2X_3. \tag{2.14}$$

For the symmetry classification corresponding to B_1, the eigenvalues already found, together with Equations (2.13) and (2.14), allow only $P = 1$. For the B_2 case, $P = -1$. However, for the B_3 case,

$$P + P' = 0, \quad P'P = -1. \tag{2.15}$$

P in this case can therefore have both values: $+1$ and -1. These results are given in Table 2.1.

TABLE 2.1
Eigenvalues of X_1, X_2, X_3 and P. There are three diffeent sets of eigenvalues resulting in three different symmetry classifications.

B	X_1	X_2	X_3	P	Symbol
B_1	1	1	1	1	S
B_2	1	−1	1	−1	A
B_3	1	0	−1/2	1	α
				−1	β

3. Construction of Symmetrized Wave Functions

We now construct three-particle wave functions having the symmetry properties summarized in Table 2.1. Let us consider three particles which can be in any of three quantum states x, y and z, with one particle in state x, another in y, and another in z. We can then write the general state for such a system as

$$\psi = a|xyz\rangle + b|yxz\rangle + c|xzy\rangle + d|zyx\rangle + f|zxy\rangle + g|yzx\rangle, \tag{3.1}$$

where $a, b, \ldots, g$ are coefficients to be determined by the symmetry property of the wave function.

A state of classification S (totally symmetric) will obey

$$X_2\psi = X_3\psi = P\psi = \psi. \tag{3.2}$$

Explicit calculation shows that this requires $a = b = c = d = f = g$, resulting in the totally symmetric wave function:

$$|S\rangle = (1/\sqrt{6})[|xyz\rangle + |yxz\rangle + |xzy\rangle + |zyx\rangle + |zxy\rangle + |yzx\rangle]. \tag{3.3}$$

If we want an A state, we must have $-X_2\psi = X_3\psi = -P\psi = \psi$, or $a = -b = -c = -d = f = g$, which results in the totally antisymmetric state:

$$|A\rangle = (1/\sqrt{6})[|xyz\rangle - |yxz\rangle - |xzy\rangle - |zyx\rangle + |zxy\rangle + |yzx\rangle], \tag{3.4}$$

If we want an α state, we need $X_2\psi = 0$, $X_3\psi = -(1/2)\psi$, and $P\psi = \psi$. These conditions will lead to

$$a + f + g = 0, \quad a = c, \quad b = g, \quad d = f. \tag{3.5}$$

Hence, there are four equations and six unknowns. This means that there will be a two-dimensional subspace of the α state. We can pick two linearly independent α states as

$$|\alpha\rangle_1 = (1/2\sqrt{3})[|xyz\rangle + |xzy\rangle + |yxz\rangle + |yzx\rangle - 2|zxy\rangle - 2|zyx\rangle],$$
$$|\alpha\rangle_2 = (1/2)[|xyz\rangle - |yzx\rangle + |xzy\rangle - |yxz\rangle]. \tag{3.6}$$

The first α state is the one given by Feynman et al. (1971), and the second is orthogonal to it.

Lastly, for a β state, the conditions from Table 2.1 are $X_2\psi = 0$, $X_3\psi = -(1/2)\psi$, $P\psi = -\psi$, or

$$a + f + g = 0, \quad a = -c, \quad b = -g, \quad d = -f. \tag{3.7}$$

Here again, we have a two-dimensional subspace of possible states. We can pick the first β state to be that given by Feynman et al. (1971), and the one orthogonal to it to be the second β state:

$$|\beta\rangle_1 = (1/2)[|xyz\rangle - |xzy\rangle + |yxz\rangle - |yzx\rangle],$$
$$|\beta\rangle_2 = (-1/2\sqrt{3})[|xyz\rangle - |yxz\rangle - |xzy\rangle + |yzx\rangle + 2|zyx\rangle - 2|zxy\rangle]. \tag{3.8}$$

We now have a complete set of six linearly independent states which are contained in the four symmetry classifications S, A, α and β.

4. Symmetrized Products of Symmetrized Wave Functions

As was stated in Section 2, we have to combine spin, unitary spin, and spatial wave functions to construct the totally symmetric overall baryonic wave function. For this purpose, we consider in this section products of two symmetrized three-particle states. We are interested in a product of wave functions with values in two separate spaces, for example, spin space and unitary spin space. Our wave function will be of the form

$$|ab\rangle = |a\rangle|b\rangle, \tag{4.1}$$

where a and b represent quantum numbers in two separate spaces. The state $|ab\rangle$ that can be made to conform to the results of Sections 2 and 3 were derived from the properties of the operators X_1, X_2, X_3, and P, without any assumptions about the form the eigenstates would take. What we would like to do is to derive a complete set of states $|ab\rangle$ which conform to the symmetry classification obtained from the symmetrized $|a\rangle$ and $|b\rangle$.

A permutation operator acting on $|ab\rangle$ will permute the particles in $|a\rangle$ and $|b\rangle$ in an identical manner. Thus we can write

$$X_1 = I = I_a I_b = X_{1a} X_{1b}, \tag{4.2}$$

$$X_2 = (1/3)[(12) + (23) + (31)]$$
$$= (1/3)[(12)_a(12)_b + (23)_a(23)_b + (31)_a(31)_b], \tag{4.3}$$

$$X_3 = (1/2)[(123) + (132)]$$
$$= (1/2)[(123)_a(123)_b + (132)_a(132)_b], \tag{4.4}$$

$$P = P_a P_b = (23)_a(23)_b. \tag{4.5}$$

We are interested here in expressing symmetrized wave functions $|ab\rangle$ in terms of symmetrized $|a\rangle$ and $|b\rangle$. The simplest way to attack this problem is to write the right-hand sides of Equations (4.2)–(4.5) in terms of the operators which are diagonal in the symmetrized a and b spaces and/or other simple operators. X_1 of Equation (4.2) and P of Equation (4.5) are already in the desired form. The remaining problem is to work out X_2 and X_3. For this purpose, we carry out first the following simple calculations:

$$X_{2a} X_{2b} = (1/3) X_2 + (2/3) X_2(X_{3b}), \tag{4.6}$$

or equivalently

$$X_{2a} X_{2b} = 1/3 X_2 + (2/3) X_2(X_{3a}), \tag{4.7}$$

and

$$X_{3a} X_{3b} = (1/2) X_3 + X_3(X_{3b}) - (1/2) X_{3a}, \tag{4.8}$$

or

$$X_{3a} X_{3b} = (1/2) X_3 + X_3(X_{3a}) - (1/2) X_{3b}, \tag{4.9}$$

Let us consider a state which is a product of an S state in the a space with an S state in the b space:

$$\psi = |a\rangle_S |b\rangle_S, \tag{4.10}$$

and look at what the relations given in Equations (4.6) and (4.9) tell us about the wave function ψ of Equation (4.10). Clearly,

$$X_{ia} \psi = X_{ib} \psi = \psi, \quad i = 1, 2, 3, \tag{4.11}$$

so that

$$(X_{2a} X_{2b}) \psi = (X_{3a} X_{3b}) \psi = \psi, \tag{4.12}$$

giving

$$\psi = [(1/3) X_2 + (2/3) X_2(X_{3b})] \psi = X_2 \psi, \tag{4.13}$$

$$\psi = [(1/2) X_3 + X_3(X_{3a}) - (1/2) X_{3b}] \psi. \tag{4.14}$$

Hence

$$X_3\psi = \psi. \tag{4.15}$$

Also, from P of Equation (4.5),

$$P\psi = \psi. \tag{4.16}$$

Thus we have established that $\psi = |a\rangle_S |b\rangle_S$ is an eigenstate of X_1, X_2, X_3 and P, and that the wave function of Equation (4.10) is an S state: $|ab\rangle_S$.

In a similar manner, we can use Equations (4.6)–(4.9) to show that the following states fall into the given symmetry classifications:

$$\begin{aligned}
&|a\rangle_S|b\rangle_S = |ab\rangle_S, \quad |a\rangle_S|b\rangle_\alpha = |ab\rangle_\alpha, \\
&|a\rangle_S|b\rangle_\beta = |ab\rangle_\beta, \quad |a\rangle_S|b\rangle_A = |ab\rangle_A, \\
&|a\rangle_A|b\rangle_S = |ab\rangle_A, \quad |a\rangle_A|b\rangle_\alpha = |ab\rangle_\beta, \\
&|a\rangle_A|b\rangle_\beta = |ab\rangle_\alpha, \quad |a\rangle_A|b\rangle_A = |ab\rangle_S.
\end{aligned} \tag{4.17}$$

However, when ψ is taken as the product of an α or β state in the a space with an α or β state in the b space, the terms $X_2\psi$ and $X_3\psi$ appear in Equations (4.6)–(4.9) with zero coefficient. These equations thus reduce to the identities from which no information can be obtained. To handle these cases, we consider the operators which simply change an α to a β state, and vice versa. For this purpose, let us introduce the operator

$$R = (1/\sqrt{3})\,[(12) - (31)]. \tag{4.18}$$

Acting on an $|a\rangle$, $|b\rangle$, or $|ab\rangle$ state, R has the following properties:

$$[X_i, R] = 0, \quad i = 1, 2, 3, \tag{4.19}$$

$$PR = -RP, \quad R^2 = (2/3)\,[1 - X_3]. \tag{4.20}$$

From Equation (4.20), we see that if a state $|a\rangle$ is an eigenstate of P_a, then $R_a|a\rangle$ will be an eigenstate of P_a with an eigenvalue opposite to that of $|a\rangle$. From Equation (4.19), we see that $R_a|a\rangle$ will have the same eigenvalues under X_{1a}, X_{2a}, and X_{3a} as $|a\rangle$. The logic is the same for the b space.

Let us first consider the action of R upon an S state:

$$P(R|\rangle_S) = -R|\rangle_S, \quad X_i(R|\rangle_S) = R|\rangle_S, \quad i = 1, 2, 3, \tag{4.21}$$

where $|\rangle$ can be $|a\rangle$, $|b\rangle$, or $|ab\rangle$. Thus $R|\rangle_S$ will be an eigenstate of X_1, X_2, X_3, and P with eigenvalues 1, 1, 1, and -1 respectively. From the argument of Section 2, it is clear that no such state can possibly exist. We conclude, therefore, that $R|\rangle_S = 0$, and from a similar argument, $R|\rangle_A = 0$. This agrees with Equation (4.20), which for S and A states reduces to $R^2 = 0$.

Next, we turn to the α and β states. R will again change the sign of the P

eigenvalues, while leaving the X_1, X_2, and X_3 eigenvalues unchanged. This means that

$$P(R|\rangle_a) = -R|\rangle_a, \quad X_1[R|\rangle_a) = R|\rangle_a,$$

and

$$X_2(R|\rangle_a) = 0, \quad X_3(R|\rangle_a) = -(1/2)R|\rangle_a. \tag{4.22}$$

Thus $R|\rangle_a$ will be a β state. Conversely, R operating on a β state will have an α state. Since $X_3 = -1/2$ for α and β state, we will have $R^2 = 1$, and

$$R^2|\rangle_\alpha = |\rangle_\alpha, \quad R^2|\rangle_\beta = |\rangle_\beta. \tag{4.23}$$

We can pick

$$|\rangle_\beta = R|\rangle_\alpha \quad \text{and} \quad R|\rangle_\beta = |\rangle_\alpha. \tag{4.24}$$

uniquely, given $|\rangle_\alpha$. In terms of the R operators acting on the a and b spaces, we can write X_2 of Equation (4.3) as

$$X_2 = (1/3)P_a P_b + R_a R_b. \tag{4.25}$$

In order to derive a similar formula for X_3, we introduce the operator

$$R' = (1/2)^{1/2}[(123)-(132)]. \tag{4.26}$$

Here again

$$[R', X_i] = 0, \quad i = 1, 2, 3, \quad \text{and} \quad R'P = -PR'. \tag{4.27}$$

As for R defined in Equation (4.18),

$$RR' = -R'R = P - X_2. \tag{4.28}$$

Thus for any α state $|\rangle_a$,

$$R(R'|\rangle_a) = (P - X_2)|\rangle_a. \tag{4.29}$$

Since $R^2 = 1$ from Equation (4.20),

$$RR'|\rangle_a = R^2|\rangle_a, \quad R'|\rangle_a = R|\rangle_a. \tag{4.30}$$

From Equation (4.28),

$$R'R|\rangle_a = -|\rangle_a. \tag{4.31}$$

From Equations (4.30) and (4.31), we derive, for $|\rangle_\alpha$ and $|\rangle_\beta$ obeying Equation (4.24):

$$R'|\rangle_\alpha = |\rangle_\beta \quad \text{and} \quad R'|\rangle_\beta = -|\rangle_\alpha. \tag{4.32}$$

In terms of the R operators acting on the a and b spaces, we can write X_3 of Equation (4.4) as

$$X_3 = (1/4)[4X_{3a}X_{3b} + 3R_a R_b]. \tag{4.33}$$

Using the X_2 and X_3 operators given in Equations (4.25) and (4.33) respectively, together with X_1 and P of Equations (4.2) and (4.5) we can show that with $|\rangle_\beta = R|\rangle_\alpha$,

$$
\begin{aligned}
|ab\rangle_s &= [|a\rangle_\alpha|b\rangle_\alpha + |a\rangle_\beta|b\rangle_\beta]/\sqrt{2}, \\
|ab\rangle_\alpha &= [-|a\rangle_\alpha|b\rangle_\alpha + |a\rangle_\beta|b\rangle_\beta]/\sqrt{2}, \\
|ab\rangle_\beta &= [|a\rangle_\alpha|b\rangle_\beta + |a\rangle_\alpha|b\rangle_\beta]/\sqrt{2}, \\
|ab\rangle_a &= [-|a\rangle_\alpha|b\rangle_\beta + |a\rangle_\beta|b\rangle_\alpha]/\sqrt{2}.
\end{aligned}
\qquad (4.34)
$$

The symmetrized wave functions given in Section 3 together with the combination formulas given in Equations (4.17) and (4.34) form the mathematical basis for constructing three-quark baryonic wave functions for spin, unitary spin, and harmonic oscillator excitations, and for constructing the total wave function by making symmetrized combinations.

5. Spin Wave Functions for the Three-Quark System

We can now apply the results of Section 3 to obtain the spin wave functions for the symmetric quark model. We observe that, as in the case of electrons, the quark spin can be either up (+) or down (−). Therefore, in constructing spin wave functions for the three-quark system, the quantum numbers x, y, and z used in Section 3 each take on the value of either + or −. As is specified in Schiff's book on quantum mechanics (1968), the totally symmetric state $|\rangle_S$ represents a spin 3/2 state, while $|\rangle_{\alpha,\beta}$ are spin-1/2 states. Because at least two of the three quantum numbers are the same, the totally antisymmetric wave function does not exist.

We write here for completeness the spin wave functions. For total spin-3/2, the totally symmetric wave functions are

$$
\begin{aligned}
|3/2, 3/2\rangle &= |+++\rangle, \\
|3/2, 1/2\rangle &= [|++-\rangle + |+-+\rangle + |-++\rangle]/\sqrt{3}, \\
|3/2, -1/2\rangle &= [|--+\rangle + |-+-\rangle + |+--\rangle]/\sqrt{3}, \\
|3/2, -3/2\rangle &= |---\rangle.
\end{aligned}
\qquad (5.1)
$$

where the first number in the ket vectors on the left-hand side is the total spin and the second number is its third component. Because at least two quarks are in the same quantum state, there are only four totally symmetric states.

The α states with total spin 1/2 are

$$
\begin{aligned}
|1/2, 1/2\rangle_\alpha &= [|++-\rangle + |+-+\rangle - 2|-++\rangle]/\sqrt{6}, \\
|1/2, -1/2\rangle_\alpha &= [|--+\rangle + |-+-\rangle - 2|+--\rangle]/\sqrt{6}.
\end{aligned}
\qquad (5.2)
$$

The β states with total spin 1/2 are

$$|1/2, 1/2\rangle_\beta = [|++-\rangle - |+-+\rangle]/\sqrt{2},$$
$$|1/2, -1/2\rangle_\beta = [|--+\rangle - |-+-\rangle]/\sqrt{2}. \tag{5.3}$$

α and β here are α_1 and β_1 of Equations (3.6) and (3.8). We noted there that there are two α states. In the spin case in which there are only two different quantum states for each particle, the second α state either vanishes or becomes dependent on other states. For this reason, we consider only the first α state. The situation is the same for the β state.

The spin wave functions given in Equations (5.1)–(5.3) are usually covered in the standard quantum mechanics course and are discussed in Chapter IV of this book. We wrote them down explicitly here in order to emphasize that they represent a particular case of the more general formulas given in Section 3, and that the unitary spin wave functions discussed in the next section represent a more general case of the same symmetry classification method.

6. Three-quark Unitary Spin and SU(6) Wave Functions

In addition to the spin, quarks carry unitary spin which is often called "flavor." When the quark model was proposed by Gell-Mann in 1964, it was believed that unitary spin could only take three different values, namely, up (u), down (d), and strange (s). It is now believed that there are three additional flavor quantum numbers which are called charm (c), bottom (b), and top (t). A summary of the currently known quarks and their quantum numbers is given in Table 1.1. As can be seen from the table, quarks carry fractional charges and baryon numbers. They are combined in such a way that the resulting hadrons have observable charges and baryonic numbers.

We shall discuss here only the unitary spin wave functions of the three traditional quarks u, d and s. It is straightforward to apply the method for these traditional quarks to states involving newer quarks. In constructing unitary spin wave functions for baryons consisting of three quarks, we can use the symmetrized forms given in Section 3.

The quark model utilizes the concept of isotopic spin, in which the proton and the neutron belong to the two-dimensional isospin multiplet. The mathematics for this isospin formalism is the same as that of $SU(2)$ which we discussed in Chapter IV. The fact that the strong interaction is invariant under rotations in this isospin space is well known and is discussed widely in textbooks in modern physics.

The conservation in strong interaction physics of isospin and strangeness has led us to believe that the three quarks should form a multiplet (Gell-Mann, 1964) under transformations of the $SU(3)$ group which is the group of

Hadronic Mass Spectra

untiary unimodular transformations in the three-dimensional complex space (Exercise 1 in Section 11). Clearly, a rotation in unitary spin space will have no effect on the exchange symmetry of a given wave function. For instance, if the wave function is totally symmetric under exchange of particles, this symmetry is not affected by rotation in unitary spin space.

If we consider all possible combinations of unitary spins for three quarks, there are ten totally symmetric states, eight α states, and eight β states. There is one totally antisymmetric state. There are therefore all together 27 ($= 3^3$) states. As is illustrated later in this section, these multiplets can be worked out explicitly. For example, the baryon Σ^{*0} belongs to the totally symmetric decuplet with the unitary spin wave function

$$|\Sigma^{*0}\rangle = |uds\rangle_S$$
$$= [|uds\rangle + |dsu\rangle + |sud\rangle + |dus\rangle + |sdu\rangle + |usd\rangle]/\sqrt{6}. \quad (6.1)$$

The proton is a member of the α or β octet, and

$$|p\rangle = |uud\rangle_\alpha$$
$$= [|uud\rangle + |udu\rangle - 2|duu\rangle]/\sqrt{6}, \quad (6.2)$$

or

$$|p\rangle = |uud\rangle_\beta$$
$$= [|uud\rangle - |udu\rangle]/\sqrt{2}. \quad (6.3)$$

Likewise, we can construct all baryonic unitary spin wave functions, and they are given in Table 6.1. The reason for the $\Sigma^0 - \Lambda^0$ degeneracy is that the (uds) states are the only ones with three different quantum numbers so that the second α and β states do not vanish as they did in the discussion of Section 5 where two of the quantum numbers are the same.

The interesting question now is how these $SU(3)$ wave functions are combined with the $SU(2)$ spin wave functions of Section 5 to generate the $SU(6)$ multiplets. We can now use the combination formulas given in Section 4, using the notation $|10\rangle_S$ for the totally symmetric decuplet, and $|8\rangle_{\alpha,\beta}$ for the α and β octets, respectively. There are 56 totally symmetric $SU(6)$ wave functions, and they are

$$|3/2\rangle_S|10\rangle_S$$

and

$$[|1/2\rangle_\alpha|8\rangle_\alpha + |1/2\rangle_\beta|8\rangle_\beta]/\sqrt{2}. \quad (6.4)$$

The $SU(6)$ 56 state consists of the spin-3/2 decuplet (with 40 states) and the spin-1/2 octet (16 states). These $SU(6)$ wave functions combine with totally symmetric spatial wave functions. Since the ground-state spatial wave function is believed to be totally symmetric, the above multiplets indeed

TABLE 6.1

Baryon multiplets in the unitary spin space consisting of (*uds*) quarks. There are ten totally symmetric and eight α or β states. When the spatial wave function is in the ground state and is totally symmetric, the decuplet combines with the spin-3/2 states, and the octets combine with the spin-1/2 states to form totally symmetric overall wave functions. The combination of spin and unitary spin states gives the $SU(6)$ multiplets. The $SU(6)$ wave functions, combined with totally symmetric ground-state spatial wave functions, form the familiar baryon octet and decuplet in a 56 $SU(6)$ multiplet. The numbers in parentheses are the observed baryonic masses measured in MeV.

Quarks	Q	I_3	S	Octet: α or β (spin = 1/2)	Decuplet: S (spin = 3/2)
uuu	2	3/2	0		N^{*++}
uud	1	1/2	0	p (938)	N^{*+} (1238)
udd	0	−1/2	0	n (939)	N^{*0} (1238)
ddd	−1	−3/2	0		N^{*-}
uus	1	1	−1	Σ^+ (1189)	Σ^{*+}
uds	0	0	−1	Λ^0 (1115), Σ^0 (1182)	Σ^{*0} (1385)
dds	−1	−1	−1	Σ^- (1179)	Σ^{*-}
uss	0	1/2	−2	Ξ^0 (1314)	Ξ^{*0} (1530)
dss	−1	−1/2	−2	Ξ^- (1320)	Ξ^{*-}
sss	0	0	−3		Ω^0 (1675)

represent the most commonly observed baryonic multiplet, and are usually called the $SU(6)$ multiplets.

There are 70 α states consisting of

$$\begin{aligned} & 32\,|3/2\rangle_S|8\rangle_\alpha, \\ & 20\,|1/2\rangle_\alpha|10\rangle_S, \\ & 16\,[-|1/2\rangle_\alpha|8\rangle_\alpha + |1/2\rangle_\beta|8\rangle_\beta]/\sqrt{2}, \\ & 2\,|1/2\rangle_\beta|1\rangle_A. \end{aligned} \qquad (6.5)$$

There are also 70 β states. They are

$$\begin{aligned} & 32\,|3/2\rangle_S|8\rangle_\beta, \\ & 20\,|1/2\rangle_\beta|10\rangle_S, \\ & 16\,[|1/2\rangle_\alpha|8\rangle_\beta + |1/2\rangle_\beta|8\rangle_\alpha]/\sqrt{2}, \\ & 2\,|1/2\rangle_\alpha|1\rangle_A. \end{aligned} \qquad (6.6)$$

The above α and β type states are combined with the α and β type spatial wave functions to generate totally symmetric overall wave functions.

In addition, there are 20 totally antisymmetric states consisting of

$$4\,|3/2\rangle_S|1\rangle_A,$$
$$16\,[-|1/2\rangle_\alpha|8\rangle_\beta + |1/2\rangle_\beta|8\rangle_\alpha]/\sqrt{2}. \tag{6.7}$$

This multiplet can be combined with totally antisymmetric spatial wave functions. However, there has not been any experimental evidence to indicate the existence of this antisymmetric multiplet. We shall therefore omit this multiplet in the following discussions.

7. Three-body Spatial Wave Functions

We are considering here the quantum relativistic bound state of three quarks in order to obtain the baryonic mass spectra. As was noted in Chapters V and VI, the harmonic oscillator potential can be made consistent with the principle of special relativity, and is mathematically simple and convenient for discussing symmetry properties. We shall therefore use the oscillator model for calculating the mass spectrum of baryons.

Let us consider that $\mathbf{x}_a$, $\mathbf{x}_b$, and $\mathbf{x}_c$ represent the coordinates of the three quarks. Then the "Hamiltonian" corresponding to the baryonic (mass)^2 can be written as

$$H = -3(\nabla_a^2 + \nabla_b^2 + \nabla_c^2) + \frac{\Omega^2}{36}\,[(\mathbf{x}_a - \mathbf{x}_b)^2 +$$
$$+ (\mathbf{x}_a - \mathbf{x}_c)^2 + (\mathbf{x}_c - \mathbf{x}_a)^2] + m_0^2, \tag{7.1}$$

where Ω is the constant specifying the strength of the oscillator force, usually measured in (GeV)^2. It should be noted that the eigenvalue equation should be written for (mass)^2 in view of the representations of the Poincaré group discussed in Chapters V and VI. The above form is totally symmetric under the exchange of quarks, and thus commutes with all six three-particle permutations discussed in Section 2.

For this three-quark system, it is more convenient to use the variables [Moshinsky (1969)]:

$$\mathbf{R} = (\sqrt{\Omega}/3)\,(\mathbf{x}_a + \mathbf{x}_b + \mathbf{x}_c),$$
$$\mathbf{r} = (\sqrt{\Omega}/6)\,(-2\mathbf{x}_a + \mathbf{x}_b + \mathbf{x}_c),$$
$$\mathbf{s} = (\sqrt{\Omega}/2\sqrt{3})\,(\mathbf{x}_c - \mathbf{x}_b). \tag{7.2}$$

The variable $\mathbf{R}$ is the overall center-of-mass coordinate for this three-quark system. The variables $\mathbf{r}$ and $\mathbf{s}$ are the internal quark coordinates and have α and β type symmetries respectively. The quantity $(r^2 + s^2)$ is totally symmetric under the exchange of quarks. In terms of the above coordinate variables, the Hamiltonian takes the form

$$H = \frac{\Omega}{2}(-\nabla_r^2 + r^2) + \frac{\Omega}{2}(-\nabla_s^2 + s^2) + m_0^2. \quad (7.3)$$

We ignore here the trivial kinetic energy term associated with the coordinate **R**. The coordinate variables u and v now have been completely separated, and the eigenvalue of the K operator is

$$H = \Omega(N + 3) + m_0^2, \quad (7.4)$$

where N is the "total" excitation number, and is the sum of the excitations in the r and s spaces:

$$N = n_r + n_s. \quad (7.5)$$

For each of the r and s coordinates, we can solve the eigenvalue problem according to the well-defined procedure for the three-dimensional isotropic harmonic oscillator, and the overall spatial solution is a product of the r and s solutions or linear combinations of them. In order to see this more clearly, let us work out the explicit forms for $N = 0, 1$, and 2.

When $N = 0$, the only possible choice for n_r and n_s is

$$n_r = n_s = 0. \quad (7.6)$$

Therefore the total spatial wave function is

$$\exp[-(r^2 + s^2)/2]. \quad (7.7)$$

We ignore here normalization constants. This form is totally symmetric under the exchange of quarks.

For $N = 1$, we have to consider the following two possibilities:

$$n_r = 1 \text{ and } n_s = 0, \quad \text{or} \quad n_r = 0 \text{ and } n_s = 1. \quad (7.8)$$

For the first case, the wave function is

$$|N = 1, L = 1\rangle_\alpha = r_i \exp[-(r^2 + s^2)/2], \quad i = 1, 2, 3. \quad (7.9)$$

This α-state solution corresponds to a total angular momentum $L = 1$ state. Likewise, the wave function for the second case is

$$|N = 1, L = 1\rangle_\beta = s_i \exp[-(r^2 + s^2)/2]. \quad (7.10)$$

Both the r_i and s_i wave functions have three spatial components. This Cartesian representation can be transformed into spherical form and can be written in terms of the spherical harmonics.

Let us next consider the $n = 2$ case, with the following three possible degeneracies.

$$n_r = 2 \text{ and } n_s = 0,$$
$$n_r = 1 \text{ and } n_s = 1, \quad (7.11)$$
$$n_r = 0 \text{ and } n_s = 2.$$

We have to combine these degenerate solutions to make the $L = 0, 1,$ and 2 states.

For the $L = 0$ states, the wave functions are rotationally invariant and take the form

$$|N = 2, L = 0\rangle_S = (r^2 + s^2) \exp[-(r^2 + s^2)/2],$$
$$|N = 2, L = 0\rangle_a = (-r^2 + s^2) \exp[-(r^2 + s^2)/2], \quad (7.12)$$
$$|N = 2, L = 0\rangle_\beta = (r_1 s_1 + r_2 s_2 + r_3 s_3) \exp[-(r^2 + s^2)/2].$$

For $L = 1$, the wave function becomes

$$|N = 2, L = 1\rangle_A = \varepsilon_{i,j,k} \, r_i s_k \, \exp[-(r^2 + s^2)/2]. \quad (7.13)$$

Because total $L = 1$, there are three degenerate states.

For $N = L = 2$, we have

$$|N = 2, L = 2\rangle_A = [r_i s_j + r_i s_j$$
$$\qquad - \delta_{ij}(r^2 + s^2)/3] \exp[-(r^2 + s^2)/2],$$
$$|N = 2, L = 2\rangle_a = [r_i s_j - r_i s_j$$
$$\qquad - \delta_{ij}(r^2 - s^2)/3] \exp[-(v^2 + s^2)/2], \quad (7.14)$$
$$|N = 2, L = 2\rangle_\beta = [r_i s_j + r_i s_j$$
$$\qquad - \tfrac{2}{3} \delta_{ij} \mathbf{r} \cdot \mathbf{s}] \exp[-(r^2 + s^2)/2].$$

The above expressions are symmetric in i and j, and thus there appears to be a six-fold degeneracy. However, owing to the δ_{ij} terms, only five of them are independent as in the case of $Y_2^m(\theta, \phi)$. The above wave functions can also be written in terms of the spherical harmonics (Problem 7 in Section 11).

8. Totally Symmetric Baryonic Wave Functions

We combine in this section the $SU(6)$ wave functions of Section 6 and the spatial wave functions of Section 7 to generate the overall wave function. As we mentioned in Section 1, the quarks carry an additional quantum number called "color." It is believed that there are three colors and that all observed baryons are in the color singlet state. Since the color wave function is totally antisymmetric, and since quarks are fermions, the rest of the wave function consisting of space, spin and flavor has to be totally symmetric. We shall outline in this section how to construct totally symmetric baryonic wave functions which do not include the color part.

We have to combine the oscillator wave functions discussed in Section 7 with the $SU(6)$ parts given in Section 6 using again the technique developed in Section 4. The useful formulas from Equation (4.17) and Equation (4.34) which lead to the totally symmetric state are

$$|ab\rangle_S = |a\rangle_S |b\rangle_S,$$
$$|ab\rangle_S = |a\rangle_A |b\rangle_A, \quad (8.1)$$
$$|ab\rangle_S = (|a\rangle_\alpha |b\rangle_\beta + |a\rangle_\beta |b\rangle_\alpha)/\sqrt{2},$$

where a and b in this case correspond to the $SU(6)$ and spatial parts respectively.

Let us start with $N = 0$. We have in this case only one spatial wave function given in Equation (7.7) which is totally symmetric. Hence it can only be combined with the totally symmetric 56 multiplet of $SU(6)$ given in Equation (6.4). Within this multiplet are the spin-3/2 decuplet containing the N^*, Σ^*, Ξ^* resonances and the Ω^- particle, and the spin-1/2 octet containing the nucleons, Σ, Λ, and Ξ hyperons. By the $SU(6)$ scheme, we usually mean this totally symmetric 56 state. This scheme has been extensively discussed in the literature, and the explicit wave functions are given in the paper of Van Royen and Weisskopf (1967).

Let us next look at the $N = 1$ states. Since u_i and v_i have the α and β type symmetries, respectively, they should be combined with the 70 α and 70 β states of Equations (6.5) and (6.6) to generate a totally symmetric overall wave function of the form

$$|\text{overall}\rangle_S = (|70\rangle_\alpha r_i + |70\rangle_\beta s_i) \exp[-(r^2 + s^2)/2]. \quad (8.2)$$

All the baryons belonging to this multiplet are contained in the $SU(3)$ octets with spins 3/2 and 1/2, and the decuplet with spin 1/2.

For $N = 2$, we have to consider each value of L separately. For $L = 0$, we have S, α, β type spatial wave functions. The s-type state is combined with the totally symmetric $SU(6)$ 56 multiplet which contains a spin-1/2 octet and spin-3/2 decuplet, just as in the $N = 0$ case. There is good experimental evidence for the existence of this $L = 0$ multiplet, and a more detailed explanation including the mass spectrum is given in Section 9.

For the $N = 2$, $L = 1$ state, we note that there is only a totally antisymmetric spatial wave function, which can be combined only with the antisymmetric $SU(6)$ wave functions given in Equation (6.7). Baryons containing the totally antisymmetric unitary spin state $|1\rangle_A$ have not yet been observed. Thus we omit the states containing this $SU(3)$ singlet from the $SU(6)$ wave functions listed in Equations (6.5) and (6.6).

For $N = 2$, $L = 2$, we have again a totally symmetric spatial wave function which combines with the 56 $SU(6)$ totally symmetric wave function yielding again a spin-1/2 octet and a spin-3/2 decuplet. As is indicated in Equation (7.14), there are also spatial wave functions with α and β type symmetries. They are combined with the $SU(6)$ 70 α and 70 β states containing spin-3/2 and $-1/2$ octets, and a spin-1/2 decuplet. As is discussed in Section 10, there is good experimental indication that this multiplet exists in nature.

In combining the spatial and $SU(6)$ wave functions, we also have to consider the addition of the ortbial angular momentum in the spatial part and the spin angular momentum in the $SU(6)$ wave function. This angular momentum addition is a standard item to be covered in the standard quantum mechanics curriculum, and was discussed in Chapter IV.

9. Baryonic Mass Spectra

If the hadronic masses were determined from the Hamiltonian given in Equation (7.1) and its counterpart for the mesons, they should depend only on the harmonic oscillator quantum numbers. This is not the case in the real world. There are indeed several perturbation terms to be added to the Hamiltonian which remove the $SU(6)$ and $SU(3)$ degeneracies. These symmetry breaking interactions are due to the mass difference between strange and nonstrange quarks, spin-spin interaction, interaction between two unitary spins, and the combined interaction of the spin and unitary spin. There are many excellent review articles on the $SU(3)$ and $SU(6)$ mass formulas. The review article of Levin and Frankfurt (1968) contains the mass formula containing the perturbation terms due to the above mentioned interactions.

Using the Levin-Frankfurt mass formula, we can predict the mass of the hadrons. For the $N = 0$ spatial state, hadronic mass spectra are those of the $SU(6)$ scheme and are in excellent agreement with the experimental observation. The mass formula indeed produces accurately the observed masses of the $N = 0$ hadrons given in Tabels 6.1 and 9.1. It is customary to use the linear mass in calculating the $SU(6)$ and $SU(3)$ symmetry breakings for the baryonic mass, while the harmonic oscillator spectrum is calculated for (mass)^2. Though this is an unfortunate convention, it shall not cause any major confusion (Problem 8 in Section 11).

As N becomes 1, 2, or 3, the experimental situation is not yet as clear as in the $N = 0$ case. Table 9.2 summarizes the accuracy of the present experimental data in relation to the quark model multiplet scheme. In general, the baryonic spectra are better understood than those of mesons, and nonstrange hadronic spectra are better than those of strange hadrons. Among the four possible mass spectra, the spectrum of nonstrange baryons offers us a unique challenge. There are in this case three clean levels for us to test the linearity in the harmonic oscillator excitations.

TABLE 9.1

Mass spectrum of nonstrange baryons. The calculated masses based on Equations (9.1) and (9.2). The experimental masses are from the 'Review of Particle Properties', *Rev. Mod. Phys.* **56**, No. 2, Part II (April, 1984). The last column contains the identification code of the pion-nucleon resonance used in *Reviews of Modern Physics*. For $N = 0$ and 1, the quark model multiplet scheme is in excellent agreement with the experimental world. For $N = 2$, the model seems to work well, but more work is needed on both the theoretical and experimental fronts. There are still very few particles in $N = 3$. Baryonic masses are measured in MeV.

N	L	SU(6)	SU(3)	Spin	J	Calculated mass	Experimental mass	RMP-ID
0	0	56	8	1/2	1/2	940	939	P_{11}****
			10	3/2	3/2	1240	1232	P_{33}****
1	1	70	8	1/2	1/2	1520	1535	S_{11}****
					3/2	1520	1520	D_{13}****
			8	3/2	1/2	1688	1650	S_{11}****
					3/2	1688	1700	D_{13}***
					5/2	1688	1675	D_{15}****
			10	1/2	1/2	1652	1620	S_{31}****
					3/2	1652	1700	D_{33}****
2	0	56	8	1/2	1/2	1480	1440	P_{11}****
			10	3/2	3/2	1780	1600	P_{33}**
		70	8	1/2	1/2	1730	1710	P_{11}***
			8	3/2	3/2	1898		
			10	1/2	1/2	1862		
2	2	56	8	1/2	3/2	1660	1720	P_{13}****
					5/2	1660	1680	F_{15}****
			10	3/2	1/2	1960	1910	P_{31}****
					3/2	1960	1920	P_{33}****
					5/2	1960	1905	F_{35}****
					7/2	1960	1950	F_{37}****
		70	8	1/2	3/2	1900		
					5/2	1900	2000	F_{15}**
			8	3/2	1/2	2078	2100	P_{11}*
					3/2	2078		
					5/2	2078	2000	F_{15}**
					7/2	2078	1990	F_{17}**
			10	1/2	3/2	2030	2160	P_{33}
					5/2	2030		
3	1	70	8	1/2	1/2	2060		
					3/2	2060	2080	D_{13}**
			8	3/2	1/2	2228	2090	S_{11}*
					3/2	2228		
					5/2	2228		

Table 9.1 (continued)

N	L	SU(6)	SU(3)	Spin	J	Calculated mass	Experimental mass	RMP-ID
			10	1/2	1/2	2192	1900	S_{31}***
					3/2	2192		
		70	8	1/2	1/2	2060		
					3/2	2060		
			8	3/2	1/2	2228		
					3/2	2228		
					5/2	2228		
			10	1/2	1/2	2192		
			10	1/2	1/2	2192		
					3/2	2192		
		56	8	1/2	1/2	1810		
					3/2	1810		
			10	3/2	1/2	2110	2150	S_{31}*
					3/2	2110	1940	D_{33}*
					5/2	2110	1930	D_{35}***
2		70	8	1/2	3/2	2180		
					5/2	2180		
			8	3/2	1/2	2348		
					3/2	2348		
					5/2	2348		
					7/2	2348	2190	G_{17}****
			10	1/2	3/2	2312		
					5/2	2360		
3		70	8	1/2	5/2	2528		
					7/2	2528		
			8	3/2	3/2	2528		
					5/2	2528	2200	D_{15}**
					7/2	2528		
					9/2	2528	2250	G_{19}****
			10		1/2	5/2	2452	
				7/2	2492	2200	G_{37}*
		56	8	1/2	5/2	2110		
					7/2	2110		
			10	3/2	3/3	2410		
					5/2	2410	2350	D_{35}*
					7/2	2410		
					9/2	2410	2400	G_{39}**

In addition, there are resonances which cannot be fitted in this table. They are $N(1540, P_{13}, 3/2, *)$, $N(2220, H_{19}, 9/2, ****)$, $N(2600, I_{1,11}, 11/2, ***)$, $N(2700, K_{1,13}, 13/2, *)$, $\Delta(1550, P_{31}, 1/2, *)$, $\Delta(2300, H_{39}, 9/2, **)$, $\Delta(2390, F_{37}, 7/2, *)$, $\Delta(2420, H_{3,11}, 11/2, ***)$, $\Delta(2750, I_{3,13}, 13/2, **)$, $\Delta(2950, K_{3,15}, 15/2, **)$. N and Δ denote the $SU(3)$ octet and decuplet respectively. The first number in parentheses is the mass of the resonance. The last number is the total angular momentum. Four stars mean that the accuracy in measurement is excellent. One star means the accuracy is poor. Since most of these resonances correspond to even-parity baryons, they presumably belong to $N = 4$ multiplet.

TABLE 9.2
Summary of the present experimental status of the quark model multiplet scheme. A means excellent, B means very good, etc. The situation is indeed excellent for $N = 0$ multiplets which are usually called the $SU(6)$ multiplets. As N increases, the agreement is less than ideal. For $N = 3$, no meaningful statement can be made at this time.

	Baryons		Mesons	
N	Nonstrange	Strange	Nonstrange	Strange
0	A	A	A	A
1	A	A^-	B^-	C
2	A^-	C	D	D
3	D	?	?	?

If we confine ourselves to nonstrange hadrons, there are only two unitary spin states for the quarks, and therefore the mathematics for the unitary spin becomes as simple as that for the spin. With this point in mind, Kim and Noz (1974) simplified the mass formula with which we can study not only the $SU(3)$ and $SU(6)$ symmetry breakings but also the spatial excitations. Their mass formula is

$$M = M_0 + aN + b\left(S - \frac{1}{2}\right) + d\left(I - \frac{1}{2}\right)$$

$$+ d\left(\frac{T - 56}{14}\right) + g L(L + 1), \tag{9.1}$$

where N, S, I, T, and L are the total oscillator quantum number, spin, isospin, $SU(6)$ number (56 or 70), and the total orbital angular momentum, respectively. We are interested in the first two terms in the above expression which measure the linearity in the mass spectrum. However, the hadronic masses we measure are perturbed values which include the remaining terms in the above mass formula. The following choice of the parameters gives a good agreement with the observed masses:

$$\begin{gathered} M_0 = 940 \text{ MeV}, \quad a = 270 \text{ MeV}, \quad b = 168 \text{ MeV}, \\ d = 132 \text{ MeV}, \quad f = 250 \text{ MeV}, \quad g = (30 \pm 10) \text{ MeV}. \end{gathered} \tag{9.2}$$

The crucial test is whether the mass spectrum exhibits the harmonic oscillator characteristics through the linearity in N in Equation (9.1). The coefficient $a = 270$ MeV given above corresponds to the spring constant of $\Omega \simeq 0.47 \text{ (GeV)}^2$.

We list the calculated and experimental masses for the nonstrange baryons

Hadronic Mass Spectra

in Table 9.1. It is quite clear from Table 9.1 that the quark model multiplet scheme works well for $N = 0$ and 1. For $N = 2$, however, some more work is needed, particularly in the 2000 MeV region. It is too early to make a meaningful statement abut the $N = 3$ states.

The study of baryonic mass spectra is one of the long-lasting programs in modern physics. If we do not use the above simplification, the mass formula becomes much more complicated. In addition, there are many other sources of perturbation. There are also reasons to believe that the force between the quarks even before the symmetry breaking perturbation is not exactly of the harmonic oscillator type (Isgur and Karl, 1979; Stanly and Robson, 1980; Maltman and Isgur, 1984).

10. Mesons

Mesons differ from baryons in that they are bound states of two particles, a quark and an antiquark which we denote by q and $\bar{q}$, respectively. Since q and $\bar{q}$ are different particles, mesonic states are not subject to the symmetry considerations. This, however, should not cause any difficulty because we are dealing here with only two constituent particles.

Since the quark can take three different unitary spin quantum numbers, and the antiquark has the quantum numbers opposite to those of the quark, there are nine $\bar{q}q$ states, such as $\bar{u}u$, $\bar{u}d$, etc. Of these combinations, all define unique values for the resulting mesons, as shown in Table 10.1, except the diagonal $\bar{u}u$, $\bar{d}d$, $\bar{s}s$ states, all of which satisfy $B = I_z = S = Q = 0$. This means that we can make suitable linear combinations of these diagonal elements in order to identify the particles observed in nature. The first combination would be to construct $(\bar{u}u - \bar{d}d)/\sqrt{2}$ in order to complete the isospin triplet with the du and ud states, as indicated in Table 10.1. We shall discuss the remaining two linear combinations after we discuss the spin wave functions.

As for the total spin of the mesons, we are considering here the addition of two spins, which is a routine procedure. The resultant total spin is either 1 or 0, with the following spin wave functions:

$$|1\rangle_S = |++\rangle, \quad [|+-\rangle + |-+\rangle]/\sqrt{2}, \quad \text{or} \quad |--\rangle, \tag{10.1}$$

and

$$|0\rangle_A = [|+-\rangle - |-+\rangle]/\sqrt{2}. \tag{10.2}$$

We can construct the combined spin-unitary spin wave functions by taking products of spin and unitary spin wave functions. As in the case of baryons, we shall call these $SU(6)$ wave functions. The unitary spin wave functions in this case are also called $SU(3)$ wave functions.

TABLE 10.1

Meson multiplets in the unitary spin space. Because all quarks have baryon number 1/3 and antiquarks −1/3, mesons have baryon number 0. The π, K ρ, and K^* mesons each form separate isospin multiplets. The η, X^0, ω, and ϕ mesons are singlets. The number in parentheses is the mass measured in MeV.

$q\bar{q}$ pair	q	I_z	S	Pseudoscalar Mesons (spin = 0)	Vector mesons (spin = 1)
$\bar{d}u$	1	1	0	π^+ (140)	ρ^+
$(\bar{u}u - \bar{d}d)/\sqrt{2}$	0	0	0	π^0 (135)	ρ^0 (780)
$\bar{u}d$	−1	−1	0	π^- (140)	ρ^-
$\bar{s}u$	1	1/2	1	K^+ (494)	K^{*+}
$\bar{s}d$	0	−1/2	1	K^0 (498)	K^{*0} (890)
$\bar{d}s$	0	1/2	−1	K^- (494)	$\bar{K}^{*0}$
$\bar{u}s$	−1	−1/2	−1	K^- (494)	K^{*-}
$(2\bar{s}s - \bar{u}u - \bar{d}d)/\sqrt{2}$	0	0	0	η(549)	
$(\bar{u}u + \bar{d}d + \bar{s}s)/\sqrt{6}$	0	0	0	X^0 (958)	
$(\bar{u}u + \bar{d}d)/\sqrt{2}$	0	0	0		ω (780)
$\bar{s}s$	0	0	0		ϕ (1019)

We should note in Table 10.1 that the $SU(3)$ wave functions for the spin-0 η and X^0 wave functons are different from those for the ω and ϕ mesons. As we noted earlier, these are linear combinations of diagonal elements with vanishing quantum numbers. Therefore, the combination is made in accordance with what we observe in the experimental world.

These $SU(6)$ wave functions are then combined with the spatial wave function. The spatial component in this case is a two-body isotropic harmonic oscillator, and its form is well known. The spin singlet should not cause any problem in coupling the orbital and spin angular momenta. The spin triplet should be coupled to the orbital wave function according to the usual angular momentum addition rule.

11. Exercises and Problems

Exercise 1. Next to the Poincaré group which contains $O(3)$ and $SU(2)$ as subgroups, $SU(3)$ has played the most important role in the development of modern elementary particle physics. What is $SU(3)$?

This is the group of unitary and unimodular transformations in the three-dimensional complex space. Its fundamental representation consists of three-by-three untiary matrices with unit determinant. This group is therefore generated by three-by-three Hermitian traceless matrices. There are eight linearly independent matrices which meet this specification. They are, in Gell-Mann's notation (Gell-Mann, 1961),

$$\lambda_1 = \begin{bmatrix} 0 & 1 & 0 \\ 1 & 0 & 0 \\ 0 & 0 & 0 \end{bmatrix}, \quad \lambda_2 = \begin{bmatrix} 0 & -i & 0 \\ i & 0 & 0 \\ 0 & 0 & 0 \end{bmatrix}, \quad \lambda_3 = \begin{bmatrix} 1 & 0 & 0 \\ 0 & -1 & 0 \\ 0 & 0 & 0 \end{bmatrix},$$

$$\lambda_4 = \begin{bmatrix} 0 & 0 & 1 \\ 0 & 0 & 0 \\ 1 & 0 & 0 \end{bmatrix}, \quad \lambda_5 = \begin{bmatrix} 0 & 0 & -i \\ 0 & 0 & 0 \\ i & 0 & 0 \end{bmatrix}, \quad \lambda_6 = \begin{bmatrix} 0 & 0 & 0 \\ 0 & 0 & 1 \\ 0 & 1 & 0 \end{bmatrix},$$

$$\lambda_7 = \begin{bmatrix} 0 & 0 & 0 \\ 0 & 0 & -i \\ 0 & 0 & 0 \end{bmatrix}, \quad \lambda_8 = \frac{1}{\sqrt{3}} \begin{bmatrix} 1 & 0 & 0 \\ 0 & 1 & 0 \\ 0 & 0 & -2 \end{bmatrix}. \quad (11.1)$$

These matrices are discussed in standard textbooks in elementary particle physics (Frazer, 1966; Huang, 1982). The above eight matrices form the closed Lie algebra:

$$[\lambda_i, \lambda_j] = 2if_{ijk}\lambda_k, \quad (11.2)$$

where f_{ijk} is antisymmetric in ijk, and $f_{123} = 1$, $f_{147} = 1/2$, $f_{156} = 1/2$, $f_{246} = 1/2$, $f_{257} = 1/2$, $f_{345} = 1/2$, $f_{367} = -1/2$, $f_{458} = \sqrt{3}/2$, $f_{678} = \sqrt{3}/2$.

The most general form of the transformation matrix is

$$U(\alpha) = \exp\left[-i\frac{1}{2}\sum_{i=1}^{8} a_i\lambda_i\right], \quad (11.3)$$

applicable to the $SU(3)$ spinors:

$$x_1 = \begin{bmatrix} 1 \\ 0 \\ 0 \end{bmatrix}, \quad x_2 = \begin{bmatrix} 0 \\ 1 \\ 0 \end{bmatrix}, \quad x_3 = \begin{bmatrix} 0 \\ 0 \\ 1 \end{bmatrix}, \quad (11.4)$$

and their linear combinations. As in the case of $SU(2)$, because the matrix $U(\alpha)$ is unitary, we can consider the Hermitian conjugates of the above column vectors:

$$x_1^\dagger = (1, 0, 0), \quad x_2^\dagger = (0, 1, 0), \quad x_3^\dagger = (0, 0, 1). \quad (11.5)$$

For the three-quark system, we take a direct product of three $SU(3)$ spinors $x_i x_j x_k$ and symmetrize the indices according to the procedures described in Section 6. For mesons consisting of a quark and antiquark, we take the product $x_i x_j^\dagger$.

Exercise 2. We defined in Equations (11.4) and (11.5) the $SU(3)$ spinors. What are the $SU(3)$ vectors?

As in the case of $SU(2)$ or $SL(2, c)$, transformation matrices applicable to the above spinors form a spinor representation. We can next consider the three-by-three matrix X whose elements are $x_i x_j^\dagger$. The transformation of this matrix is carried out through

$$X' = U(\alpha) \, X \, U^\dagger(\alpha). \tag{11.6}$$

This is very similar to the procedure of recovering vectors from the spinors in $SU(2)$ and $SL(2, c)$ discussed in Chapter IV. The Hermitian matrix X is an *SU(3) vector*, and the above transformation forms a vector representation. Furthermore, X can be written as a linear combination of the eight λ matrices given in Equation (11.1) plus λ_0, where

$$\lambda_0 = \left(\frac{2}{3}\right)^{1/2} \begin{bmatrix} 1 & 0 & 0 \\ 0 & 1 & 0 \\ 0 & 0 & 1 \end{bmatrix}. \tag{11.7}$$

If we use the nine λ matrices as the basis vectors and define their inner product as

$$(\lambda_i, \lambda_j) = \tfrac{1}{2} \operatorname{Tr}(\lambda_i^\dagger \lambda_j) = \delta_{ij}, \tag{11.8}$$

Then, for

$$X = \sum_{i=0}^{8} \lambda_i a_i, \tag{11.9}$$

the inner product is

$$(X', X) = \frac{1}{2} \operatorname{Tr}[(X')^\dagger X] = \sum_{i=0}^{8} (a_i')^\dagger a_i. \tag{11.10}$$

Exercise 3. Discuss the little groups of $SU(3)$ using the vector representation.

The $U(2)$ subgroup generated by $\lambda_1, \lambda_2, \lambda_3$ and λ_8 leaves λ_8 invariant, when the transformation is performed according to Equation (11.6). This is a little group. The subgroup generated by λ_2 and λ_8 leaves $(a\lambda_3 + b\lambda_8)$ invariant, where $a \neq \pm\sqrt{3}$. This subgroup is also a little group. Since λ_3 and λ_8 comute with each other, this little group is a direct product of $U(1)$ and $U(1)$ (Kim and Markley, 1969).

Problem 1. We noticed in Exercise 2 that there is an $SU(2)$ subgroup generated by λ_1, λ_2 and λ_3. Show that there are two other $SU(2)$-like subgroups in $SU(3)$ and that they are unitarily equivalent.

Problem 2. In $SU(2)$, we can write the transformation matrix in the form of $U = \exp[-i(\alpha/2)\mathbf{n} \cdot \boldsymbol{\sigma}]$. This can be expanded to

$$U = \left(\cos\frac{\alpha}{2}\right) I - i\left(\sin\frac{\alpha}{2}\right) \mathbf{n} \cdot \boldsymbol{\sigma}. \tag{11.11}$$

Is it possible to expand the expression in Equation (11.3) in a finite polynomial in the λ matrices? See Kim and Markley (1969).

Problem 3. Very often in physics, particularly in gauge theory (Huang, 1982), we deal with an eight-component field or wave function of the form

$$\psi(x) = \sum_{i=1}^{8} \lambda_i \psi_i(x). \tag{11.12}$$

Show that the above expression can be decomposed into a polar form

$$\psi(x) = e^{-i\theta(x)} \rho(x) e^{i\theta(x)}, \tag{11.13}$$

where

$$\theta(x) = \sum_{i=1}^{8} \lambda_i \theta_i(x), \quad \text{and} \quad \rho(x) = \sum_{i=1}^{8} \lambda_i \theta_i(x) \cdot .$$

Show in particular that the generators of the little group mentioned in Exercise 3 can remain in $\rho(x)$, while others are exponentiated so that

$$\rho(x) = \lambda_1 \rho_2(x) + \lambda_2 \rho_2(x) + \lambda_3 \rho_3(x) + \lambda_8 \rho_8(x),$$
$$\theta(x) = \lambda_4 \theta_4(x) + \lambda_5 \theta_5(x) + \lambda_6 \theta_6(x) + \lambda_7 \theta_7(x), \tag{11.14}$$

for the $U(2)$ little group. Likewise, for the little group generated by λ_3 and λ_8,

$$\rho(x) = \lambda_8 \rho_8(x) + \lambda_3 \rho_3(x),$$
$$\theta(x) = \lambda_1 \rho_1(x) + \lambda_2 \rho_2(x) + \lambda_4 \rho_4(x) + \lambda_5 \rho_5(x) + \lambda_6 \rho_6(x) + \lambda_7 \rho_7(x). \tag{11.15}$$

See Kim and Markley (1969).

Problem 4. In Equations (11.14) and (11.15), the λ matrices are separated into two distinctive sets. Show that this separation remains invariant under transformations of their respective little groups.

Problem 5. Repeat the calculations from Equation (2.6) to Equation (2.10) with $B = X_1 + 2X_2 + 3X_3$, instead of the form given in Equation (2.6), to obtain the result given in Table 2.1.

Problem 6. Construct the Casimir operators for $SU(3)$ (Okubo, 1962).

Problem 7. Express the wave functions given in Section 7 in terms of the spherical harmonics (De *et al.*, 1973).

Problem 8. Estimate the numerical difference in calculating the mass versus the (mass)^2 for baryons. See Kim and Noz (1974).

Problem 9. Prove that the quantity in Equation (11.9) transforms under Equation (11.6) as an $SU(3)$ scalar (singlet) plus an $SU(3)$ vector (octet), and that no $SU(3)$ transformation mixes them.

Problem 10. Since λ_3 and λ_8 are diagonal, they commute with each other. Let us call them the charge matrices. λ_3 and λ_8 are proportional to I_3 (third component of isospin) and Y (hypercharge) respectively.

(a) Compute the charges of the three spinor components of Equations (11.4) and (11.5) by

$$Q x_i = q_i x_i, \quad i = 1, 2, 3. \tag{11.16}$$

where $Q = I_3, Y$.

(b) Compute the charges of the eight vector components of Equation (11.1) by using the condition:

$$Q Q \lambda_m - \lambda_m Q = q \lambda_m, \tag{11.17}$$

where $m = 1, 2, \ldots, 8$.

(c) Use the result of part (b) to show that $SU(3)$-spinor (triplet) can be decomposed into an $SU_I(2)$-spinor (doublet) + an $SU_I(2)$ scalar (singlet). Show also that $SU(3)$-vector (octet) decomposes into an $SU_I(2)$-vector (triplet) + an $SU_I(2)$-scalar (singlet), both with zero hypercharge and two $SU_I(2)$-spinors (doublet) with opposite hypercharges. Note that, since the $SU_I(2)$-vector and scalar generate the $U(2)$ of Exercise 3, the two $SU_I(2)$-spinors represent the $SU(3)/U(2)$ coset space.

(d) Verify that the direct product $x_i x_j^\dagger$ contains all the states found in (c) by explicit matrix multiplication and expansion in terms of Equation (11.1) as well as by computing the I_3 and Y charges.

Problem 11

(a) Do parts (a) and (b) of Problem 10 for the $SU(2)$-like subgroup found in Problem 1.

(b) Show that the $SU(3)$-spinor (triplet) can also serve as an $SO(3)$-vector. Show also that an $SU(3)$-vector (octet) decomposes into $SO(3)$-vector (triplet) + an $SO(3)$-second-rank tensor (quintet). Note that, while the $SO(3)$-vector components found in the $SU(3)$ vector generate $SO(3)$, the $SO(3)$-second-rank tensor represents the $SU(3)/SO(3)$ coset space.

Problem 12. Show that
(a)

$$x^a x_b^\dagger = \sum_{i=0}^{\infty} a_i \, (\lambda_i)_b^a = a_0 \, \delta_b^a + \sum_{i=1}^{\infty} a_i \, (\lambda_i)_b^a. \tag{11.18}$$

where a and b denote row and column indices of the three-by-three λ_i matrices.

(b) Note that λ_1, λ_2, and λ_3 operate only on the first two indices and that λ_8 is an identity matrix on them. Using this observation on splitting the set of indices a, b, and c into (1, 2) and (3), carry out the following decomposition.

x^a and $x_b^\dagger$,

$$\begin{aligned} x^{[a} x^{b]} &= \tfrac{1}{2} (x^a x^b - x^b x^a), \\ x^{\{a} x^{b\}} &= \tfrac{1}{2} (x^a x^b + x^b x^a), \\ x^a x_b^\dagger \ &(\text{with } a_0 = 0). \end{aligned} \tag{11.19}$$

For an extensive use of this method for the symmetry breaking of $SU(N)$, see Hubsch *et al.* (1985).

(c) Show that $\varepsilon_{abc} \, x^b x^c$ has the same charge as $x_a^\dagger$.

Chapter XII

Lorentz-Dirac Deformation in High-Energy Physics

In Chapter XI, we studied the problem of hadronic mass spectra and concluded that it is safe to believe that hadrons are quantum bound states of quarks having localized probability distribution. As in all bound state cases, this localization condition is reponsible for the existence of discrete mass spectra. Therefore the hadronic mass spectra plays a very important role in demonstrating that hadrons are quantum bound states of quarks.

However, the above-mentioned picture of bound states is applicable only to observers in the Lorentz frame in which the hadron is at rest. How would hadrons appear to observers in other Lorentz frames? More specifically, can we use Dirac's picture of Lorentz deformation described in Chapter VI to tackle this problem. The purpose of this chapter is to examine whether this picture is consistent with what we observe in high-energy laboratories. We are particularly interested in studying the nucleon form factors, Feynman's parton picture, and the hadronic jet phenomenon.

The size of the proton is 10^{-5} of that of the hydrogen atom. Therefore, it is not unnatural to assume that the proton has a point charge in atomic physics. However, while carrying out experiments on electron scattering from proton targets, Hofstadter in 1955 observed that the proton charge is spread out. In this experiment, an electron emits a virtual photon, which then interacts with the proton. If the proton consists of quarks distributed within a finite space-time region, the virtual photon will interact with quarks which carry fractional charges. The scattering amplitude will depend on the way in which quarks are distributed within the proton.

The portion of the scattering amplitude which describes the interaction between the virtual photon and the proton is called the form factor. Although there have been many attempts to explain this phenomenon within the framework of quantum field theory, it is quite natural to expect that the wave function in the quark model will describe the charge distribution. In high-energy experiments, we are dealing with the situation in which the momentum transfer in the scattering process is large. We therefore have to know how to describe the quark-model wave functions for rapidly moving protons. The first application of the Lorentz-Dirac deformation described in Chapter VI should be made on this form factor problem.

While the form factor is the quantity which can be extracted from the elastic scattering, it is important to realize that in high-energy processes, many particles are produced in the final state. They are called inelastic processes. While the elastic process is described by the total energy and momentum transfer in the center-of-mass coordinate system, there is, in addition, the energy transfer in inelastic scattering. Therefore, we would expect that the scattering cross section would depend on the energy, momentum transfer, and energy transfer. However, one prominent feature in inelastic scattering is that the cross section remains nearly constant for a fixed value of the momentum-transfer/energy-transfer ratio. This phenomenon is called "scaling."

In order to explain the scaling behavior in inelastic scattering, Feynman in 1969 observed that a fast-moving hadron can be regarded as a collection of many "partons" whose properties do not appear to be identical to those of quarks. For example, the number of quarks inside a static proton is three, while the number of partons in a rapidly moving proton appears to be infinite. The question then is how the proton looking like a bound state of quarks to one observer can appear different to an observer in a different Lorentz frame? Can the Lorentz-deformation picture described in Chapter VI explain this puzzle? We shall deal with this problem in the present chapter.

Another peculiar behavior in high-energy processes is the jet phenomenon. In high-energy collisions, many hadrons are produced in the final state. Their momenta are not randomly distributed. They are bunched together to form a "jet." The question then is whether the Lorentz deformation of hadronic wave functions is responsible for hadrons coming out in the same direction. We shall also study this problem.

The Lorentz-Dirac deformation was extensively discussed in Chapter VI. In Section 1 of the present chapter, we use the technique of space-time diagrams to illustrate the Lorentz deformation property of hadrons which are regarded as bound states of quarks. Section 2 explains the connection between the nucleon form factors and the development of the concept of extended hadrons having non-zero space-time extension which is shared by the quark model. In Section 3, we discuss how the Lorentz-Dirac deformation can explain the relativistic behavior of the form factors for large momentum transfer.

Section 4 explains what Feynman's parton picture is. It is shown that, as the hadron moves very fast, the Lorentz deformation leads to widespred longitudinal spatial and momentum distributions for the constituent quarks. It is pointed out that these widespread distributions are responsible for those parton properties of the quarks in which they appear as free independent particles with a widespread momentum distribution.

In Section 5, we resolve the apparent paradox associated with simultaneous widespread distribution of the spatial and momentum wave functions. It is shown that this phenomenon, while consistent with Lorentz invariance of Planck's constant, is responsible for the infinite number of partons. In Section 6, we discuss the present status of the efforts being made to determine quantitatively the distribution of partons inside the proton.

In Section 7, it is shown that the Lorentz-Dirac deformation property can explain the formation of hadronic jets. Section 8 contains exercises and problems connected with the derivation of some of the well-known formulas in high-energy physics.

1. Lorentz-Dirac Deformation of Hadronic Wave Functions

The covariant harmonic oscillator formalism and its Lorentz-Dirac deformation property have been discussed in Chapters V and VI using the language of the Poincaré group. Since the purpose of this chapter is to see whether the deformation propertry can be observed in high-energy laboratories, we translate the group-theoretical treatment of the Lorentz-Dirac deformation into a language suitable for studying experimental data.

As before, we consider two quarks bound together inside a hadron. Let x_a and x_b be space-time coordinates for the first and second quarks respectively. Then it is more convenient to use the variables X and x defined as

$$X = (x_a + x_b)/2,$$
$$x = (x_a - x_b)/2\sqrt{2}. \tag{1.1}$$

The coordinate variable X specifies the space-time position of the hadron, and the x coordinate is for the space-time separation between the quarks. The spatial component of the four vector X specifies where the hadron is, and its time component tells how old the hadron and the quarks become. The spatial components of the four vector x specify the relative spatial separation between the quarks. Its time component is the time interval or separation between the quarks.

Let us start with a hadron at rest. As far as the spatial components are concerned, the relative quark motion is dictated by the attractive force between the quarks and by Heisenberg's uncertainty relation. If the attractive force is that of a harmonic oscillator, we can write down the wave function as

$$\psi(x, y, z) = H_k(\sqrt{\Omega}\, x)\, H_m(\sqrt{\Omega}\, y)\, H_n(\sqrt{\Omega}\, z)\, \exp[-\Omega(x^2 + y^2 + z^2)/2], \tag{1.2}$$

where Ω is the spring constant, and $H_n(z)$ is the Hermite polynomial of order n. We are not worrying about the normalization constant. This simple wave function is in fact the starting point for the quark model analysis of the observed mass spectra, as is shown in Chapter XI.

When the hadron moves, we have to boost the above wave function. Before carrying out the boost, we have to complete the wave function by taking into consideration the time-dependence. In the case of the oscillator formalism, we can accomplish this by multiplying the above nonrelativistic wave function by $\exp(-\Omega t^2/2)$. As was discussed thoroughly in Chapter VI, this procedure is consistent with the existence of the time-energy uncertainty relation and with the non-existence of time-like excitations. If the hadron moves along the x direction, the transverse components are not affected, and can therefore be ignored. Furthermore, spatial excitations do not play any major role in discussion of the Lorentz-Dirac deformation property. For this reason, we shall drop all Hermite polynomials in Equation (1.2). The relevant portion of the wave function is

$$\psi(z, t) = \exp[-\Omega(z^2 + t^2)/2]. \tag{1.3}$$

We emphasize again that t is the time-separation variable and is not the time variable appearing in the time-dependent Schrödinger equation. The above expression is localized within the region around $z = t = 0$, and vanishes rapidly as $(z^2 + t^2)$ becomes large. This circular region can be written as

$$(z^2 + t^2) < 1/\Omega. \tag{1.4}$$

If the hadron moves along the z direction with speed parameter, β, z and t in Equation (1.3) should be replaced by

$$\begin{aligned} z' &= (z - \beta t)/(1 - \beta^2)^{1/2}, \\ t' &= (t - \beta z)/(1 - \beta^2)^{1/2}. \end{aligned} \tag{1.5}$$

so that

$$\psi_\beta(z, t) = \psi(z', t') = \exp[-\Omega(z'^2 + t'^2)/2]. \tag{1.6}$$

This boost procedure becomes very convenient if we introduce the light-cone variables:

$$\begin{aligned} u &= (t + z)/\sqrt{2}, \\ v &= (t - z)/\sqrt{2}. \end{aligned} \tag{1.7}$$

In terms of these variables, the Lorentz boost of Equation (1.5) takes a simple form:

$$\begin{aligned} u' &= [(1 - \beta)/(1 + \beta)]^{1/2}\, u, \\ v' &= [(1 + \beta)/(1 - \beta)]^{1/2}\, v. \end{aligned} \tag{1.8}$$

The light-cone coordinate system consisting of the orthogonal variables u and v is only a rotation of the zt system by 45°. Thus

$$z^2 + t^2 = u^2 + v^2. \tag{1.9}$$

In terms of the light-cone variable, the oscillator wave function for the moving hadron is

$$\psi_\beta(z, t) = \exp[-\Omega(u'^2 + v'^2)/2]$$

$$= \exp\left[-\left(\frac{\Omega}{2}\right)\left(\frac{1-\beta}{1+\beta}u^2 + \frac{1+\beta}{1-\beta}v^2\right)\right]. \tag{1.10}$$

The above expression becomes that of Equation (1.3) when $\beta = 0$. When $\beta \to 1$, the wave function becomes concentrated along one of the light-cone axes.

As for the momentum-energy wave function, let p_a and p_b be the four-momenta of the first and second quarks respectively. Then the four-momenta conjugate to X and x of Equation (1.1) are

$$P = (p_a + p_b), \quad q = \sqrt{2}(p_a - p_b), \tag{1.11}$$

respectively. The momentum-energy wave function is identical to the space-time wave function if Ω is replaced by $1/\Omega$, and its form is given in Section 3 of Chapter VI.

While it is much easier to discuss the problem using a two-body bound state, hadrons of experimental interest, such as the nucleon, are bound-states of three quarks. If we let x_a, x_b, and x_c be the space-time coordinates of the quarks, it is more convenient to use the variables:

$$X = \tfrac{1}{3}(x_a + x_b + x_c),$$
$$r = \tfrac{1}{6}(x_a + x_b - 2x_c), \tag{1.12}$$
$$s = \tfrac{1}{2}(x_b - x_a),$$

and their conjugate variables.

$$P = p_a + p_b + p_c,$$
$$q = p_a + p_b - 2p_c, \tag{1.13}$$
$$k = \sqrt{3}\,(p_b - p_a).$$

The above formulas are the inversions of the equations given in Equations (7.7) and (7.12) respectively of Chapter V.

In terms of these variables, the covariant harmonic oscillator wave function for the three-particle bound system takes the form

$$\psi_\beta(r, s) = \exp\left[-\frac{\Omega}{2}(r_z'^2 + r_0'^2 + s_z'^2 + s_0'^2)\right], \tag{1.14}$$

where, as in Equation (1.10), the transverse components have been ignored. The momentum-energy wave function is

$$\phi_\beta(q, k) = \exp\left[-\frac{1}{2\Omega}(q_z'^2 + q_0'^2 + k_z'^2 + k_0'^2)\right]. \quad (1.15)$$

The Lorentz-Dirac deformation property of r and s and of their conjugate coordinate system are the same as x and q for the two-body system.

We should emphasize here that the above-mentioned Lorentz-Dirac deformation property is not restricted to the harmonic oscillator wave functions, and that it is not difficult to find the same deformation property in other theoretical and phenomenological models (Gribov et al., 1965; Ioffe, 1969; Drell and Yan, 1971; Kim and Zaoui, 1971; Preparata and Craigie, 1976). One important advantage of using the harmonic oscillator formalism is that it is mathematically simple. According to Dirac (1970), a theory possessing mathematical beauty is more likely to survive than those which do not.

2. Form Factors of Nucleons

The quark model is intimately tied to the idea that hadrons, such as nucleons, are bound states of more fundamental particles with non-zero spatial extension. The idea that nucleons have space-time extension has a much longer history than the quark model. As was noted in Chapter V, Yukawa developed this idea as early as 1953. As Yukawa pointed out in his 1953 paper, his work was also based on earlier ideas. Thus it is somewhat endless to trace the origin of concept of extended model of elementary particles. Furthermore, this idea will persist whenever we find new particles, such as quarks. It is not difficult these days to find articles on composite models of quarks.

However, experimental discoveries of such an effect are rather rare. The first experimental discovery of the non-zero size of the proton was made by Hofstadter in 1955, who used electron-proton scattering to measure the charge distribution inside the proton. Since then the study of form factors has been an important branch of physics. For instance, we use this method to distinguish newly discovered leptons from hadrons. Leptons are still point particles while hadrons are not.

For this reason, the study of form factors cannot be separated from the quark model. In Chapter XI, we studied the hadronic mass spectra to believe that there is a spatial wave function whose localization condition provides discrete mass spectra. The question then is whether the same wave function can describe the charge distribution observed in form factor experiments. The purpose of this section is to study whether we can explain the behavior of form factors in terms of the Lorentz-Dirac deformation properties of the nucleon.

Let us first see what effect the charge distribution has on the scattering amplitude, using nonrelativistic scattering in Born approximation. If we scatter electrons from a fixed charge distribution whose density is $e\rho(r)$, the scattering amplitude is

$$f(\theta) = -\frac{e^2 m}{2\pi} \int d^3x \, d^3x' \, \frac{\rho(r')}{R} \, e^{-i\mathbf{Q}\cdot\mathbf{x}}, \tag{2.1}$$

where $r = |\mathbf{x}|$, $R = |\mathbf{r} - \mathbf{r}'|$, and $\mathbf{Q} = \mathbf{K}_f - \mathbf{K}_i$ which is the momentum transfer. This amplitude can be reduced to (Problem 3 in Section 8)

$$f(\theta) = [2me^2/Q^2] \, F(Q^2). \tag{2.2}$$

$F(Q^2)$ is the Fourier transform of the density function

$$F(Q^2) = \int d^3x \, \rho(r) \, e^{-i\mathbf{Q}\cdot\mathbf{x}}. \tag{2.3}$$

The above quantity is called the *form factor* and describes the charge distribution in terms of the momentum transfer. The charge density is normalized:

$$\int \rho(r) \, d^3x = 1. \tag{2.4}$$

Then $F(0) = 1$ from Equation (2.3). If the density function is a delta function $\delta(\mathbf{r})$ corresponding to a point charge, $F(Q^2) = 1$ for all values of Q^2, and the scattering amplitude of Equation (2.1) becomes the Rutherford formula for Coulomb scattering. The deviations from Rutherford scattering for increasing values of Q^2 give a measure of the charge distribution (Hofstadter and McAllister, 1955; Hofstadter, 1963). This was precisely what Hofstadter found in the scattering of electrons from a proton target.

As the energy of incoming electrons becomes higher, we have to take into account the recoil effect of target protons, and formulate the problem relativistically. It is generally agreed that electrons and their electromagnetic interaction can be described by quantum electrodynamics, in which the method of pertubation theory using Feynman diagrams is often employed for practical calculations (Schweber, 1961; Itzykson and Zuber, 1980). In this pertubation approach the scattering amplitude is expanded in power series of the fine structure constant $\alpha = e^2/4\pi$. Therefore in lowest order in α, we can describe the scattering of an electron by a proton using the diagram given in Figure 2.1. The corresponding matrix element is given in many textbooks on elementary particle physics (Frazer, 1966). It is proportional to

$$\bar{U}(P_f) \, \Gamma_\mu(P_f, P_i) \, U(P_i) \, (1/Q^2) \, \bar{U}(K_f) \, \gamma^\mu \, U(K_i), \tag{2.5}$$

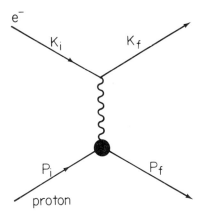

Fig. 2.1. Elastic electron-proton scattering. It is assumed that the electron behaves like a point charge. However, the proton has its hadronic structure, and its charge distribution is spread out.

where P_i, P_f, K_i and K_f are the initial and final four-momenta of the proton and electron respectively. $U(P_i)$ is the Dirac spinor for the initial proton. Q^2 is (four-momentum transfer)^2 and is

$$Q^2 = (P_f - P_i)^2 = (K_i - K_f)^2. \tag{2.6}$$

The $(1/Q^2)$ factor in Equation (2.5) comes from the virtual photon being exchanged between the electron and the proton. It is customary to use the letter t for Q^2 in form factor studies, and this t should not be confused with the time separation variable used in Section 1. In the metric we use, this quantity is positive for physical values of the four-momenta for the particles involved in the scattering process.

The unknown function $\Gamma_\mu(P_f, P_i)$ in Equation (2.5) represents the shaded circle in Figure 2.1 and carries the effect of the nucleon structure. If the proton were a point charge, we would have $\Gamma_\mu = \gamma_\mu$. If the proton has an extended charge structure, it will be modified by a factor, so that $\Gamma_\mu = \gamma_\mu F(t)$. This is not enough to explain the *anomalous magnetic moments* whose values are 2.79 and -1.91 in units of $e/2M$ for the proton and neutron respectively. We shall ignore the proton-neutron mass difference, and use M as the nucleon mass throughout this chapter. If we include these obesrved anomalous magnetic moments, Γ_μ should be written as

$$\Gamma_\mu = \gamma_\mu F_1(t) + i[\sigma_{\mu\nu} Q^\nu/2M] F_2(t). \tag{2.7}$$

At $t = 0$,

$$\begin{aligned} F_1^p(0) &= 1, \quad F_1^n(0) = 0, \\ F_2^p(0) &= 1.79, \quad F_2^n(0) = -1.91. \end{aligned} \tag{2.8}$$

The above form factors are Lorentz-invariant functions in the Lorentz-invariant variable t.

When we study the quark model, it is more convenient to use the following linear combinations:

$$G_M(t) = F_1(t) + F_2(t),$$
$$G_E(t) = F_1(t) + (t/4M^2) F_2(t). \tag{2.9}$$

The scattering cross section takes a relatively simple form in terms of these form factors (Problem 2 in Section 8). When $t = 0$,

$$G_{pM}(0) = \mu_p = 2.79, \quad G_{pE}(0) = 1 \text{ for proton},$$
$$G_{nM}(0) = \mu_n = -1.91, \quad G_{nE}(0) = 0 \text{ for neutron.} \tag{2.10}$$

These numbers are the magnetic moments of the proton and neutron respectively. The measured values of these moments were known for a long time (Blatt and Weisskopf, 1954). There were many unsuccessful attempts to calculate these numbers, before the invention of the quark model. Indeed, one of the early successes of the quark model was the calculation of the magnetic moment ratio (Problem 4 in Section 8):

$$\mu_n/\mu_p = -2/3. \tag{2.11}$$

However, the real challenge in modern physics is to calculate the form factors for increasing values of t.

At present, we can make the following experimental observations. Of the four form factors in the nucleonic system given in Equation (2.10), the neutron charge form factor is zero at $t = 0$, and remains small (not zero) for all values of t. The three remaining form factors decrease like $1/t^2$ as t increases beyond the value of (nucleon mass)^2. This behavior is usually called the *dipole fit*.

The question then is whether it is possible to calculate the above-mentioned dipole behavior using the wave function obtained from the static quark model, such as the harmonic oscillator wave function discussed in Chapter XI. If we are interested only in the overall behavior of the above-mentioned form factors, each of them can be derived for the single form factor $G(t)$ by multiplication of a constant factor, where $G(t)$ is normalized as

$$G(0) = 1, \tag{2.12}$$

and is proportional to $1/t^2$ for large values t, as is illustrated in Figure 2.2. In the case of the neutron charge form factor, we have to multiply the above $G(t)$ by zero within the framework of the approximation we make.

If we assume that the wave function for the nucleon in the quark model is Gaussian as is indicated in Chapter XI, the straightforward Fourier transformation of the charge distribution function, according to Equation (2.3), will lead to the form factor of the form

$$G(t) = \exp(-t/4\Omega). \tag{2.13}$$

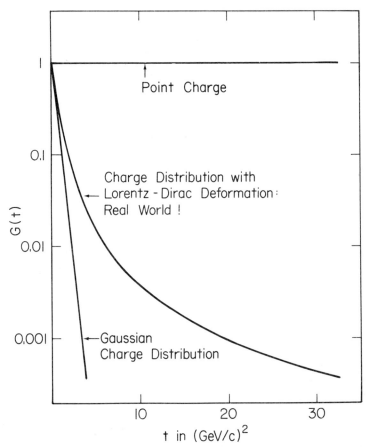

Fig. 2.2. Form factor behaviors for increasing values of $t = Q^2$. If the proton is a point charge, the form factor should be independent of t. If the charge distribution is Gaussian, the nonrelativistic calculation leads to an exponential decrease for increasing t. The relativistic calculation gives a dipole behavior. We observe this dipole behavior in the real world. See Stanley and Robson (1982) for the curves summarizing experimental data.

This expression is not consistent with the above-mentioned dipole fit. Indeed, it is a challenge to use the relativistic effect to convert the above expression into a dipole fit.

Before calculating the form factor behavior using relativistic wave functions, we should point out that there have been and there are still many attempts to do the same using other theoretical approaches. Perhaps, the single most important calculation which enhanced the development of the quark model is the work of Frazer and Fulco (1960) based on dispersion relations. The concept of the ρ meson, which played the key role in the $SU(3)$ formulation of the mesonic multiplets, was initially invented to explain the form factor behavior for moderate values of t.

3. Calculation of the Form Factors

In order to make a relativistic calculation of the form factor, let us go back to the definition of the form factor given in Equation (2.3). The density function $\rho(r)$ depends only on the target atom, and is proportional to $|\psi(\mathbf{r})|^2$, where $\psi(\mathbf{r})$ is the wave function for electrons in the target atom. This expression is a special case of the more general case:

$$\rho(\mathbf{r}) = \psi_f^\dagger(\mathbf{r}) \, \psi_i(\mathbf{r}), \tag{3.1}$$

where ψ_i and ψ_f are the initial and final wave function of the target atom. Indeed, the form factor of Equation (2.3) can be written in the form

$$F(q^2) = (\psi_f, \, e^{-i\mathbf{Q} \cdot \mathbf{r}} \, \psi_i). \tag{3.2}$$

Starting from this expression, we can make the required Lorentz generalization using the relativistic wave functions for hadrons. It is possible to obtain a Lorentz-invariant expression for the above quantity starting from a theoretical model based on the covariant Lagrangian formalism (Probelm 5 in Section 8).

However, in order to see the details of the transition to relativistic physics, we would be able to replace each quantity in the expression of Equation (3.2) by its relativistic counterpart. Let us go to the Lorentz frame in which the momenta of the incoming and outgoing nucleons have equal magnitude but opposite signs.

$$\mathbf{p}_f + \mathbf{p}_i = 0. \tag{3.3}$$

The Lorentz frame in which the above condition holds is usually called the *Breit frame*. We assume without loss of generality that the proton comes along the z direction before the collision and goes out along the negative z direction after the scattering process, as is illustrated in Figure 3.1. In this frame, the four vector $Q = (K_f - K_i) = (P_i - P_f)$ has no time-like component. Thus the exponential factor $\exp[-i\mathbf{Q} \cdot \mathbf{x}]$ can be replaced by the Lorentz-invariant form $\exp[-iQ \cdot x]$. As for the wave functions for the protons, we can use the covariant harmonic oscillator wave functions discussed in Chapters V and VI assuming that the nucleons are in the ground state. Then the only difference between the nonrelativistic and relativistic case is that the integral in the evaluation of Equation (3.2) is four-dimensional, including that for the time-like direction. This integral in the time-separation variable does not interfere with the exponential factor which does not depend on the time-separation variable.

Let us now write down the integral:

$$g(t) = \int d^4x \, \psi^\dagger_{-\beta}(x) \, \psi_\beta(x) \, e^{-iQ \cdot x}, \tag{3.4}$$

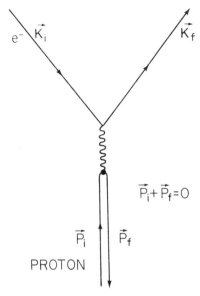

Fig. 3.1. Breit frame for electron-nucleon scattering. The momentum of outgoing nucleon is equal in magnitude but opposite in sign to that of the incoming nucleon.

where β is the velocity paramter for the incoming proton, and the wave function $\psi_{\pm\beta}(x)$ takes the form:

$$\psi_{\pm\beta}(x) = \frac{\Omega}{\pi}(\exp[-\Omega(x^2+y^2)/2])$$

$$\times \exp\left(-\frac{\Omega}{4}\left[\left(\frac{1\mp\beta}{1\pm\beta}\right)(z\pm t)^2 + \left(\frac{1\pm\beta}{1\mp\beta}\right)(z\mp t)^2\right]\right). \quad (3.5)$$

After the above decomposition of the wave functions, we can perform the integrations in the x and y variables trivially. After dropping these trivial factors, we can write the product of the two wave functions as

$$\psi_{-\beta}^\dagger(x)\psi_\beta(x) = \frac{\Omega}{\pi}\exp(-\Omega[(1+\beta^2)/(1-\beta^2)](z^2+t^2)). \quad (3.6)$$

Thus the z and t variables have been separated. Since the exponential factor in Equation (3.2) does not depend on t, the t integral in Equation (3.4) can also be trivially performed, and the integral of Equation (3.4) can be written as

$$g(t) = \left[\frac{\Omega}{\pi}(1-\beta^2)/(1+\beta^2)\right]^{1/2}\int dz\, e^{-2iPz}\exp\left(-\Omega\frac{1+\beta^2}{1-\beta^2}z^2\right), \quad (3.7)$$

where P is the z component of the momentum of the incoming nucleon. The (momentum transfer)^2 variable Q^2 is $4P^2$. Indeed, the distribution of the hadronic material along the longitudinal direction became contracted (Licht and Pagnamenta, 1970), as is illustrated in Figure 3.2. If we use the relation between β and t:

$$\beta^2 = t/(t + 4M^2), \tag{3.8}$$

the evaluation of the above integral for $g(t)$ in Equation (3.7) leads to

$$g(t) = (2M^2/(t + 2M^2)) \exp(-M^2 t/[2\Omega(t + 2M^2)]). \tag{3.9}$$

At $t = 0$, the above expression becomes $= 1$. It decreases as $1/t$ for large values of t.

We have so far carried out the calculation for an oscillator bound state of two quarks. The proton consists of two u and one d quarks, as we discussed in Chapter XI. There are one u and two d quarks in the neutron. If there are three quarks, there are two oscillator modes as was pointed out in Section 1. The generalization of the above calculation to the three-quark system is straightforward, and the result is that the form factor $G(t)$ takes the form (Problem 6 in Section 8):

$$G(t) = (2M^2/(t + 2M^2))^2 \exp(-2M^2 t/[\Omega(t + 2M^2)]), \tag{3.10}$$

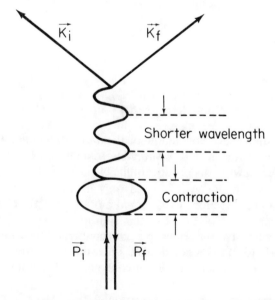

Fig. 3.2. Lorentz contraction of the hadronic density in the form factor calculation. As the momentum transfer increases, the width of the longitudinal distribution decreases. The wavelength of the photons also decreases.

which is 1 at $t = 0$, and decreases as $1/t^2$ for large values of t. Indeed, this function satisfies the requirement of the dipole behavior discussed in Section 2.

Let us re-examine the above calculation. If we replace β by zero in Equation (3.7), $g(t)$ will become the nonrelativistic result of Equation (2.13). The above-mentioned dipole behavior is due to the relativistic effect. Indeed, the longitudinal hadronic distribution becomes contracted as β increases, as is illustrated in Figure 3.2, while the size remains constant in nonrelativistic calculation. However, β is correlated to P in Equation (3.7) or to the momentum transfer variable by Equation (3.8). As the energy P or the momentum transfer increases, both the longitudinal size and the wavelength of the virtual photon decrease. This correlation prevents a harsh high-frequency cut-off which happens in the nonrelativistic case.

In order to gain a deeper understanding of the above-mentioned correlation, let us study the case using the momentum-energy wave functions:

$$\phi_\beta(q) = \left(\frac{1}{2\pi}\right)^2 \int d^4x\, e^{-iq\cdot x}\, \psi_\beta(x). \tag{3.11}$$

As before, we can ignore the transverse components. Then $g(t)$ can be written as

$$\int dq_0\, dq_z\, \phi^+_{-\beta}(q_0, q_z - P)\phi_\beta(q_0, q_z + P). \tag{3.12}$$

We have sketched the above overlap integral in Figure 3.3. When $t = 0$ or $P = 0$, the two wave functions overlap completely in the $q_z q_0$ plane. As P increases, the wave functions become separated. However, they maintain a small overlapping region due to the Lorentz-Dirac deformation. In the nonrelativistic case, where the deformation is not taken into account, there is no overlapping region. This is precisely why the relativistic calculation gives a slower decrease in t than in the nonrelativistic case.

We have so far been interested only in the space-time behavior of the hadronic wave function. We should not forget the fact that quarks are spin-1/2 particles. The effect of this spin manifests itself prominently in the baryonic mass spectra. Since we are concerned here with the relativistic effects, we have to construct relativistic spin wave function for the quarks. This quark wave function should give the hadronic spin wave function. In the case of nucleons, the quark spins should be combined in a manner to generate the vertex function of Equation (2.7).

The most naive approach to this problem is to use free Dirac spinors for the quarks (Feynman et al., 1971). However, it was shown by Lipes (1972)

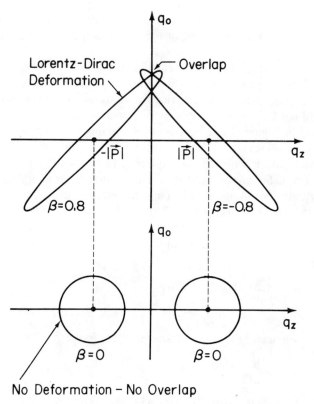

Fig. 3.3. Lorentz-Dirac deformation of the momentum-energy wave functions in the form factor calculation. As the momentum transfer increases, the two wave functions become separated. In the relativistic case, the wave functions maintain an overlapping region. Wave functions become completely separated in the nonrelativistic calculation. This lack of overlapping region leads to an unacceptable behavior of the form factor.

that the use of free-particle Dirac spinors leads to a wrong form factor behavior. Since quarks in a hadron are not free particles, Lipes's result does not alarm us. The difficult problem is to find a suitable mechanism in which quark spins are coupled to orbital motion in a relativistic manner. This is a non-trivial research problem. It is interesting to note that there has been progress at least in the study of form factors (Problem 7 in Section 8).

4. Scaling Phenomenon and the Parton Picture

The parton picture was developed originally by Feynman to explain the scaling behavior in deep inelastic electron-proton scattering. Let us consider the process $e + p \to e +$ hadrons, as is described in Figure 4.1. The matrix element for this process can be written as

$$M_{fi} = (e^2/Q^2)\, U(K_f)\gamma_\mu U(K_i)\langle P', \text{all}|J^\mu|P\rangle, \tag{4.1}$$

where $Q = K_i - K_f$ is the four-momentum transfer of the electron. We use P for the four-momentum of the initial nucleon. P' is the total four-momentum of the final-state hadrons. Here again, the electron-photon vertex is point-like. If we do not measure spins, the rate of transition is

$$\text{Rate} = \left(\frac{e^2}{4\pi}\right)^2 \int d^3K_f \frac{m^2}{EE'}\, (1/Q^2)^2\, \eta^{\mu\nu} W_{\mu\nu}, \tag{4.2}$$

where m is the electron mass, and

$$\eta^{\mu\nu} = \left(\frac{1}{2m}\right)^2 \text{Tr}[(\gamma\cdot K_f - m)\gamma^\mu(\gamma\cdot K_i - m)\gamma^\nu], \tag{4.3}$$

The non-trivial part of physics is contained in

$$W_{\mu\nu} = (2\pi)^3 \sum_{\text{all}} \langle P|J_\mu|\text{all}\rangle\langle\text{all}|J_\nu|P\rangle \delta(P' - P - Q). \tag{4.4}$$

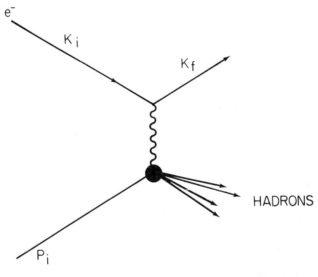

Fig. 4.1. Inelastic electron-proton scattering. The electron emits a virtual photon. After absorbing the virtual photon, the proton becomes many hadrons. Unlike the case of the elastic scattering described in Figure 2.1, there is an energy transfer through the virtual photon, in addition to the momentum transfer.

The summation is over all hadrons in the final state. The above expression can be reduced to

$$W_{\mu\nu} = W_1(Q^2, \nu)(g_{\mu\nu} - Q_\mu Q_\nu/Q^2)$$
$$+ W_2(Q^2, \nu)\left(\frac{1}{M}\right)^2 (P_\mu - (P \cdot Q/Q^2)Q_\mu)$$
$$\times (P_\nu - (P \cdot Q/Q^2)Q_\nu), \qquad (4.5)$$

where

$$\nu = -Q \cdot P/M.$$

W_1 and W_2 are now the Lorentz-invariant functions of the two Lorentz-invariant variables.

Unlike the case of elastic scattering, the cross section depends also on the energy transfer. If the proton and electron spins are not measured, the cross section in the laboratory frame takes the form (Problem 8 in Section 8):

$$\frac{d^2\sigma}{d\Omega dE'} = (\alpha^2/4E^2\sin^4(\theta/2))(W_2(Q^2, \nu)\cos^2(\theta/2)$$
$$+ 2W_1(Q^2, \nu)\sin^2(\theta/2)), \qquad (4.6)$$

where E and E' are the energies of the initial and final electrons, and Ω is the solid angle of the scattered electron. ν is the energy transfer, and is thus

$$\nu = E - E',$$
$$Q^2 = (4EE')\sin\frac{\theta}{2}. \qquad (4.7)$$

It should be emphasized here that W_1 and W_2 in Equation (4.6) are basically different from the form factors G_1 and G_2 discussed in Section 2. The W functions depend not only on Q^2, but also on the energy transfer variable. The fundamental difference is that the W_1 and W_2 are proportional to the cross section, while the form factors are proportional to the scattering amplitude.

We call the above inelastic scattering *deep inelastic scattering* when both the momentum and energy transfer variables become very large. In this case, we can define the ratio:

$$\omega = \frac{-Q^2}{2Q \cdot P}. \qquad (4.8)$$

It was observed (Bjorken, 1969) that both W_1 and νW_2 become functions of only one variable in the deep inelastic limit:

$$W_1 \to F_1(\omega),$$
$$\nu W_2 \to F_2(\omega). \tag{4.9}$$

This is called the *scaling phenomenon*.

There are several different theoretical approaches to explain this scaling behavior in deep inelastic scattering. One of the first explanations was that of Feynman's parton picture, and the parton model still plays a major role in high-energy physics. In 1969, Feynman observed that a rapidly moving proton, which is now believed to be a bound state of three quarks, can be regarded as a collection of particles called *partons* which exhibit the following peculiar features.

(a) The picture is valid only for hadrons moving with velocity close to that of light
(b) The interaction time between the quarks becomes dilated, and partons behave as free independent particles.
(c) The momentum distribution of partons becomes widespread as the hadron moves fast.
(d) The number of partons seems to be much larger than that of quarks, as is discussed in Figure 4.2.

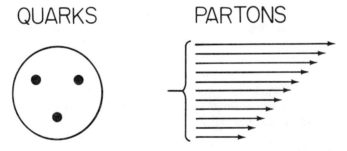

Fig. 4.2. Proton as it appears in the quark and parton models. Suppose that a proton is sitting quietly on the desk. According to the quark model, it appears like a bound state of three quarks to an observer who is sitting on the chair. However, to an observer who is on a jet plane with its speed close to that of light, the proton would look like a collection of free particles with a wide momentum distribution. This is called Feynman's parton picture.

Because the hadron is believed to be a bound state of two or three quarks, each of the above phenomena appears as a paradox, particularly (b) and (c) together. This paradox will be resolved in Section 5. In the meantime, let us see how the parton picture will explain the scaling phenomenon.

We ignore for simplicity spins and assume that all the particles are scalar particles. Then $W_{\mu\nu}$ of Equation (4.4) becomes

$$W = (2\pi)^3 \sum_{\text{all}} \langle P|J(0)|\text{all}\rangle\langle\text{all}|J(0)|P\rangle \delta(P' - P - Q). \tag{4.10}$$

If we use the expression:

$$\delta(P' - P - Q) = \left(\frac{1}{2\pi}\right)^4 \int e^{-i(P'-P-Q)\cdot x} d^4x, \tag{4.11}$$

the W function can be transformed into

$$W = \frac{1}{2\pi} \int d^4x \, e^{iQ\cdot x} \langle P|[J(0), J(x)]|P\rangle. \tag{4.12}$$

Indeed, the "summation over all" in Equation (4.10) drastically simplifies the problem. All we have to do is to compute the commutator in the above expression. It is not difficult to generalize the above expression to the case where the spins are considered (Problem 8 in Section 8).

The calculation of the commutator requires the knowledge of dynamics, and a simple calculation is possible only for free point particles, and the nucleon is not a point particle. However, in the limit of large Q^2, the integrand makes a significant contribution to the integral only in the small-x^2 region near the light-cones. This is called the light-cone dominance. There is no reason to expect that the commutator will behave like a free-field commutator in this region. If we use the free-field commutators for the current in Equation (4.12), we are still making a drastic assumption. Indeed, this is the place where the parton picture plays the decisive role. Let us go to the Lorentz frame in which the nucleon momentum is very large. Since partons are free particles, we can evaluate the integral of Equation (4.12) using the free-field approximation. As is shown in Exercise 2 in Section 8, the evaluation of the integral leads to

$$\nu W(Q^2, \nu) = (1/2M^2) \, \delta(Q^2/2Q\cdot P - 1). \tag{4.13}$$

If the nucleon is a collection of free partons, the relevant momentum variable in the above expression is the four-momentum of one of the partons. Therefore Equation (4.13) should be replaced by

$$\nu W(Q^2, \nu) = (1/2M^2) \sum_{\substack{\text{all}\\ \text{partons}}} C_i^2 \, \delta(1 - Q^2/Q\cdot p_i), \tag{4.14}$$

where p_i and C_i^2 are the four-momentum and the (charge in unit of e)^2 of the i-th parton respectively. If we parameterize the parton four-momentum by variable x_i:

$$p_i = x_i P, \tag{4.15}$$

then the formula in Equation (4.13) becomes

$$\nu W(Q^2, \nu) = (1/2M^2) \sum_i C_i^2 \, x_i \, \delta(x_i - \omega), \tag{4.16}$$

where ω is given in Equation (4.8). Indeed νW is a function only of the variable ω.

In order to calculate the above quantity, we have to know the charge and momentum distributions of partons. Before starting this calculation, we are led to the question of whether partons are quarks. The question is then whether the quark model and the parton model are only two different manifestations of one physical entity which is a relativistic bound state of quarks. We shall study this problem in Section 6 using the covariant harmonic oscillator formalism.

5. Covariant Harmonic Oscillators and the Parton Picture

We examine in this section how the Lorentz-Dirac deformation leads to the parton picture using for simplicity the covariant harmonic oscillator wave function for a two-body bound state (Kim and Noz, 1977a). By taking the Fourier transformation of Equation (1.10), we get the momentum-energy wave function:

$$\phi(q) = \exp\left[-\left(\frac{1}{2\Omega}\right)\left(\frac{1+\beta}{1-\beta}\, q_u^2 + \frac{1-\beta}{1+\beta}\, q_v^2\right)\right], \tag{5.1}$$

with

$$q_u = (q_0 - q_z)/\sqrt{2}, \quad q_v = (q_0 + q_z)/\sqrt{2}. \tag{5.2}$$

Because we are using here the harmonic oscillator, the mathematical form of the above momentum-energy wave function is identical to that of the space-time wave function. The Lorentz deformation properties of these wave functions are given in Figure 5.1. When $\beta = 0$, both wave functions behave like those for the static bound states of quarks. As β increases, the wave functions become continuously Lorentz-deformed until they become concentrated along their respective positive light-cone axes. When $\beta \to 1$, $v \to 0$ and $q_u \to 0$. Therefore, in this limit,

$$u = \sqrt{2}\, z = \sqrt{2}\, t, \quad q_v = \sqrt{2}\, q_z = \sqrt{2}\, q_0. \tag{5.3}$$

If we use the approximation $1 + \beta = 2$, and write the wave functions of Equations (1.10) and (5.1) in terms of the z and q_z variables only,

$$\begin{aligned}\psi_\beta(z) &\to \exp[-\Omega(1-\beta)z^2/2], \\ \phi_\beta(q_z) &\to \exp[-(1-\beta)q_z^2/2\Omega].\end{aligned} \tag{5.4}$$

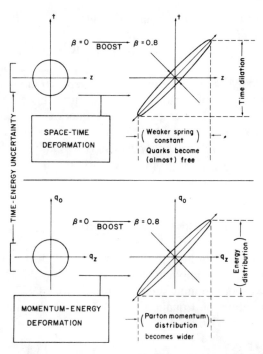

Fig. 5.1. Pictorial explanation of Feynman's parton picture based on the Lorentz-Dirac deformations in both the space-time and momentum-energy coordinate systems. In the harmonic oscillator model, the Lorentz deformation property of the momentum-energy wave function is identical to that of the space-time wave function. As the hadron moves very fast, both the space-time and the momentum-energy distribution become concentrated and elongated along their respective positive light-cone axes.

Both of the above wave functions become widespread. Pictorially, we can obtain these distributions by taking the longitudinal projections of the elliptic deformations given in Figure 5.1.

Let us consider first the longitudinal distribution. From Equation (1.1), we can write the longitudinal coordinate for each quark as

$$z_a = Z + \sqrt{2}\, z,$$
$$z_b = Z - \sqrt{2}\, z.$$
(5.5)

where Z is the longitudinal coordinate of the hadron. The distribution in z given in Equations (5.3) and (5.4) and in Figure 5.1 tells us that the position of each quark appears widespread to the observer in the laboratory frame, and that the quarks appear like free particles as the hadronic velocity parameter β approaches 1. This effect, first noted by Feynman, is universally observed in high-energy hadronic experiments.

Because the quarks appear almost free, we would normally expect that the momentum of each quark will appear sharply defined. However, let us consider the longitudinal momentum distribution derivable from Equation (5.5). We write the momentum of each quark as

$$p_{az} = P_z/2 + q_z/2\sqrt{2},$$
$$p_{bz} = P_z/2 - q_z/2\sqrt{2}. \tag{5.6}$$

When the hadron moves very fast, we can use the approximation

$$(1 - \beta) = (P_0/M)^2/2, \tag{5.7}$$

in the expression given in Equation (5.4) where M and P_0 are the hadronic mass and energy respectively. Then q_z is spread over the region

$$-\sqrt{2\Omega}\, P_0/M < q_z < \sqrt{2\Omega}\, P_0/M, \tag{5.8}$$

which becomes widespread as P_0 becomes large. Consequently, the longitudinal momentum of the first quark given in Equation (5.6) mostly lies within the interval between

$$p_z^{max} = P_0(1/2 + \sqrt{\Omega}/2M),$$

and

$$p_z^{min} = P_0(1/2 - \sqrt{\Omega}/2M), \tag{5.9}$$

where we assumed that $P_z = P_0$. If we assume that M is the proton mass, and use the value $\Omega = 1$ (GeV)^2, which is commonly used for the quark model analysis, then the quantity $\sqrt{\Omega}/2M$ is of the same order of magnitude as $1/2$. For this reason, the (almost) light-like four-momentum $p_{1\mu}$ can be written as

$$p_{1\mu} = x P_\mu, \tag{5.10}$$

where the parameter x ranges approximately between zero and 1. This type of distribution, which was postulated first by Feynman, is commonly observed in high-energy hadronic experiments.

We can make similar projections to the vertical axes in Figure 5.1. When $\beta \to 1$, the relative quark motion is an oscillation along the positive light-cone axis. Figure 5.1 clearly indicates that the oscillation time is dilated. Classically, the quarks make two "collisions" with each other within one period of oscillation. Therefore, the interquark collision time is one half of the oscillation time. If the hadron is at rest with $\beta = 0$, the collision time is $1/2\sqrt{\Omega}$. If the hadronic velocity is close to that of light, and if we make the same approximation as the one used for the derivation of Equation (5.4) the collision time becomes $1/[2\sqrt{\Omega}(1-\beta)]$ to an observer in the laboratory frame. If we use the expression of Equation (5.7) the collision time ratio becomes

$$\frac{\text{Collision time for } \beta \to 1}{\text{Collision time for } \beta = 0} = P_0/2M. \tag{5.11}$$

As the hadron moves very fast, the above ratio becomes very large. This is also consistent with what we observe in the real world, and also with our previous observation that quarks appear as almost free particles to observers in the laboratory frame. This time dilation allows us to make an incoherent sum of cross sections due to each parton.

The widespread momentum distribution shown in Figure 5.1 appears to contradict our initial expectation that, based upon the widespread spatial distribution, the quark be a free particle with a sharply defined momentum. This apparent contradiction presents to us the following two fundamental questions:

(a) If both the spatial and momentum distributions become widespread as the hadron moves, and if we insist on Heisenberg's uncertainty relation, is Planck's constant dependent on the hadronic velocity?

(b) Is this apparent contradiction related to another apparent contradiction that the number of partons is infinite while there are only two or three quarks inside the hadron?

The answer to the first question is "No", and that for the second question is "Yes". The purpose of this section is to explain these answers.

Let us answer the first question which is related to the Lorentz invariance of Planck's constant. If we take the product of the width of the longitudinal momentum distribution and that of the spatial distribution, we end up with the relation

$$\langle z^2 \rangle \langle q_z^2 \rangle = (1/4) \, [(1 + \beta)/(1 - \beta)]^2. \tag{5.12}$$

The right-hand side increases as the velocity parameter increases. This could lead us to an erroneous conclusion that Planck's constant becomes velocity dependent. This is not correct, because the product of two longitudinal coordinates is not Lorentz invariant.

In order to maintain the Lorentz-invariance of the uncertainty product, we have to work with a conjugate pair of variables whose product does not depend on the velocity parameter. We noted in Section 3 of Chapter VI that the use of the light-cone variables will solve this problem. The content of the argument given there is contained in Figure 5.1. The major axis of the space-time ellipse in Figure 5.2 is conjugate to the minor axis of the momentum-energy ellipse, as is illustrated in Figure 5.1. Therefore, the exact calculation gives

$$\langle u^2 \rangle \langle q_u^2 \rangle = 1/4, \quad \langle v^2 \rangle \langle q_v^2 \rangle = 1/4. \tag{5.13}$$

Planck's constant is indeed Lorentz invariant. It was noted in Section 5 of Chapter VI the manner in which the above uncertainty products can be stated in terms of the conventional longitudinal and time-like variables.

Let us next resolve the puzzle of why the number of partons appears to be infinite while there are only a finite number of quarks inside the hadron. It is

Lorentz-Dirac Deformation in High-Energy Physics

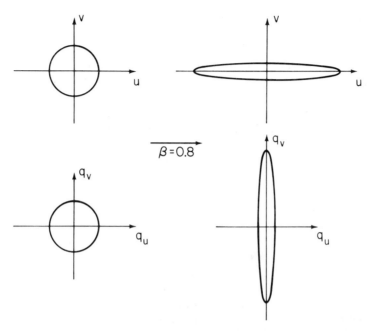

Fig. 5.2. The Lorentz-Dirac deformation in the light-cone coordinate system. The major (minor) axis in the uv coordinate system is conjugate to the minor (major) axis in the $q_u q_v$ coordinate system. If we rotate these figures by 45°, they become those in Figure 5.1.

not difficult to see from Figure 5.1 that both the x and q distributions become concentrated along the positive light-cone axis, as the hadronic velocity approaches the speed of light. This means that the quarks also move with velocity very close to that of light. We have in fact shown that the quark four-momentum is proportional to that of the hadron in Equation (5.10). Quarks in this case behave like massless particles.

We then know from statistical mechanics that the number of massless particles is not a conserved quantity. For instance, in black-body radiation, free light-like particles have a widespread momentum distribution. However, this does not contradict the known principles of quantum mechanics, because the massless photons can be divided into infinitely many massless particles with a continuous momentum distribution.

Likewise, in the parton picture, massless free quarks have a widespread momentum distribution. They can appear as a distribution of an infinite-number of free particles. These free massless particles are the partons. We shall see in Section 6 whether the widespread momentum distribution can be measured experimentally.

6. Calculation of the Parton Distribution Function for the Proton

We studied in Sections 4 and 5 the kinematics through which the peculiarities of Feynman's parton picture can be derived from the Lorentz-Dirac deformation of hadronic wave functions. A more interesting problem is to calculate the parton momentum distribution function in the proton and compare it with the distribution measured in high-energy experiments.

Before carrying out this program, let us see what we do in the Lamb-shift calculation. Because calculations in QED (quantum electrodynamics) are much more complicated than those in nonrelativistic quantum mechanics, we often forget the fact that the starting point in the Lamb shift calculation is the Rydberg formula for unperturbed hydrogen energy levels. The Rydberg formula is obtained from the solution of the Schrödinger equation for the hydrogen atom. The discreteness of the energy levels is a consequence of the boundary condition on the wave function requiring that the probability distribution be localized around the proton. QED then makes corrections to the Rydberg formula by taking into account the detailed interaction mechanism between photons and electrons.

Likewise, in calculating the parton distribution function, we have to know first how the partons are distributed. After that, we may make corrections due to the detailed interaction between the free partons and probing particles. Like QED in the Lamb-shift calculation, there is at present a field theoretic method called quantum chromodynamics (QCD) for calculating the corrections to the distribution function.

The purpose of this section is to provide a calculation of the parton distribution function using the covariant harmonic oscillator wave function. The nucleon in its rest frame is regarded as the ground state of the three-quark bound state. As before, we name the three quarks a, b, c, respectively, and use three independent four-momentum variables P, q and k given in Equation (1.13). p represents the four-momentum of the hadron, and is on its mass shell. q and k are the relative internal momenta. Here again, we ignore the transverse components. In terms of the q and k variables, the ground state wave function takes the form of Equation (1.15). The distribution function is the square of the wave function.

In the parton model calculation, only one of the quarks interacts with the virtual photon. Since the wave function is totally symmetric under the exchange of quarks, we can pick quark a as the one interacting with the virtual photon. As can be seen from Equation (1.13), the four-momentum of this quark does not depend on k. Thus we can integrate over the k variables. After this integration, the probability distribution function for the ground state takes the form

$$\left(\frac{1}{\pi\Omega}\right) \exp\left\{-\left(\frac{1}{\Omega}\right)\left[\left(\frac{1-\beta}{1+\beta}\right) q_+^2 + \left(\frac{1+\beta}{1-\beta}\right) q_-^2\right]\right\}, \qquad (6.1)$$

where

$$q_\pm = q_z \pm q_0. \tag{6.2}$$

In the limit $\beta \to 1$, the q_- integral can be performed as in the case of the two-quark system. If we introduce the x variable defined as

$$P_{a0} = xP_0, \tag{6.3}$$

then the parton distribution function becomes

$$\rho(x) = \frac{3M}{\sqrt{2\pi\Omega}} \exp[-(M^2/2\Omega)(3x-1)^2]. \tag{6.4}$$

The above distribution function is normalized as

$$\int_{-\infty}^{\infty} \rho(x)\,dx = 1. \tag{6.5}$$

This normalization is of course that of the three-quark wave function in the hadronic rest frame.

We are now interested in relating the above distribution function to the measurable quantity $\nu W(Q^2, \nu)$ given in Equation (4.16). Since this function depends only on one variable, and since the scaling variable ω is equal to x, this quantity is usually written as $F_2(x)$ which is called the *structure function*. If we assume that partons interact with the virtual photons like an electron, then the definition given in Equation (4.16) and the distribution given in Equation (6.4) leads to the following form of the proton structure function:

$$F_2^{ep}(x) = \langle C^2 \rangle \{x\rho(x)\} = \frac{1}{3}\left(\frac{3Mx}{\sqrt{2\pi\Omega}}\right) \exp\left(\frac{-M^2}{2\Omega}(3x-1)^2\right). \tag{6.6}$$

$\langle C^2 \rangle$ is the average of (quark charge)^2 and is $1/3$ for the proton.

The numerical value of the proton mass M is approximately 940 MeV. Thus the only parameter in the above expression is the spring constant Ω. This number can be estimated from the mass spectrum discussed in Section 9 of Chapter XI. A reasonable estimate for this number is approximately $M^2/2$. This simplifies the expressions of Equations (6.4) and Equation (6.6) to

$$\rho(x) = \frac{3}{\sqrt{\pi}} e^{-(3x-1)^2}, \tag{6.7}$$

$$F_2^{ep}(x) = (x/\sqrt{\pi}) e^{-(3x-1)^2}, \tag{6.8}$$

respectively. If we calculate the above form of $F_2^{ep}(x)$ and compare with the measured structure function, the agreement is only qualitative. The calculated structure function is somewhat smaller than the measured curve (Kim and Noz, 1979a).

What is the source of this disagreement? The answer to this question is very simple. As we saw in the case of hadronic mass spectra in Chapter XI, there is a significant perturbation between the exact harmonic oscillator and the real world. The structure function given in Equation (6.8) does not take into account the effect of perturbation. We therefore have to retreat to the parton distribution function of Equation (6.7) and consider all necessary perturbative effects before making an attempt to compare quantitatively with the experimental curve.

The existence of this perturbative effect is manifested in the experimental result that the exact scaling is violated and that the structure function, in addition to x (or ω), depends on the momentum transfer variable Q^2. Fortunately there is an effective theoretical tool to calculate the perturbative corrections to the distribution or structure function. This method is based on a field theoretical approach called "quantum chromodynamics" or simply QCD. Since it will require a separate book to explain QCD, and since perturbative QCD is discussed exhaustively in the literature (Altarelli and Parisi, 1977; Marciano and Pagels, 1978; Burcas, 1980), we shall not go into this subject.

The strength of perturbative QCD is its ability to calculate the departure from exact scaling. On the other hand, the weakness of QCD is its inability to produce the parton distribution function. For this reason, there have been models of QCD which start from an assumed parton distribution which does not depend on Q^2. The model of Hwa is such a model (Hwa, 1980; Hwa and Zahir, 1981). Hwa's distribution function has been compared with the expression given in Equation (6.7) (Hussar, 1981), as is shown in Figure 6.1. The agreement is not good enough for us to say that the above-mentioned calculation constitutes the final word on this subject. On the other hand, the agreement is resonable enough to continue research along this line.

There are of course many related problems. For simplicity, we discussed only one of the structure functions for electron proton scattering. As we can see, there are two independent functions in Equation (4.6). If we use heavy vector mesons, instead of photons, in inelastic neutrino-nucleon scattering, there are three structure functions. We can also measure the neutron structure function using a deuteron target (Hanlon et al., 1980).

The analysis given in this section indicates that the neutron structure function, if appropriately renormalized, would be identical to that of the proton structure function. But this is not the case in the real world. The ratio of the neutron structure function to that of the proton is not constant (Bodek et al., 1973). There has been an attempt to explain this ratio using the departure from the exact $SU(6)$ scheme and the accompanying level mixing in the harmonic oscillator system and to relate this mixing parameter to the observed neutron form factor. According to the published result, the sign of the mixing parameter is not consistent with the neutron form factor (Le Yaouanc et al., 1978). This discrepancy in sign could be due to the fact that we do not yet completely understand quark spins.

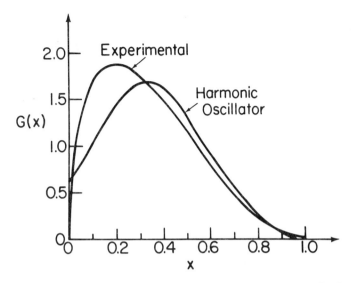

Fig. 6.1. Parton momentum distribution inside a rapidly moving proton in the harmonic oscillator model. The x variable is the ratio of the parton momentum to that of the proton. The experimental distribution is obtained from the measured structure function after the QCD correction. For $x > 0.3$, the calculated curve is very close to the experimental distribution. For $x < 0.3$, the agreement does not appear to be impressive. However, it should be noted that the error bars are still very large in this region, and that strong-interaction dynamics is not yet completely understood for small values of x. See Hussar (1981) for a detailed discussion.

The determination of the structure functions, both experimental and theoretical, will remain as one of the main branches of high energy physics for many years to come.

7. Jet Phenomenon

We shall study another universally observed high-energy phenomenon. In high-energy collisions, many hadrons are produced in the final state. A peculiar phenomenon is that these hadrons are bunched together to form two or three jets. This is called the *jet phenomenon*. What makes these hadronic momenta to line up?

Let us consider for simplicity a high-energy e^+e^- collision producing many hadrons in the final state. In the center-of-mass system where the total momentum is zero, we can consider a plane containing the initial e^+e^- beams. This plane divides the entire space into two subspaces which we call L and R respectively. As is illustrated in Figure 7.1a, the total momentum carried by hadrons in L should be equal in magnitude and opposite in sign to that carried by hadrons in R.

Since we are interested only in the qualitative behavior, we shall consider the case where only two hadrons are produced in R, and restrict ourselves to

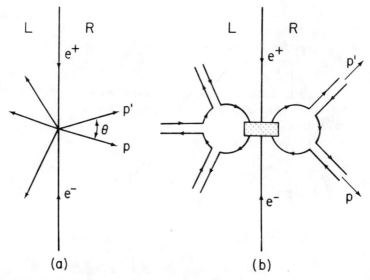

Fig. 7.1. Hadronic production in e^+e^- collision in the center-of-mass system. (a) Energy-momentum conservation alone does not make the angle between two hadronic momenta small. (b) Hadrons are produced from quark loops.

this region. Energy-momentum conservation laws do not put any restriction on the magnitude of the total momentum is either region. Therefore, there is no restriction on the angle between the hadronic momenta from these conservation laws. We shall see however that the Lorentz-Dirac deformation makes this angle smaller as the hadrons move faster, and is thus consistent with the formation of jets in hadronic production.

The idea that the Lorentz deformation may be responsible for the origin of jets was made originally by Kitazoe and Hama (1979) who used the Bethe-Salpeter wave functions, and followed up by Kim *et al.* (1979c) and by Kitazoe and Morii (1980a, b). We use here the harmonic oscillator wave functions to study this problem. For simplicity, we assume that all hadrons are mesons which are regarded as bound states of a quark and antiquark, and use the ground-state harmonic oscillator wave function which contains all essential features of the Lorentz-Dirac deformation. One important difference between the calculations of the present section and those in earlier sections is that we have to consider here a hadron moving not necessarily along the z direction. We therefore need a more general expression for the wave function than those for the hadron moving along the z direction. Indeed, the generalization to an arbitrary direction of the hadronic wave function given in Equation (1.6) or Equation (1.10) is (Problem 9 in Section 8)

$$\psi(x, P) = \frac{1}{\pi\Omega} \exp\left\{-\frac{\Omega}{2}\left[x \cdot x + 2\left(\frac{P \cdot x}{M}\right)^2\right]\right\}, \qquad (7.1)$$

where x is, as defined before, the space-time separation between the quarks, and P is the four-momentum of the hadron.

Since our primary purpose is to extract the kinematical effect of the Lorentz-Dirac deformation, we work with the simplest dynamical model, in which a quark-antiquark pair is initially created for the hadrons in region R and then this pair is combined with a newly created pair to form two hadrons as is indicated in Figure 7.1b. The simplest possible dynamics for creation of the original pair, as indicated in the black box in Figure (7.1b) is a point interaction. Then the dependence of the production amplitude on the angle between the two hadronic momenta becomes:

$$F(\theta) = \int d^4x \, \psi(x, P) \, \psi(x, P') \exp[-i(P + P') \cdot x], \tag{7.2}$$

where P and P' are the four-momenta of the first and second hadrons respectively.

The above form is not unlike the expression for the electromagnetic form factor discussed in Section 3. Except for the sign of P' in the exponential factor, the integrand of the above expression is identical to that of Equation (3.4). Here again the integral can be illustrated by the two overlapping momentum wave functions. The overlap is maximal when the two hadronic momenta are lined up, as is illustrated in Figure 7.2. After the integration, $F(\theta)$ becomes

$$F(\theta) = \frac{-M^2}{P \cdot P'} \exp\left(\frac{-M^2(P + P')^2}{4\Omega P \cdot P'}\right). \tag{7.3}$$

In order to extract the θ dependence, we consider the symmetric case where $|\mathbf{P}| = |\mathbf{P}'|$. Then

$$F(\theta) = \frac{1}{1 + 2b^2 \sin^2(\theta/2)} \exp\left[-\frac{M^2}{\Omega}\left(1 + \frac{1}{1 + 2b^2 \sin^2(\theta/2)}\right)\right], \tag{7.4}$$

where $b = |\mathbf{P}|/M$. The first factor in the expression of Equation (7.4) is like a

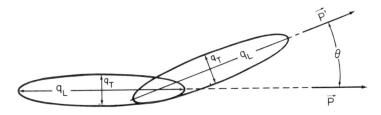

Fig. 7.2. Two overlapping Lorentz-deformed wave functions in momentum space. As the hadronic momenta becomes large, the wave functions become more eccentric and the overlap becomes a rapidly decreasing function of θ.

Coulomb scattering amplitude and has a sharp peak at $\theta = 0$ as b becomes very large. On the other hand the exponential factor has its minimum value at $\theta = 0$, and increases to a finite asymptotic value as θ becomes large. This exponential factor therefore tempers the sharp Coulomb-like peak at $\theta = 0$.

This means that the scattering amplitude is maximum when the two momenta are parallel and decreases rapidly when the angle between them increases. This explains qualitatively why the hadrons in the final state have to form a jet.

8. Exercises and Problems

Exercise 1. We discussed in Section 3 how to effect a transition from a nonrelativistic calculation of the form factor into a relativistic expression. In scattering theory also, such a transition exists. Show how the Born approximation amplitude for a Yukawa potential scattering can be generalized to the Feynman amplitude for the scattering with one-meson exchange.

The Born approximation amplitude for the Yukawa potential:

$$V(r) = (G^2/2m^2) \frac{e^{-\lambda r}}{r} \tag{8.1}$$

is

$$f(\theta) = \frac{-G^2}{Q^2 + \lambda^2}, \tag{8.2}$$

where $Q^2 = (\mathbf{K}_f - \mathbf{K}_i)^2$. If we calculate the lowest-order Feynman amplitude for the scattering of two spinless nucleons through exchange of one meson with mass λ, in the center-of-mass frame, the expression becomes identical to the above formula. In the center-of-mass system, the energy transfer for this scattering process is zero. Therefore Q^2 is also the (four-momentum transfer)^2. This is the way in which Yukawa was originally led to formulate the idea of a meson being exchanged to generate the force between nucleons.

Exercise 2. Show that the expression of Equation (4.13) for W leads to scaling if partons are free and independent particles, by deriving Equation (4.14) from Equation (4.13). Assume for simplicity that the proton is a scalar particle.

For the scalar field $\phi(x)$, we use for the current $J(x)$ the normal-ordered product

$$J(x) = :\phi^\dagger(x)\phi(x): . \tag{8.3}$$

Then

$$[J(x), J(0)] = i\Delta(x)\Delta_1(x)$$
$$+ i\Delta(x)\{:\phi^\dagger(x)\phi(0): + :\phi^\dagger(0)\phi(x):\}, \tag{8.4}$$

with

$$\Delta(x) = -i(1/2\pi)^3 \int d^4q \, e^{-iq \cdot x} \, \varepsilon(q_0) \, \delta(q^2 + M^2),$$

(8.5)

$$\Delta_1(x) = (1/2\pi)^3 \int d^4p \, e^{-iq \cdot x} \, \delta(q^2 + M^2),$$

For space-like Q, it can be shown that

$$\int d^4x \, e^{-iQ \cdot x} \Delta(x) \Delta_1(x) = 0.$$

Thus

$$W(Q^2, \nu) = (1/2\pi)^2 \int d^4x \, d^4q \, e^{-i(Q-q) \cdot x} \, \varepsilon(q_0) \, \delta(q^2 + M^2)$$

$$\times \langle P | \{ :\phi^\dagger(x)\phi(0): + :\phi^\dagger(0)\phi(x): \} | P \rangle,$$

(8.6)

The evaluation of this integral is straightforward, and the result is

$$W(Q^2, \nu) = \frac{1}{M} \varepsilon(Q_0 \pm P_0) \, \delta[(Q \pm P)^2 + M^2]$$

$$= (-1/2MQ \cdot p) \, \delta(\omega \pm 1).$$

(8.7)

Thus

$$\nu W(Q^2, \nu) = (1/2M^2) \, \delta(\omega - 1).$$

(8.8)

Problem 1. The Lamb shift is the energy difference between the $2s$ and $2p$ state due to the delta function potential at the origin caused by the effect of quantum electrodynamics. Since the behavior of wave function at the origin is different for $2s$ and $2p$ states, the perturbing potential near origin will give rise to the energy splitting between the two states. We now know that the proton charge is spread out. Assuming that the proton charge distribution is of the form $\exp(-r/r_0)$ with $r_0 = 2.3 \times 10^{-14}$ cm, calculate the contribution to the Lamb shift due to this charge distribution. Would this significantly change the measured value of the Lamb shift which is approximately 1058 megacycles per second?

Problem 2. Show that, in terms of the form factor $G_M(t)$ and $G_E(t)$ given in Equation (2.9), the electron proton elastic cross section in the laboratory frame (in which the initial proton is at rest) takes the form:

$$\frac{d\sigma}{d\Omega} = \sigma_0 \left(\frac{G_E^2(t) - G_M^2(t)/2m^2}{1 - t/4m^2} - \frac{t}{2m^2} G_M^2(t) \tan^2\left(\frac{\theta}{2}\right) \right),$$

(8.9)

with

$$\sigma_0 = \frac{\alpha^2}{4E^2\sin^2(\theta/2)} \left(\frac{\cos^2(\theta/2)}{1 + (2E/M)\sin^2(\theta/2)} \right), \tag{8.10}$$

where E is the incident electron energy, and θ is the scattering angle.

Problem 3. Show that the Rutherford scattering amplitude in the nonrelativistic Born approximation is $f(\theta) = 2m^2/Q^2$ if the electron is scattered by a point charge. Show next that, when the electron is scattered by a charge whose distribution is $e\rho(\mathbf{x})$, the scattering amplitude becomes

$$f(\theta) = [2m^2/Q^2] \int \rho(\mathbf{x}) \, e^{i\mathbf{Q}\cdot\mathbf{x}} \, d^3x. \tag{8.11}$$

See the discussion given in Section 2.

Problem 4. Assuming that the electromagnetic field couples directly and minimally to the quarks, calculate the neutron/proton magnetic moment ratio. See Beg *et al.* (1964).

Problem 5. It is possible to derive formally the expression for the factor given in Equation (3.2) from the Lagrangian formalism. Carry out this derivation using the minimal electromagnetic coupling to the quark charge. See Lipes (1972).

Problem 6. Show that the form factor for the three-quark nucleon takes the form of Equation (3.10). See Fujimura *et al.* (1970) and Lipes (1972).

Problem 7. Show that the use of free Dirac spinors for the quarks leads to a wrong asymptotic behavior for the nucleon form factor (Lipes, 1972). Show however that it is possible to construct spinor models in which the quarks do not but the nucleon does satisfy the free Dirac equation (Henriques *et al.*, 1975; Ishida *et al.*, 1979; Haberman, 1984).

Problem 8. Carry out the calculation from (4.1) to (4.12), and repeat the calculation of Exercise 2 for spin-1/2 partons.

Problem 9. Show that the wave function given in Equation (7.1) is the covariant form for the oscillator wave function of Equation (1.10) for the hadron moving along the z direction.

Problem 10. Show that the weak interaction coupling ratio $-g_A/g_V$ is 5/3 in the quark model if there is no interaction between the quarks. Show however that the internal quark motion, such as that of the harmonic oscillator, brings down this number to an acceptable value of approximately 1.16 (Ruiz, 1975).

·*Problem 11.* This chapter is concerned with the relativistic bound-state problem. This has been and still is regarded as one of the most difficult problems in physics. For this reason, there has been an attempt to understand this bound-state problem using the S matrix which can be formulated covariantly. In this approach, bound-states correspond to poles in the complex energy plane. Show that approximations in the S matrix approach do not guarantee the localization of bound-state wave functions. Show however that it is possible to construct a nonrelativistic bound-state wave function from the S matrix. See Kim and Noz (1978c).

References

Abraham, R. and J. E. Marsden, *Foundations of Mechanics*, 2nd ed. (Benjamin/Cummings, Reading, Mass., 1978).
Aharonov, Y. and D. Bohm, *Phys. Rev.* **122**, 1649 (1961).
Aharonov, Y. and D. Bohm, *Phys. Rev.* **134**, 1417 (1964).
Aldinger, R. R., A. Bohm, P. Kielanowski, M. Loewe, P. Magnollay, N. Mukunda, W. Drechsler, and S. R. Kony, *Phys. Rev. D* **28**, 3020 (1983).
Ali, S. T., in *Lecture Notes in Physics*, Vol. 139, ed. by H. D. Doebner (Springer-Verlag, Berlin, 1981).
Allcock, J. R., *Ann. Phys. (NY)* **53**, 253, 286, 311 (1969).
Altarelli, G. and G. Parisi, *Nucl. Phys.* **B126**, 298 (1977).
Arfken, G. B., *Mathematical Methods for Physicists*, 3rd ed. (Academic, New York, 1985).
Aricks, F., J. Broeckhove, and E. Deumens, *Nucl. Phys. A* **318**, 269 (1979).
Arnold, V. I., *Mathematical Methods of Classical Mechanics* (Springer-Verlag, Berlin and New York, 1978).
Ashtekar, A. and M. Strawbel, *Proc. Roy. Soc. (London) A* **376**, 585 (1981).
Bacry, H. and M. Cadilhac, *Phys. Rev A* **23**, 2533 (1981).
Balazs, N. L. and G. G. Zippel, *Ann. Phys. (NY)* **77**, 139 (1973).
Bargmann, V., *Ann. Math.* **48**, 568 (1947).
Bargmann, V., *Ann. Math.* **59**, 1 (1954).
Bargmann, V. and E. P. Wigner, *Proc. Natl. Acad. Sci. (U.S.A.)* **34**, 211 (1946).
Bargmann, V., E. P. Wigner, R. Blankenbecler, M. L. Goldberger, S. B. Treiman, and A. S. Wightman, *Rev. Mod. Phys.* **34**, 587 (1962).
Bartlett, M. S. and J. E. Moyal, *Proc. Cambridge Phil. Soc.* **45**, 545 (1949).
Basu, D. and K. B. Wolf, *J. Math. Phys.* **23**, 189 (1982).
Barut, A. O. and C. Fronsdal, *Proc. Roy. Soc. (London)* **A287**, 532 (1965).
Barut, A. O. and R. Raczka, *Theory of Group Representations and Applications* (PWN, Warsaw, 1977).
Bauer, M. and P. A. Mello, *Proc. Natl. Acad. Sci. (U.S.A.)* **73**, 283 (1976).
Bauer, M. and P. A. Mello, *Ann. Phys.* **111**, 38 (1978).
Beg, M. B., B. W. Lee, and A. Pais, *Phys. Rev. Lett.* **13**, 514 (1964).
Biedenharn, L. C., *J. Math. Phys.* **2**, 433 (1961).
Biedenharn, L. C. and H. van Dam, *Phys. Rev. D* **9**, 471 (1974).
Bjorken, J. D. and S. D. Drell, *Relativistic Quantum Mechanics and Relativistic Quantum Fields* (McGraw-Hill, New York, 1964).
Bjorken, J. D. and S. D. Drell, *Phys. Rev.* **179**, 1546 (1969).
Bjorken, J. D. and E. A. Paschos, *Phys. Rev.* **185**, 1975 (1969).
Blaker, J. W., *Geometric Optics — The Matrix Theory* (Marcel Dekker, New York, 1971).
Blanchard, C. H., *Am. J. Phys.* **50**, 642 (1982).

Blankenbecler, R. and M. L. Goldberger, *Phys. Rev.* **126**, 766 (1962).
Blatt, J. M. and V. Weisskopf, *Theoretical Nuclear Physics* (Wiley, New York, 1954).
Bodek, A. *et al., Phys. Rev. Lett.* **30**, 1087 (1973).
Boerner, H., *Representations of Groups* (North-Holland, Amsterdam, 1970).
Bogoliubov, N. N., A. A. Logunov, and I. T. Todorov, *Introduction to Axiomatic Quantum Field Theory* (Benjamin, Reading, Mass., 1975).
Bohm, A., *Quantum Mechanics* (Springer-Verlag, Berlin, 1979).
Bohm, A., M. Loewe, L. C. Biedenharn, and H. van Dam, *Phys. Rev. D* **28**, 3032 (1983).
Bohr, A. and B. R. Mottelson, *Nuclear Structure, Vol. II: Nuclear Deformations* (Benjamin, New York, 1969).
Bohr, N., *Atomic Physics and Human Knowledge* (Wiley, New York, 1958).
Bowen, M. and J. Coster, *Am. J. Phys.* **49**, 80 (1981).
Brooke, J. A., *J. Math. Phys.* **19**, 952 (1978).
Brooke, J. A., *J. Math. Phys.* **21**, 617 (1980).
Brown, L., D. I. Fivel, B. W. Lee, and R. F. Sawyer, *Ann. Phys. (NY)* **23**, 187 (1963).
Buras, A. J., *Rev. Mod. Phys.* **52**, 199 (1980).
Byers, N. and C. N. Yang, *Phys. Rev.* **142**, 976 (1966).
Cartan, Elie, *The Theory of Spinors* (Dover, New York, 1966).
Carruthers, P. and M. M. Nieto, *Phys. Rev. Lett.* **14**, 387 (1965).
Carruthers, P. and C. C. Shih, *Phys. Lett.* **127B**, 242 (1983).
Carruthers, P. and F. Zachariasen, *Rev. Mod. Phys.* **55**, 245 (1983).
Chew, G. F., *Rev. Mod. Phys.* **34**, 394 (1962).
Chou, T. T. and C. N. Yang, *Phys. Rev.* **170**, 1591 (1968).
Collins, P. D. B. and E. J. Squires, *Regge Poles and Particle Physics* (Springer-Verlag, Berlin, 1968).
Critchfield, C. L., *J. Math. Phys.* **17**, 261 (1976).
Davies, R. W. and K. T. R. Davies, *Ann. Phys.* **89**, 261 (1975).
De, T., Y. S. Kim, and M. E. Noz, *Nuovo Cimento* **13A**, 1089 (1973).
Deenan, D and C. Quesne, *J. Math. Phys.* **23**, 878 (1982).
Dirac, P. A. M., *Proc. Roy. Soc. (London)* **A114**, 243, 710 (1927).
Dirac, P. A. M., *Quantum Electrodynamics* (Comm. Dublin Inst. Adv. Studies, 1943).
Dirac, P. A. M., *Proc. Roy. Soc. (London)* **A183**, 284 (1945).
Dirac, P. A. M., *Rev. Mod. Phys.* **21**, 392 (1949).
Dirac, P. A. M., *The Principles of Quantum Mechanics*, 4th ed. (Oxford Univ. Press, London, 1958).
Dirac, P. A. M., *Physics Today* **23**, No. 4 (April), 29 (1970).
Dirac, P. A. M., *Development of Quantum Theory* (Gordon and Breach, New York, 1972).
Dominici, D. and G. Longhi, *Nuovo Cimento* **42A**, 235 (1977).
Domokos, G. and P. Suranyi, *Nucl. Phys.* **54**, 529 (1964).
Drell, S. D. and T. M. Yan, *Ann. Phys. (NY)* **60**, 578 (1971).
Dyson, F. J., *J. Math. Phys.* **3**, 140 (1962).
Eberly, J. A. and L. P. S. Singh, *Phys. Rev. D* **7**, 359 (1973).
Edmonds, A. R., *Angular Momentum in Quantum Mechanics* (Princeton Univ. Press, Princeton, New Jersey, 1957).
Faddeev, L. D., *Act Congress Int. Math.* **3**, 35 (1970).
Feinberg, G., *Phys. Rev.* **159**, 1089 (1967).
Feynman, R. P., in *High Energy Collisions*, Proceeding of the Third International Conference, Stony Brook, New York, ed. by C. N. Yang *et al.* (Gordon and Breach, New York, 1969).
Feynman, R. P., M. Kislinger, and F. Ravndal, *Phys. Rev. D* **3**, 2706 (1971).
Fleming, G. N., *Phys. Rev.* **137**, B188 (1965).
Fleming, G. N., *J. Math. Phys.* **11**, 1959 (1966).
Fock, V. A., *J. Exptl. Theoret. Phys. (U.S.S.R.)* **42**, 1135 (1962); *Soviet Phys. JETP* **15**, 784 (1962).

Foldy, L. L. and S. A. Wouthusen, *Phys. Rev.* **78**, 29 (1950).
Frazer, W. and J. Fulco, *Phys. Rev. Lett.* **2**, 365 (1960).
Frazer, W. and J. Fulco, *Elementary Particle Physics* (Prentice Hall, Englewood Cliffs, New Jersey, 1966).
Fujimura, K., T. Kobayashi, and M. Namiki, *Prog. Theor. Phys.* **43**, 73 (1970).
Fujiwara, T., K. Wakita, and H. Yoro, *Prog. Theor. Phys.* **64**, 363 (1980).
Gel'fand, I. M., M. I. Graev, and N. Ya. Vilenkin, *Generalized Functions* (Academic, New York, 1966), Vol. 5, Chapt. II.
Gell-Mann, M., Calif. Inst. of Tech. Report CTSL-20, 1961 (unpublished).
Gell-Mann, M., *Phys. Rev.* **125**, 1067 (19962).
Gell-Mann, M., *Phys. Lett.* **13**, 598 (1964).
Gilmore, R., *Lie Groups and Lie Algebras, and Some of their Applications* (Wiley, New York, 1974).
Glauber, R. J., *Phys. Rev.* **131**, 2766 (1963).
Glauber, R. J., *Phys. Lett.* **21**, 650 (1966).
Goldhaber, M., L. Grodzins, and A. W. Sunyar, *Phys. Rev.* **109**, 1015 (1958).
Goldin, E., *Waves and Photons* (Wiley & Sons, New York, 1982).
Goto, T., *Prog. Theor. Phys.* **58**, 1635 (1977).
Grebogi, C. and A. N. Kaufman, in *Long-Time Prediction in Dynamics*, ed. by C. W. Holton *et al.* (Wiley, New York, 1983).
Greenberg, O. W., *Phys. Rev. Lett.* *13*, 598 (1964).
Greenberg, O. W., *Am. J. Phys.* **50**, 1074 (1982).
Greenberg, O. W. and C. A. Nelson, *Physics Reports* **33C**, 71 (1977).
Greenberg, O. W. and M. Resnikoff, *Phys. Rev.* **163**, 1844 (1967).
Gribov, V. N., B. L. Ioffe, and I. Ya. Pomeranchuk, *J. Nucl. Phys. (U.S.S.R.)* **2**, 768 (1965) or *Soviet J. Nucl. Phys.* **2**, 549 (1966).
Gursey, F. and S. Orfanidis, *Nuovo Cimento* **11A**, 225 (1972).
Haberman, M. L., *Phys. Rev. D* **29**, 1412 (1984).
Halpern, F. R. and E. Branscomb, *Wigner's Analysis of the Unitary Representations of the Poincaré Group*, Univ. of Calif., Lawrence Radiation Lab. Report UCRL 12359 (1965).
Halpern, F. R., *Relativistic Quantum Mechanics* (Prentice Hall, Englewood Cliffs, New Jersey, 1968).
Hamermesh, M., *Group Theory* (Addison Wesley, Reading, MA, 1962).
Han, D. and Y. S. Kim, *Prog. Theor. Phys.* **64**, 1852 (1980).
Han, D. and Y. S. Kim, *Am. J. Phys.* **39**, 348 (1981a).
Han, D. and Y. S. Kim, *Am. J. Phys.* **49**, 1157 (1981b).
Han, D., Y. S. Kim, and M. E. Noz, *Found. of Phys.* **11**, 895 (1981c).
Han, D., Y. S. Kim, and D. Son, *Phys. Rev. D* **24**, 461 (1982a).
Han, D., M. E. Noz, Y. S. Kim, and D. Son, *Phys. Rev. D* **25**, 1740 (1982b).
Han, D., Y. S. Kim, and D. Son, *Phys. Rev. D* **26**, 3717 (1982c).
Han, D., Y. S. Kim, and D. Son, *Phys. Rev. D* **27**, 2384 (1983a).
Han, D., M. E. Noz, Y. S. Kim, and D. Son, *Phys. Rev. D* **27**, 3032 (1983b).
Han, D., Y. S. Kim, and D. Son, *Phys. Lett.* **131V**, 327 (1983c).
Han, D., Y. S. Kim, M. E. Noz, and D. Son, *Am. J. Phys.* **52**, 1037 (1984).
Han, D., Y. S. Kim, D. Son, *Bull. Am. Phys. Soc.* **30**, 706 (1985).
Hanlon, J. *et al.*, IIT-Maryland-Stony Brook-Tohoku-Tufts Collaboration, *Phys. Rev. Lett.* **45**, 1817 (1980).
Hanson, A. J. and T. Regge, *Ann. Phys. (NY)* **87**, 498 (1974).
Harish-Chandra, *Proc. Roy. Soc. (London) A* **189**, 372 (1947).
Heisenberg, W., *Z. Phys.* **45**, 172 (1927).
Heisenberg, W., *Am. J. Phys.* **43**, 389 (1975).
Heitler, W., *The Quantum Theory of Radiation*, 3rd ed. (Oxford Univ. Press, London, 1954).
Hendry, A. W. and D. B. Lichtenberg, *Rep. Prog. Phys.* **41**, 113 (1978).

Henriques, A. B., B. H. Kellet, and R. G. Moorehouse, *Ann. Phys. (NY)* **93**, 125 (1975).
Hermann, R., *Lie Groups for Physicists* (Benjamin, New York, 1966).
Hofstadter, R. and R. W. McAllister, *Phys. Rev.* **98**, 217 (1955).
Hofstadter, R. and R. W. McAllister, *Nuclear and Nucleon Structure* (Benjamin, New York, 1963).
Holman, W. J., *Ann. Phys. (NY)* **52**, 176 (1969).
Holman, W. J. and L. C. Biedenharn, *Ann. Phys. (NY)* **39**, 1 (1966).
Holman, W. J. and L. C. Biedenharn, *Ann. Phys. (NY)* **47**, 205 (1968).
Horwitz, L. P. and C. Piron, *Helv. Phys. Acta* **46**, 316 (1973).
Huang, K., *Quarks, Leptons and Gauge Fields* (World Scientific, Singapore, 1982).
Hubsch, T., S. Meljanic, and S. Pallua, *Phys. Rev. D* **31**, 352 (1985).
Hussar, P. E., *Phys. Rev. D* **23**, 2781 (1981).
Hussar, P. E., Y. S. Kim, and M. E. Noz, *Am. J. Phys.* **48**, 1038 (1980a).
Hussar, P. E., Y. S. Kim, and M. E. Noz, *Am. J. Phys.* **48**, 1043 (1980b).
Hussar, P. E., Y. S. Kim, and M. E. Noz, *Am. J. Phys.* **53**, 142 (1985).
Hwa, R. C., *Phys. Rev. D* **22**, 759 (1980).
Hwa, R. C. and M. S. Zahir, *ibid.* **23**, 2539 (1981).
Inonu, E. and E. P. Wigner, *Nuovo Cimento* **9**, 705 (1952).
Inonu, E. and E. P. Wigner, *Proc. Natl. Acad. Sci. (US)* **39**, 510 (1953).
Ioffe, B. L., *Phys. Lett. B.* **30**, 123 (1969).
Isgur, N. and G. Karl, *Phys. Rev. D* **19**, 2653 (1979).
Ishida, S., *Prog. Theor. Phys.* **46**, 1570, 1905 (1971).
Ishida, S., A. Matsuda, and M. Namiki, *Prog. Theor. Phys.* **57**, 210 (1977).
Ishida, S., K. Takeuchi; S. Tsuruta, M. Watanabe, and M. Oda, *Phys. Rev.* **D20**, 2906 (1979).
Itzykson and J. B. Zuber, *Quantum Field Theory* (McGraw-Hill, New York, 1980).
Iwasawa, K., *Ann. Math.* **50**, 507 (1949).
Jacob, M. and G. C. Wick, *Ann. Phys. (NY)* **40**, 149 (1959).
Janner, A. and T. Janssen, *Physica* **60**, 292 (1972).
Jersak, J. and D. Rein, *Z. Phys. C* **3**, 339 (1980).
Kadanoff, L. P. and G. Baym, *Quantum Statistical Mechanics* (Benjamin, New York, 1962).
Kalnins, E. G. and W. Miller, *J. Math. Phys.* **15**, 1263 (1974).
Kalnins, E. G. and W. Miller, *J. Math. Phys.* **18**, 1 (1977).
Kim, Y. S. and F. L. Markley, *Nuovo Cimento* **63A**, 60 (1969).
Kim, Y. S., F. L. Markley, and R. Zaoui, *Phys. Rev. D* **4**, 1738 (1971).
Kim, Y. S. and M. E. Noz, *Nuovo Cimento* **11A**, 513 (1972).
Kim, Y. S. and M. E. Noz, *Phys. Rev. D* **8**, 3521 (1973).
Kim, Y. S. and M. E. Noz, *Nuovo Cimento* **19A**, 657 (1974).
Kim, Y. S. and M. E. Noz, *Phys. Rev. D* **15**, 335 (1977a).
Kim, Y. S. and M. E. Noz, *Prog. Theor. Phys.* **57**, 1373 (1977b).
Kim, Y. S. and M. E. Noz, *Phys. Rev. D* **15**, 3032 (1977c).
Kim, Y. S. and M. E. Noz, *Prog. Theor. Phys.* **60**, 801 (1978a).
Kim, Y. S. and M. E. Noz, *Am. J. Phys.* **46**, 480 (1978b).
Kim, Y. S. and M. E. Noz, *Am. J. Phys.* **46**, 484 (1978c).
Kim, Y. S. and M. E. Noz, *Found of Phys.* **9**, 375 (1979a).
Kim, Y. S. M. E. Noz, and S. H. Oh, *Am. J. Phys.* **47**, 892 (1979b).
Kim, Y. S. M. E. Noz, and S. H. Oh, *Found. of Phys.* **9**, 947 (1979c).
Kim, Y. S. M. E. Noz, and S. H. Oh, *J. Math. Phys.* **20**, 1341 (1979d).
Kim, Y. S. M. E. Noz, and S. H. Oh, *J. Math. Phys.* **21**, 1224 (1980).
Kim, Y. S. and M. E. Noz, *J. Math. Phys.* **22**, 2289 (1981).
Kim, Y. S. and M. E. Noz, *Am. J. Phys.* **50**, 721 (1982).
Kim, Y. S. and M. E. Noz, *Am. J. Phys.* **51**, 368 (1983).
Kitazoe, T. and S. Hama, *Phys. Rev. D* **19**, 2006 (1979).
Kitazoe, T.and T. Morii, *Phys. Rev. D* **21**, 685 (1980a).

Kitazoe, T. and T. Morii, *Nucl. Phys. B* **164**, 76 (1980b).
Klauder, J. R. and E. C. G. Sudarshan, *Quantum Optics* (Benjamin, New York, 1968).
Klauder, J. R. and E. C. G. Sudarshan, *Phys. Rev. D* **19**, 2349 (1979).
Kokkedee, J. J. J., *The Quark Model* (Benjamin, New York, 1969).
Kuriyan, J. G., N. Mukunda, and E. C. G. Sudarshan, *J. Math. Phys.* **9**, 2100 (1968).
Kuriyan, J. G., N. Mukunda, and E. C. G. Sudarshan, *Commun. Math. Phys.* **8**, 204 (1968).
Landau, L. D. and E. M. Lifschitz, *Quantum Mechanics*, 2nd ed. (Pergamon, New York, 1958).
Lang, S., $SL(2,r)$ (Addison-Wesley, Reading, Mass., 1975).
Le Yaouanc, A., L. Oliver, O. Pene, and J. C. Raynal, *Phys. Rev. D* **12**, 2137 (1975).
Le Yaouanc, A., L. Oliver, O. Pene, and J. C. Raynal, *Phys. Rev. D* **15**, 844 (1977).
Le Yaouanc, A., L. Oliver, O. Pene, and J. C. Raynal, *Phys. Rev. D* **18**, 1733 (1978).
Lee, T. D. and C. N. Yang, *Phys. Rev.* **105**, 1671 (1957).
Leutwyler, H. and J. Stern, *Phys. Lett.* **73B**, 75 (1978).
Levin, E. M., and L. L. Frankfurt, *Sov. Phys. USP.* **11**, 106 (1968), which is the translation of the original article published in Russian in *Usp. Fiz. Nauk.* **92**, 243 (1968).
Licht, A. L. and A. Pagnamenta, *Phys. Rev. D* **2**, 1150, 1156 (1970).
Lichtenberg, D. B., *Unitary Symmetry and Elementary Particles* (Academic, New York, 1970).
Lipes, R. G., *Phys. Rev. D* **5**, 2849 (1972).
Lipsman, R. L., *Group Representations* (Springer-Verlag, Berlin, 1974).
Littlejohn, R. G., *Phys. Fluids* **24**, 1730 (1981).
Lukierski, J. and M. Oziewics, *Phys. Lett.* **69B**, 339 (1977).
Mackey, G. W., in *Theoretical Problems of Relativistic Physics*, ed. by I. E. Segal (Am. Math. Soc., Providence, R. I., 1963).
Mackey, G. W., *Induced Representations of Groups and Quantum Mechanics* (Benjamin, New York, 1968).
Mackey, G. W., *Unitary Group Representations in Physics, Probability, and Number Theory* (Benjamin/Cummings, Reading, Mass., 1978).
Maltman, K. and N. Isgur, *Phys. Rev. D* **29**, 952 (1984).
Marciano, W. and H. Pagels, *Phys. Rep, C* **36**, 138 (1978).
Markov, M., *Suppl. Nuovo Cimento* **3**, 760 (1956).
Magnus, W. and F. Oberhettinger, *Formulas and Theorems for the Functions of Mathematical Physics* (Chelsea, New York, 1949).
Merzbacher, E., *Quantum Mechanics*, 2nd ed. (Wiley, New York, 1970).
Michel, L., in *Group Theoretical Concepts and Methods in Elementary Particle Physics*, ed. by Feza Gursey (Gordon and Breach, New York, 1962).
Miller, W., *Lie Theory and Special Functions* (Academic, New York, 1968).
Miller, W., *Symmetry Groups and Their Applications* (Academic, New York, 1972).
Miller, W., *Symmetry and Separation of Variables* (Addison and Wesley, Reading, Mass., 1977).
Misra, S. P. and J. Maharana, *Phys. Rev. D* **14**, 133 (1976).
Moncrief, V., *Ann. Phys. (NY)* **114**, 201 (1978).
Morrison, P. J. and J. M. Greene, *Phys. Rev. Lett.* **46**, 790 (1981).
Morse, P. and H. Feshbach, *Methods of Theoretical Physics*, I and II (McGraw-Hill, New York, 1953).
Moshinsky, M., *Phys. Rev.* **81**, 347 (1951a).
Moshinsky, M., *Phys. Rev.* **84**, 525, 533 (1951b).
Moshinsky, M., *Rev. Mex. Fis.* **1**, 28 (1952a).
Moshinsky, M., *Phys. Rev.* **88**, 625 (1952b).
Moshinsky, M., *Harmonic Oscillators in Modern Physics* (Gordon and Breach, New York, 1969).
Moshinsky, M. and C. Quesne, *J. Math. Phys.* **12**, 1772 (1971).
Moshinsky, M. and P. Winternitz, *J. Math. Phys.* **21**, 1667 (1980).

Moyal, L. E., *Proc. Cambridge Phil. Soc.* **45**, 99 (1949).
Mukunda, N., *J. Math. Phys.* **8**, 2210 (1967).
Mukunda, N., *J. Math. Phys.* **9**, 50, 417 (1968).
Mukunda, N., *J. Math. Phys.* **10**, 2068, 2092 (1973).
Naimark, M. A., *Am. Math. Soc. Transl.* **6**, 379 (1957).
Naimark, M. A., *Linear Representations of the Lorentz Group* (Pergamon, New York, 1964).
Nakanishi, N., *Prog. Theor. Phys. Suppl.* **51**, 1 (1972).
Ne'eman, Y., *Nucl. Phys.* **26**, 222 (1961).
Nix, J. R., *Nucl. Phys. A* **130**, 241 (1969).
Novozhilov, Yu. V., *Introduction to Elementary Particle Theory* (Pergamon Press, Oxford, 1975).
O'Connell, R. F. and A. K. Rajagopal, *Phys. Rev. Lett.* **47**, 1029 (1982).
Okubo, S., *Prog. Theor. Phys.* **27**, 469 (1962).
Oneda, S. and E. Takasugi, in: *Int'l Symposium on Mathematical Physics*, ed. by A. Bohm (Univ. of Mexico, Mexico, 1976).
Papp, E. W. R., in: *The Uncertainty Principle and Quantum Mechanics*, ed. by W. C. Price and S. S. Chissick (John Wiley & Sons, New York, 1977).
Parker, L. P. and G. M. Schmieg, *Am. J. Phys.* **38**, 218, 1298 (1970).
Pauli, W., *Ann. Inst. Henri Poincaré* **6**, 137 (1936).
Pauri, M., in: *Group Theoretical Methods in Physics*, Proc. 9th Int'l Colloquium, Cocoyoc, Mexico, ed. by K. B. Wolf (Springer-Verlag, Berlin, Heidelberg, 1980).
Pontryagin, L. S., *Topological Groups*, 2nd ed. (Gordon and Breach, New York, 1966).
Preparata, G. and N. S. Craigie, *Nucl. Phys.* **B102**, 478 (1976).
Prugovečki, E., *Quantum Mechanics in Hilbert Space*, 2nd ed. (Academic, New York, 1981).
Pukanszky, L., *Trans. Am. Math. Soc.* **100**, 116 (1961).
Pukanszky, L., *Math. Annal.* **156**, 96 (1964).
Quesne, C. and M. Moshinsky, *J. Math. Phys.* **12**, 1780 (1971).
Rayski, J and J. M. Rayski, Jr. in: *The Uncertainty Principle and Quantum Mechanics*, ed. by W. C. Price and S. S. Chissick (John Wiley & Sons, New York, 1977).
Recami, R., in: *The Uncertainty Principles and Quantum Mechanics*, ed. by W. C. Price and S. S. Chissick (John Wiley & Sons, New York, 1977).
Regge, T., *Nuovo Cimento* **14**, 951 (1959).
Rno, J. S., *J. Math. Phys.* **26**, 675 (1985)
Roman, P. and J. Haavisto, *J. Math. Phys.* **22**, 403 (1981).
Rosensteel, G. and D. J. Rowe, *Ann. Phys.* **126**, 198 (1980).
Robertson, H. P., *Phys. Rev.* **34**, 163 (1929).
Rose, M. E., *Elementary Theory of Angular Momentum* (Wiley, New York, 1957).
Rotbart, F. C., *Phys. Rev. D* **23**, 3078 (1981).
Ruiz, M. J., *Phys. Rev. D* **10**, 4306 (1974).
Ruiz, M. J., *Phys. Rev. D* **12**, 2922 (1975).
Sally, P. J., *Bull. Am. Math. Soc.* **72**, 269 (1966).
Schiff, L. I., *Quantum Mechanics*, 3rd ed. (McGraw-Hil, New York, 1968).
Schweber, S. S., *An Introduction to Relativistic Quantum Field Theory* (Row, Peterson & Co., Elmsford, NY, 1961).
Sciarrino, A. and M. Toller, *J. Math. Phys.* **8**, 1252 (1967).
Schwartz, C., *Phys. Rev. D* **25**, 1159 (1982).
Segal, I. E., *Duke Math. J.* **18**, 221 (1951).
Segel, I. E., *Theoretical Problems of Relativistic Physics* (Am. Math. Soc., Providence, RI, 1963).
Serterio, L. and M. Toller, *Nuovo Cimento* **33**, 413 (1964).
Shapiro, J. A., *Ann Phys. (NY)* **43**, 439 (1968).
Shirley, J., *Phys. Rev.* **138**, B972 (1965).
Smith, F. T., *Phys. Rev.* **118**, 349 (1960).

Smirnov, V. I., *Linear Algebra and Group Theory* (McGraw-Hill, New York, 1961).
Sogami, I., *Prog. Theor. Phys.* **46**, 1352 (1969).
Sogami, I. and H. Yabuki, *Phys. Lett.* **94B**, 157 (1980).
Sollani, G. and M. Toller, *Nuovo Cimento* **15A**, 430 (1973).
Sommerfeld, A., *Partial Differential Equations in Physics* (Academic, New York, 1949).
Stanley, D. P. and D. Robson, *Phys. Rev. Lett.* **45**, 235 (1980).
Stanley, D. P. and D. Robson, *Phys. Rev. D* **26**, 223 (1982).
Streater, R. F. and A. S. Wightman, *CPT, Spin and Statistics, and All That* (Benjamin, 1964).
Takabayasi, T., *Nuovo Cimento* **33**, 668 (1964).
Takabayasi, T., *Prog. Theor. Phys. Suppl.* **67**, 1 (1979).
Talman, J. D., *Special Functions, a Group Theoretical Approach* (Benjamin, New York, 1968).
Toller, M., *Nuovo Cimento* **37**, 631 (1968).
Van Royen, R. and V. Weisskopf, *Nuovo Cimento A* **50**, 617 (1967).
Vilenkin, Ya., *Special Functions and the Theory of Group Representations*, Trans. Am. Math. Soc. **22** (1968).
Wallace, S. J., *Phys. Rev. D* **8**, 1846 (1973).
Wallace, S. J., *Phys. Rev. D* **9**, 406 (1974).
Weyl, H., *Classical Groups*, 2nd ed. (Princeton Univ. Press, Princeton, New Jersey, 1946).
Weinberg, S., *Phys. Rev.* **134**, B882 (1964a).
Weinberg, S., *Phys. Rev.* **135**, B1049 (1964b).
Weinberg, S., *Gravitation and Cosmology* (Wiley, New York, 1972).
Weisskopf, V. and E. P. Wigner, *Z. Physik* **63**, 54 (1930); **65**, 18 (1930).
Whittaker, E. T. and G. N. Watson, *A Course on Modern Analysis* (Cambridge Univ. Press, Cambridge, England, 1927).
Wick, G. C., *Phys. Rev.* **96**, 1124 (1954).
Wightman, A. S., in: *Dispersion Relations and Elementary Particles*, by C. De Witt and R. Omnes (Hermann, Paris, 1960).
Wigner, E. P., *Phys. Rev.* **40**, 749 (1932).
Wigner, E. P., *Phys. Rev.* **51**, 106 (1937).
Wigner, E. P., *Ann. Math.* **40**, 149 (1939).
Wigner, E. P., *Z. Phys.* **124**, 665 (1948).
Wigner, E. P., *Group Theory, and Its Applications to the Quantum Theory of Atomic Spectra* (Academic, New York, 1959).
Wigner, E. P., in: *Group Theoretical Concepts and Methods in Elementary Particle Physics*, ed. by Feza Gursey (Gordon and Breach, New York, 1962a).
Wigner, E. P., in: *Theoretical Physics*, ed. by A. Salam (Int'l Atomic Energy Agency, Vienna, 1962b).
Wigner, E. P., in: *Perspective in Quantum Theory*, ed. by W. Yourgrau and A. van der Merwe (MIT Press, Cambridge, Mass., 1971).
Wigner, E. P., in: *Aspects of Quantum Theory, in Honour of P. A. M. Dirac's 70th Birthday*, ed. by A. Salam and E. P. Wigner (Cambridge Univ. Press, London, 1972).
Winternitz, P. and I. Fris, *J. Nucl. Phys. (U.S.S.R.)* **1**, 889 (1965) or *Soviet J. Nucl. Phys.* **1**, 636 (1965).
Wolf, K. B., *J. Math. Phys.* **15**, 1295, 2102 (1974).
Wolf, K. B., *Integral Transforms in Science and Technology* (Plenum, New York, 1979).
Wu, C. S., E. Ambler, R. W. Hayward, D. D. Hoppes, and R. P. Hudson, *Phys. Rev.* **105**, 1413 (1957).
Yukawa, H., *Phys. Rev.* **91**, 416 (1953).
Zweig, G., CERN Report Nos. TH401 and TH412, 1964 (unpublished).

Index

Abelian group, 3
Abelian invariant subgroup, 5, 44, 176, 206
Active transformation, 29, 45
Anomalous magnetic moments, 293
Associated Laguerre function, 117, 124, 181, 247
Associated Legendre function, 207

Baker—Campbell—Hausdorff formula, 48
Baryons and baryonic mass spectra, 275
Basis vectors, 12
Bessel function, 163, 185, 207
Bethe—Salpeter equation, 107, 134, 314
Breit frame, 196

Cartan's criterion, 44, 78, 106, 209
Casimir operators, 39, 78, 61
   of homogeneous Lorentz group, 78, 238
   of Poincaré group, 61, 116
   $SE(2)$, 39
   $SL(2, c)$, 47
   $SO(3)$, 39
Clebsch—Gordan coefficients, 10, 86
C-number time-energy uncertainty relation, 137
Coherent state representation, 134
Color, 255, 259, 273
Compact group, 41
Complex angular momentum, 225
Constraint condition, 135
Coset, 4
Coset space, 5
Coulomb gauge, 176
Covariant harmonic oscillators, 107
   differential equations for, 109
   for massless composite particles, 178
   for unitary representations of Poincaré group, 115
   in four-dimensional Euclidean space, 122, 126
   in light-cone coordinate system, 151
   in Lorentz coordinate system, 245
   in moving $O(4)$ coordinate system, 126
   normalizable relativistic wave functions, 111
   orthogonality relation, 114
   three-body, 129, 290

Deep inelastic scattering, 302
DeSitter group, 212
Dimensionality of representation, 44, 66
Dirac equation, 69
   Dirac representation, 106
   Foldy—Wouthuysen transformation, 105
   large-momentum/zero-mass limit of, 196
   Lorentz transformation of, 97
   pseudo-unitary representation, 244
   symmetry of, 95
   Weyl representation, 69, 95
Dirac's form of relativistic quantum mechanics, 135
   constraint condition, 135
   light-cone coordinate system, 147, 151
   instant form, 135, 144, 145
   front form, 135, 144
   point form, 135, 148
   Poisson brackets, 144
   representations of Poincaré group, 50, 60, 144
   subsidiary condition, 145
Direct product, 7, 10, 86, 178, 237
   of groups, 7, 237
   spinors, 10, 86, 178, 243
   of vector spaces, 10
Discrete space-time symmetry, 71

$E(2)$ and $SE(2)$, 4, 31, 45, 161, 189
  finite-dimensional representation of $SE(2)$, 164, 187, 198
  infinite-dimensional unitary representation of $SE(2)$, 163, 175, 183, 198, 207
  $SE(2)$ as contraction of $SO(3)$, 190
$E(2)$-like little group, 58, 166
  as limiting case of $O(3)$-like little group, 193
  finite-dimensional representation, 166
  for integer spin, 166, 198
  for photons, 166
  for gravitons, 199
  for neutrinos, 177
  trivial representation, 176
Electromagnetic fields, 187, 241
Equivalence classes, 7, 193, 200
Euclidean group, 4, 31, 45, 161, 188, 189
  $E(2)$ and $E(2)$-like little group, 168
  $E(2)$ as contraction of $O(3)$, 190
  in two-dimensional space, 4, 31, 45, 161
  in three-dimensional space, 188
Exponentiation, 11, 15

Flavor, 255, 268
Form factors, 291, 296
  dipole fit, 294
  nonrelativistic, 292
  of nucleons, 291, 296
Four group, 3
Four potential, 62
Fourier relation for time and energy, 139, 150, 156
Front form, 136, 144

Galilei group and transformations, 204, 212
Gauge degrees of freedom, 170, 177, 239
Gauge parameters, 62
Gauge transformation, 170, 177
Gegenbauer polynomial, 124
Generators of Lie groups, 29
  differential form of, 65
  of $E(2)$, 29
  of Lorentz group, 55
  of Poincaré group, 61
Geometric optics, 209
Geometry of $SL(2, r)$ and $Sp(2)$, 216
  of $SO(2, 1)$, 220
$GL(n, c)$ and $GL(n, r)$, 9
Gramm—Schmidt orthogonalization, 27

Graviton, 199
Group, 2
  Abelian, 3
  compact, 41
  continuous, 25
  contraction, 189
  index of, 5
  Lie group, 26
  little group, 3
  non-compact, 41
  order of, 3
  pseudo-unitary, 9
  quotient, 5
  semi-simple, 5
  simple, 5
  subgroup, 4
  unimodular, 9
  universal covering, 41
Group contraction, 189
  $E(2)$-like little group as contraction of $O(3)$-like little group, 193
  Galilei group as contraction of Lorentz group, 204
  large-momentum/zero-mass limit, 193, 196
  $SE(2)$ as contraction of $SO(3)$, 190

Hadronic distribution, 141
Hadronic mass spectra, 255
  of baryons, 275
  of mesons, 279
Hadronic wave functions, 106, 145, 262, 263, 267, 268, 271, 288
Harmonic oscillators, 106, 133, 234, 245, 271
  application to coherent states, 134
  covariant, 106, 145
  four-dimensional, 122
  three-body, 129, 271
Harmonic oscillator group, 133
Helicity, 96, 116, 171, 174
Helicity gauge, 171, 174
Helicity preserving transformation, 174, 187
Hermite polynomial, 112
  generating function of, 130
Homomorphism, 3
Hydrogen atom, 132

Index, 5
Induced representation, 64
Inhomogeneous Lorentz group, 63, see Poincaré group

Instant form, 135
Invariant subgroup, 5, 144, 187
Isomorphism, 3
    local, 36
Isospin and isotopic spin, 257
Iwasawa decomposition, 20

Jacobi identity, 24
Jet phenomenon, 313

Kalnins—Miller coordinate, 179

Large-momentum/zero-mass limit, 193, 196
Legendre function, 212, 225
Lens matrix, 210
Lie algebra, 35, 36, 212
    contraction of, 192, 195
    exponentiation, 35
    invariant subalgebra, 35
    regular representation, 37
    semi-simple, 36
    simple 36
    structure constants of, 34
    subalgebra of, 36
Lie group, 26
    compact, 41
    non-compact, 41
    simply connected, 41
    theorems of, 32, 44
Lie's theorems, 32
Light-cone coordinate system, 136, 147, 151
Linear transformation, 9
    homogeneous, 9
    inhomogeneous, 9, 28
Little group, 7, 55, 64
    $E(2)$-like little group, 57, 159, 166, 193
    $O(2, 1)$-like little group, 58, 214
    $O(3)$-like little group, 57, 117
    of $E(2)$, 7
    of Lorentz group, 55, 135, 166
Local isomorphism, 36
Lorentz condition, 168, 171
Lorentz contraction, 114, 154, 298
Lorentz coordinate system, 245
Lorentz—Dirac deformation, 149, 155, 288, 292, 305, 315
Lorentz group, 51, 237
    Casimir operators of homogeneous, 78, 238
    generators of, 55
    homogeneous, 51, 237, 238
    inhomogeneous, 60
    little groups of, 55, 135, 166
    orbits of, 55
    proper, 52
    subgroups of, 55
Lorentz-invariant uncertainty relation, 150, 152
Lorentz transformation, 51
    in light-cone coordinate system, 147, 155
    matrices, 53, 60, 75, 76
    of electric and magnetic fields, 241
    of electromagnetic four-potentials, 170, 174
    of fields, 67
    of harmonic oscillator wave functions, 113, 117, 288
    of Maxwell fields, 241
    of photon polarization vectors, 170, 174
    of spinors, 96, 105, 186, 241
    of uncertainty relations, 154, 156
    of wave functions, 64, 113, 117

Manifold, 41
Massless particles, 159
    $E(2)$-like internal space-time symmetry, 75, 166, 174
    gravitons, 203
    massless composite particles, 178
    neutrinos, 176
    photons, 166, 170, 174, 238, 241
    with integer spin, 166, 170, 174, 202
    with spin 1/2, 176
Matrix, 9, 13
    anti-Hermitian, 13
    exponentiation of, 11, 15
    Hermitian, 13
    nilpotent, 13
    triangular, 14, 20, 23
    unimodular, 9
    unitary, 13
Mesons and mesonic mass spectra, 279
Multiplication table, 3, 19

Neutrino polarization, 63, 101, 177
Nilpotent matrix, 39
Non-compact group, 41, 44
Non-unitary representation, 44, 67, 170, 179, 187, 221
    of free fields, 67
    of four-vectors, 170

of harmonic oscillators, 178, 245
of Maxwell fields, 241
of $SE(2)$, 164, 184, 198
of $SO(3, 1)$, 238
of spinors, 184, 241

$O(2, 1)$-like little group, 58
$O(3)$-like little group, 57, 60, 135
$O(3, 1)$, 51, 237, 238, 245, 249
$O(4)$, 122, 126, 132
Orbit, 8, 55, 64, 236
  of $E(2)$, 8
  of Lorentz group, 55
Orbit completion, 50, 64, 236

Parton, 303, 305, 310
  Feynman's picture, 305
  scaling phenomenon, 303
  parton distribution function, 305, 310
  structure function, 311
Passive transformation, 29, 45
Permutation group, 3, 17, 260, 262, 263
  Dirac's symmetry classification scheme, 260
  subgroups of, 5
  invariant subgroup of, 5
  multiplication table, 19
  Schur's lemma for, 19
  semi-direct product of, 7
Photons, 166, 170, 174
  four-potential, 166, 170
  Maxwell fields, 241
  polarization vectors, 170, 174
  unitary transformation of, 174
Poincaré group or inhomogeneous Lorentz group, 50, 60, 144, 159
  as semidirect product of Lorentz and translation groups, 60
  Casimir operators, 61, 116
  finite-dimensional representations, 67, 166
  generators of, 61, 116
  Lorentz subgroup, 60, see Lorentz group
  translation subgroup, 60
  unitary representations, 64, 115, 174
Point form, 135
Poisson brackets, 144
Pseudo-orthogonal and pseudo-unitary representation, 9, 66, 244

Quantum chromodynamics (QCD), 312
Quantum electrodynamics (QED), 292, 310

Quark model, 256, 286
Quotient group, 5, 237

Regge poles, 226
Regular representation, 37
Relativistic quantum mechanics, see Dirac's form of relativistic quantum mechanics
Representation, 9
  dimensionality, 44
  finite-dimensional, 37, 44
  homogeneous Lorentz group, 51, 237, 238
  infinite-dimensional, 37, 44, 175
  induced, 64
  irreducible, 10
  multiplier, 104, 231
  non-unitary, 44, 67, 170, 179, 187
  of $E(2)$-like little group, 159
  of $O(3)$-like little group, 135
  of $SE(2)$, 161, 184, 187
  of Poincaré group, 60
  pseudo-unitary, 9, 66, 244
  reducible, 11
  regular, 37
  trivial, 176
  unitary, 9, 44, 64, 112, 115, 163, 174, 183, 198, 207
Representation space, 11

Scaling phenomenon, 303
Schur's Lemma, 15, 19
$SE(2)$, 41, 29, 5, 161, 190
  as contraction of $SO(3)$, 190
  finite-dimensional representation, 164, 187, 198
  infinite-dimensional unitary representation, 163, 175, 183, 198, 207
Semi-direct product, 6, 60, 188, 206
Semi-simple group, 5, 36
Similarity transformation, 8, 13, 195, 196
Simple group, 5, 36
Simply connected Lie group, 41
$SL(2, c)$, 63, 78, 80, 91, 239
  as covering group of Lorentz group, 80
  spinors of, 91, 178
  spinors and four-vectors, 91
  subgroups of, 82
$SL(2, r)$, 84, 214, 216
$SO(2, 1)$, 58, 61, 84, 221
$SO(3)$, 3, 41, 86, 193
$SO(3, 1)$, 52, 237, 238, 245, 249
$SO(4)$, 47, 122, 126, 132
$Sp(2)$ or $Sp(2, r)$, 84, 214

Spinors, 79
  direct products of, 10, 86, 178, 238, 241
  of $SL(2, c)$, 81
  of $SU(2)$, 86
  tensor product of, 239
Step-up and step-down operators, 142
Structure constants of Lie groups, 34
Structure function, 311
  QCD corrections to, 312
$SU(2)$, 86
$SU(1, 1)$, 84, 214, 228
  unitary representation of, 228
$SU(3)$, 255, 268, 280
$SU(4)$, 49
$SU(6)$, 268
Subalgebra, 36
Subgroups, 4, 64
  invariant, 5, 44, 187
  of S_3, 4, 5
  of $SE(2)$, 5
  of $SL(2, c)$, 82
  of Lorentz group, 55
  of Poincaré group, 55, 237
Subsidiary condition, 112, 145
$SW(2)$, 37, 42, 85, 198
Symmetrized wave functions, 262, 263
  Spin wave functions, 267
  $SU(2)$ wave functions, 267
  $SU(3)$ wave functions, 268
  $SU(6)$ wave functions, 268
  Symmetrized products of, 263
  Three-body spatial wave functions, 271
  Totally symmetric baryonic wave functions, 273
Symplectic group, 84, 214

Tangent plane, 204

Tensor product, 239
Three-particle symmetry classification, 260
Time-dilation, 303, 306
Time-energy uncertainty relation, 137
Time-like excitation, 142
Time-separation variable, 139
Thomas precession, 233
Transversality condition, 170
Triangular matrix, 14, 20, 23, 197
Trivial representation, 176, 254

Uncertainty relations, 137
  covariant commutator form, 143, 152, 157
  Lorentz-invariant form, 153
  Lorentz transformations of, 156
  time-energy, 156
Unitary representation, 9, 13, 44, 64, 115, 174, 183, 198, 207
  dimensionality of, 44
  of Bargmann and Wigner, 64
  of Poincaré group, 115
  of $SE(2)$, 184, 207
  of $SO(2, 1)$ and $SU(1, 1)$, 228
  of $SO(3, 1)$, 249
  trivial, 176
Unitary spin, 168
Universal covering group, 41

Vector space, 10
  invariant, 10
  direct product of, 10

Wick rotation, 134
Wigner function, 233
Watson—Sommerfeld transformation, 226

Yukawa potential, 316